Springer Series in Solid-State Sciences

Volumes 1–89 are listed at the end of the book

90 **Earlier and Recent Aspects of Superconductivity**
Editors: J. G. Bednorz and K. A. Müller

91 **Electronic Properties of Conjugated Polymers III**
Basic Models and Applications
Editors: H. Kuzmany, M. Mehring, and S. Roth

92 **Physics and Engineering Applications of Magnetism**
Editors: Y. Ishikawa and N. Miura

93 **Quasicrystals**
Editors: T. Fujiwara and T. Ogawa

94 **Electronic Conduction in Oxides**
By N. Tsuda, K. Nasu, A. Yanase, and K. Siratori

95 **Electronic Materials** A New Era in Materials Science
Editors: J. R. Chelikowsky and A. Franciosi

96 **Electron Liquids**
By A. Isihara

97 **Localization and Confinement of Electrons in Semiconductors**
Editors: F. Kuchar, H. Heinrich, and G. Bauer

98 **Magnetism and the Electronic Structure of Crystals**
By V. A. Gubanov, A. I. Liechtenstein, and A.V. Postnikov

99 **Electronic Properties of High-T_c Superconductors and Related Compounds**
Editors: H. Kuzmany, M. Mehring, and J. Fink

97 Springer Series in Solid-State Sciences

Edited by Klaus von Klitzing

F. Kuchar H. Heinrich
G. Bauer (Eds.)

Localization and Confinement of Electrons in Semiconductors

Proceedings of the
Sixth International Winter School,
Mauterndorf, Austria, February 19–23, 1990

With 224 Figures

Springer-Verlag
Berlin Heidelberg New York London
Paris Tokyo Hong Kong Barcelona

Professor Dr. Friedemar Kuchar
Institut für Festkörperphysik, Universität Wien, A-1090 Wien, Austria

Professor Dr. Helmut Heinrich
Institut für Experimentalphysik, Universität Linz, A-4040 Linz, Austria

Professor Dr. Günther Bauer
Institut für Physik, Montanuniversität Leoben, A-8700 Leoben, Austria

Series Editors:
Professor Dr., Dres. h. c. Manuel Cardona
Professor Dr., Dr. h. c. Peter Fulde
Professor Dr., Dr. h. c. Klaus von Klitzing
Professor Dr. Hans-Joachim Queisser

Max-Planck-Institut für Festkörperforschung, Heisenbergstrasse 1,
D-7000 Stuttgart 80, Fed. Rep. of Germany

Managing Editor:
Dr. Helmut K. V. Lotsch

Springer-Verlag, Tiergartenstrasse 17, D-6900 Heidelberg, Fed. Rep. of Germany

ISBN-13:978-3-642-84274-0 e-ISBN-13:978-3-642-84272-6
DOI: 10.1007/978-3-642-84272-6

This work is related to Department of the Navy Grant N. 00014-90-J-1491 issued by the Office of Naval Research. The US Government has a royalty-free license throughout the world in all copyrightable material contained herein.

Softcover reprint of the hardcover 1st edition 1990

2154/3140-543210 – Printed on acid-free paper

Preface

The winter school on *Localization and Confinement of Electrons in Semiconductors* was the sixth of a series dealing with new developments in solid-state physics organized by the Austrian Physical Society. The school, which was held in the castle of Mauterndorf, in the federal province of Salzburg, Austria, February 19–23, 1990, was attended by about 200 participants. This is the fourth such winter school to have the proceedings published in the Springer Series in Solid-State Sciences, the earlier ones appearing as Vols. 53, 67 and 83. The contributions of the invited speakers are arranged in seven parts according to topic: quantum boxes and quantum wires; localization and the metal–insulator transition; the integer and fractional quantum Hall effects and electron–electron interactions; tunneling and localization in low-dimensional structures and superlattices; optical properties of confined systems; impurities in quantum well structures; and optical and magnetic properties of dilute magnetic semiconductor quantum well structures.

A particular intention of this book is to present relevant introductory articles alongside the most recent research reports, all written by experts in the fields of localization and condensation of electrons in semiconductors, the quantum Hall effect, lateral superlattices, ballistic transport, etc.

The fascinating subject of quantum boxes (dots) and quantum wires is treated in considerable detail. Experimental techniques for producing laterally structured two-dimensional electron gas systems are described together with methods for studying their electrical and optical properties. Also included are the latest results on coupled quantum dots, which enable the formation of a one-dimensional mini-band structure to be studied. The role of disorder and of the metal–insulator transition in the quantum Hall effect is discussed, along with the role of quasiparticles and spin-reversal in the fractional quantum Hall effect.

Coherence and localization in periodic and quasi-periodic superlattices are discussed in detail, as are magnetic interactions between confined electrons in wells and localized magnetic species in barriers. The possible impact of confined electron systems on device physics is discussed and the current state of low-dimensional device developments is reviewed.

The organizers thank the invited speakers for agreeing to prepare manuscripts for inclusion in this volume.

The winter school was supported by generous contributions from the following institutions and companies:

Austrian Physical Society (ÖPG)
Bundesministerium für Wissenschaft und Forschung
Salzburger Landesregierung
European Office of Aerospace Research and Development (EOARD), London
Office of Naval Research (ONR), Arlington
Gesellschaft für Mikroelektronik
IBM–Austria
Bruker GmbH
Alcatel
Cryogenic Consultants
BOMEM – Dipl.-Ing. Haider
Keithley
Oxford Instruments
Karl Süss
VTS – Joachim Schwarz

Mauterndorf
February 1990

F. Kuchar
H. Heinrich
G. Bauer

Contents

Part I Quantum Boxes and Quantum Wires

Part II Localization and the Metal–Insulator Transition

Part III The Integer and Fractional Quantum Hall Effects and Electron–Electron Interactions

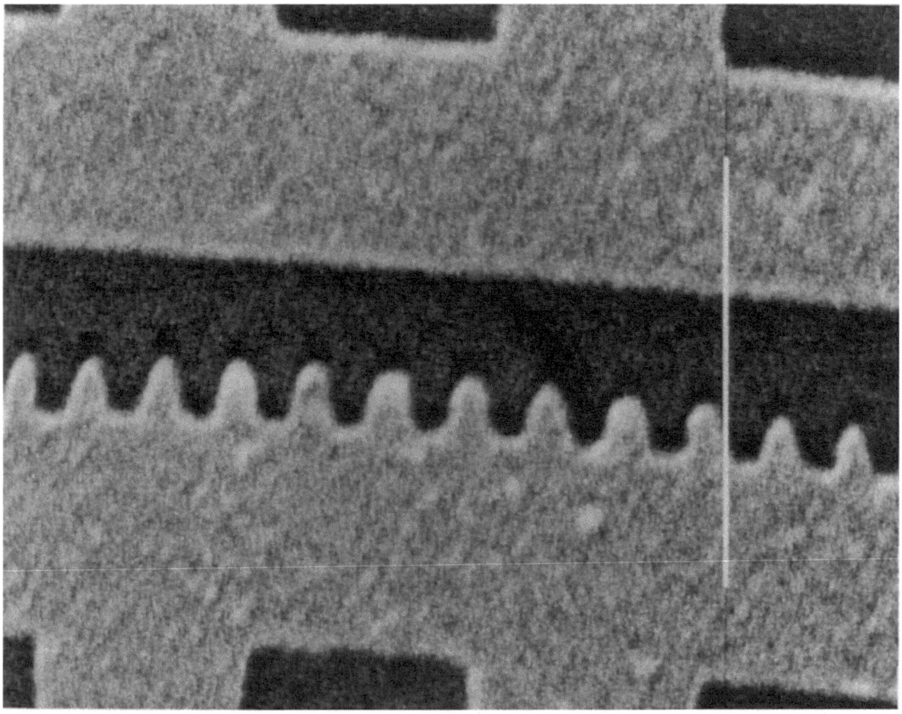

Scanning electron micrograph of a periodically corrugated gate geometry (containing 16 fingers in total), which is used to define 15 quantum dots in series in the 2DEG. Such a structure is used to study the formation of a mini-band structure in this periodic 1D crystal device. (The white marker indicates 1 μm.) Courtesy of Léo P. Kouwenhoven et al., Delft University of Technology, The Netherlands.

Part I

Quantum Boxes and Quantum Wires

Quantum Boxes, Quantum Wires, and In-Plane Superlattices: Their Impact in Device Physics and Required Breakthroughs in Material Science

H. Sakaki

Research Center for Advanced Science and Technology (RCAST), University of Tokyo, Komaba 4-6-1, Meguro-ku, Tokyo 153, Japan

The introduction of an in-plane potential V(x,y), with characteristic dimension comparable with the de Broglie wavelength of electrons, into otherwise uniform two-dimensional electron systems allows, in principle, the formation of various unique quantum states ; they include truly-discrete states confined in quantum box (QB) structures, truly-collimated propagating wave states guided in quantum wire (QWI) structures and various coupled states in coupled QB and QWI arrays, all of which are drastically different from those in more-conventional layered quantum structures. In this paper, we examine the significance of these lateral super-structures (LASS) in device physics by clarifying what particular electronic processes should be controlled to create new properties desirable for advanced electronics and opto-electronics. We discuss, in particular, the scattering processes in quantum wire and coupled QB arrays and their significance both on conventional (non-coherent) transistor action (based on charge modulation) and on novel coherent transistor action based on quantum interference (such as gate-controlled Bragg interference and the Aharonov-Bohm effect). We discuss also features of opto-electronic processes particularly in QB structures in connection with laser and other opto-electronic device applications.

We then point out the presence of stringent requirements on the geometrical features of these lateral superstructures for their device applications and emphasize the need to establish material fabrication methods with feature sizes of 200Å or less. The current status of efforts along this line will be reviewed.

1.INTRODUCTION

The introduction of an artificial potential V(x) or V(x,y) of microscopic scale (~100Å) or mesoscopic scale (~1000Å) into the (x-y) plane of two-dimensional electron systems (2DES) was first proposed by Sakaki in 1975 (Sakaki et al. 1975, 1976). The main objective was to provide means to create a variety of entirely new electronic states and bring forth various transport and/or optical properties that are found neither in bulk nor in multi-layered quantum structures (Esaki and Tsu, 1970). Examples of such systems span a wide spectrum of states, ranging from isolated quantum boxes (QBs) and quantum wires (QWIs) to quantum-mechanically coupled QWI/QB arrays (or in-plane superlattices (IP-SLs)) where a series of minibands and minigaps are expected to be formed (See Fig. 1). These systems are unique also

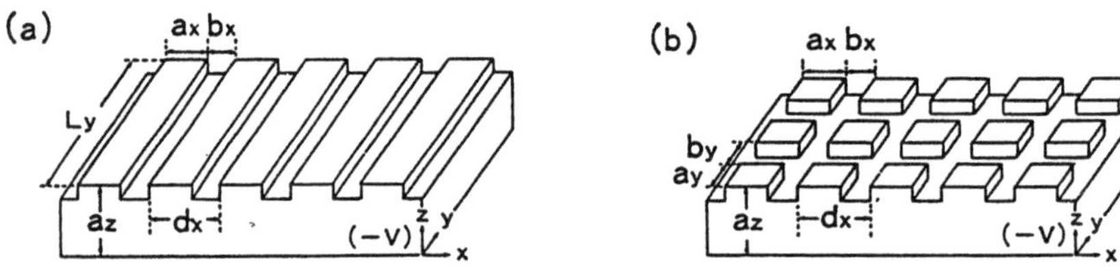

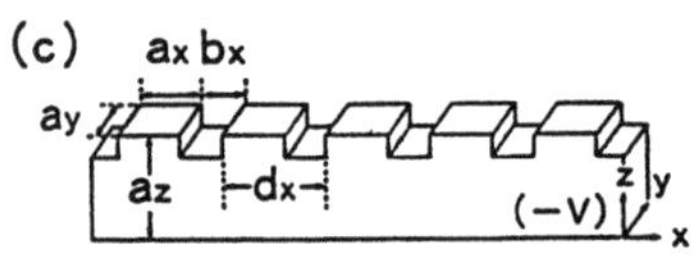

Fig. 1 Various forms of laterally-defined superstructures (LASSs): quantum wire (arrays)(a) and quantum box (arrays)(b) and (c). The coupling between adjacent elements leads to the miniband formation.

Springer Series in Solid-State Sciences, Vol. 97
Localization and Confinement of Electrons in Semiconductors
Editors: F. Kuchar · H. Heinrich · G. Bauer

Despite the attractive features of these lateral superstructures (LASSs), there existed in the early phase a strong skepticism on both their feasibility and relevance. Since 1980, however, we have witnessed growing interest and persistent effort in the exploration of this new physical system. As a result, a number of new phenomena have been predicted and /or demonstrated. In particular, a good number of experiments have successfully demonstrated the formation of QB, QWI, and IP-SL states in relatively large LASSs at low temperatures where the coherence length of electrons in high quality samples gets quite long (~1μm). In this paper, we discuss the physical basis of these unique phenomena. We examine in particular which of these phenomena will remain as subjects of academic curiosity and which of these phenomena may have an impact on device physics. In this context, we emphasize the need for breakthroughs in material processing technology to allow the preparation of LASSs with microscopic feature sizes (typically less than 200Å).

2. PHYSICAL PHENOMENA USABLE IN TRANSPORT AND OPTOELECTRONIC DEVICES AND REQUIREMENTS ON ELECTRONIC STATES

(a) Physical Requirements of Electronic Switches and Transistors.

For any electronic switches to be useful, they must satisfy the following three conditions :
(e-1) the (ON/OFF) ratio of conductance should be sufficiently high,
(e-2) the switch in the ON-state must support a sufficiently high current to drive the load with a reasonable speed,
(e-3) the switch must be able to sustain a sufficiently high voltage (particularly in the OFF state) and,
equivalently, (e-4) the switch must control the flow of more energy than the input energy required for switching.

Although quite a number of novel switching schemes have been proposed, one must examine whether each one of these satisfies the above conditions. In particular, requirements (e-2) and (e-3) force one to choose only those physical phenomena that persist at sufficiently high current (i.e. high electron concentration and/or high electron velocity) and high voltage (i.e. high electron energy and/or electron temperature). We discuss this issue in Sects.3,4, and 5 and show that only a small number of proposed switching schemes can fulfill those requirements.

(b) Physical Requirements of Lasers and Optoelectronic Devices

As will be discussed in Sect.6 in detail, most of the predicted benefits of using QB/QWI structures as optoelectronic materials result from the peaked features of the electronic density of states, in which most carriers are accommodated in a narrow range of energy. This effective reduction of energy distribution of carriers is possible only when (o-1) the majority of carriers can be accommodated in the ground level of QB/QWI and, at the same time, (o-2) the energy width of the ground level is much narrower than the average kinetic energy of carriers.

(c) General Requirements on the Possible Device Applications of Lateral Superstructures

From the considerations given above and from the discussion which will be elaborated later, it should be clear that requirements for the fabrication of relevant LASS devices are much more stringent than those for the mere observability of quantum state formation. This is because laterally-defined quantum states are formed as long as the following condition (C-1) is satisfied:
(C-1) The lateral dimension L is small enough that the energy level separation $E(2) - E(1) = 3(h^2/2m)(\pi/L)^2$ is greater than the level broadening ΔE.

In contrast, the most drastic phenomena applicable to novel LASS devices appear only when the following two conditions are additionally fulfilled:
(C-2) The level separation E(2)-E(1) should be comparable or greater than the average Fermi energy E_F (or thermal energy kT_e) of carriers so that the majority of carriers are accommodated in the ground level.

(C-3) The quantum level must be sharp with $\Delta E <<$ carrier energy E_f or kT_e so that the energy distribution of carriers n(E) gets much sharper in QWI/QB systems, while carriers remain unlocalized in QWI systems.

To indicate to what extent these conditions are satisfied by various LASS systems, we use the term LASS class-1 system when the conditions C-1,2,3 are all satisfied. If either C-2 or C-3 is violated, we refer to such systems as class-2 or class-3 LASS systems. Note that it is a class-1 system that is needed in most device applications. For GaAs LASSs to be class-1 at room temperature, it is generally necessary to set the lateral dimension L to be less than 200Å. In Sect.7, recent work toward such required breakthroughs in material sciences is described.

3. ELECTRON TRANSPORT THROUGH WELL-SEPARATED TRANSVERSE MODES IN QUANTUM WIRES, SUPPRESSION OF IONIZED IMPURITY SCATTERING AND THEIR SIGNIFICANCE

The confinement of electrons in a quantum wire with a cross-section of $L_y \times L_z$ leads to the formation of a series of discrete transverse states or modes ($k_y=n_y\pi/L_y$, $k_z=n_z\pi/L_z$), leaving the free electron motion only in the x-direction along the wire. When L_y and L_z are set sufficiently small so that $(2\pi/L_y)$ and $(2\pi/L_z)>k_f$ (Fermi wavenumber), then nearly all the electrons will be only in the ground transverse mode ($n_y=1$, $n_z=1$).

In 1980 it was theoretically predicted (Sakaki 1980) that the ionized impurity scattering of electrons in quantum wires of Fig.2 can be drastically suppressed if a large number n_e of electrons ($\geq 10^6$ / cm) are all accommodated in the sharply-defined ground subband or mode. This is because incident electrons travelling along the wire with the wave number k_i cannot be scattered elastically into any states in phase space except a small region in the vicinity of $-k_i$. Since this backward scattering requires a large exchange of momentum $\Delta k = 2k_i$, its probability is calculated to be quite small. For example, the low temperature mobility limited by bare ionized-impurities is approximately expressed for GaAs wire as 341 $cm^2/Vs \times (n_e d)$ $\exp(2\pi n_e d)$ where n_e is the electron concentration and d the impurity-wire distance. As shown in Fig.3 the mobility reaches as high as $3\times10^8 cm^2/Vs$ if n_e is increased to $1 \sim 2\times10^6$ /cm while d is around 200 ~ 100Å. Qualitatively similar enhancement of electron mobility is expected even when some electrons are accommodated in higher transverse modes (n_y and/or n_z=2,3) as long as the separation Δk of quantum states in the k-space is large enough to reduce the scattering probability.

This suppression of Coulomb and other elastic scattering processes in class-1 quantum wires (or their equivalent) implies that both the mobility and the mean free path of electrons can be substantially enhanced at low temperatures in a quantum wire with well-separated transverse modes. The recent work of Ismail et al. has indeed shown that the electron mobility in a carefully-prepared quantum wire can indeed be higher than that in a 2D system (Ismail et al. 1989), suggesting that the predicted mechanism has played some role.

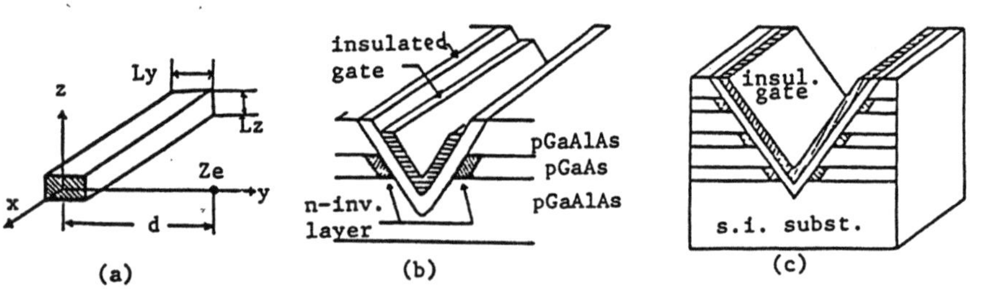

Fig. 2 Various forms of ultrafine semiconductor wires (USWs) : a basic structure (a), an insulated-gate FET-type wire (b), which utilizes n-type inversion layers on the edge of layered substrates (INVELS), and the multiple INVELS structures for parallelly-connected operations (c).

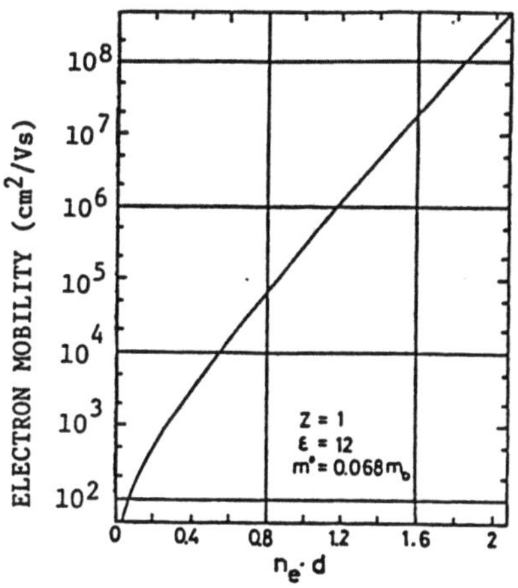

Fig. 3 Impurity dominated mobility of electrons in a class-1 GaAs quantum wire vs electron concentration n_e times d, the impurity-wire distance.

The most apparent and straightforward use of high-mobility features is to construct high-speed FETs with these QWI systems. Figure 2(b) shows one example of such FETs. Of course, in such applications, one should be well aware that the maximum current that each single-mode wire can sustain is given by qn_ev_s and is estimated to be $3\mu A$ when $n_e=10^6$/cm and the saturation velocity v_s is 2×10^7cm/s. Hence, one should construct a channel consisting of a few hundreds of parallel wires to supply a current of the order of mA. Then, one can expect some advantages of QWI FETs over the conventional modulation-doped FETs. For example, the saturation velocity of electrons of properly designed wires can be substantially higher than in the 2D system (as suggested in the work of Yamada and Sone 1989). In addition, there is a possibility that the reduced degree of free motion may contribute to some reduction of noise figures.

Aside from the straightforward application of QWIs to standard FETs, there exist other advantages of using this system since the enhancement in coherence of electron-waves and/or the single-mode (SM) feature of QWI systems can play greater roles particularly in quantum interference devices. This will be described in sect.4 and 5. The recent discovery of Van Wees et al. (1988) of the resistance quantization of ballistic electrons in ultrashort QWI systems might benefit also from the mobility enhancement since the finiteness of the number of subbands in their QWI systems may also contribute to the reduction of scattering rate to some extent.

4. ELECTRON TRANSPORT THROUGH A FINITE MINIBAND IN QUANTUM BOX ARRAYS AND QUANTUM WIRE SUPERLATTICES AND POSSIBLE SUPPRESSION OF OPTICAL PHONON SCATTERING

The discreteness of transverse modes of electrons in quantum wires is shown to result in the substantial suppression of elastic scattering processes which dominate the mobility at low temperatures. At high temperatures, however, the dominant scattering mechanism of electrons is by lattice vibrations, particularly optical phonons for polar semiconductors.

It has recently been pointed out by Sakaki (1989) that QWI superlattices or QB arrays of Fig.1(c) can provide a unique method to suppress optical phonon scattering in semiconductors. This unique feature may arise when the miniband structure of such systems satisfies the following three conditions:

(S-1) The first minigap is much wider than the average kinetic energy (E_f or kT_e) of electrons so that electron transport takes place only within the ground miniband.

(S-2) The width of the ground miniband is narrower than the optical phonon energy E_{op}(~30meV) so that neither the optical phonon emission nor its absorption is allowed within the miniband.

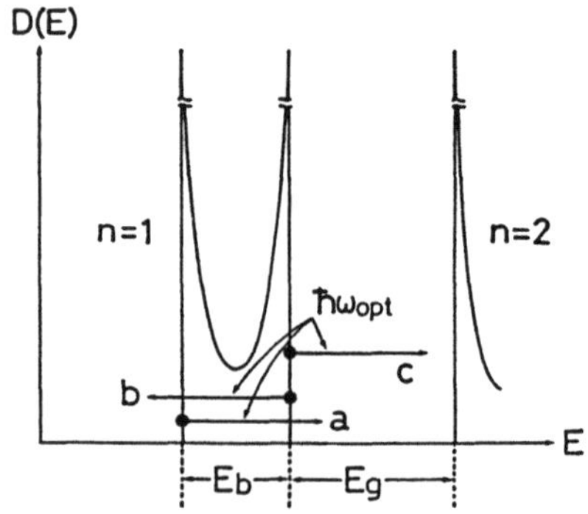

Fig. 4 Schematic illustration of state density *D(E)* for the ground miniband of a quantum wire superlattice *(QWISL)* as a function of electron energy. Note that neither the absorption (a) nor the emission (b) of optical phonons can take place within the miniband when its width is smaller than the optical phonon energy, which is represented as the length of the arrow. Similarly, the interminiband transition by the absorption of single optical phonon is forbidden when the minigap is wider than the phonon energy (c).

(S-3) The first minigap is wider than E_{op} so that no inter-miniband scattering is allowed (except through a multiple phonon process).

Figure 4 shows the density of states of electrons in QWI superlattices for this situation, where no pairs of states can be communicated by the absorption or emission of optical phonons. If the optical phonon scattering is indeed suppressed by this scheme, electrons may exhibit very high mobility($>10^5 cm^2/Vs$) even at high temperatures (Te = 200 ~ 300K) and at high electric fields . The transport in such a miniband will be quite nonlinear and may well lead to Bloch-oscillation and other coherent phenomena even at high temperatures. Note that the when the miniband width is 20 meV and the period of a superlattice is 100Å, the maximum group velocity of electrons in the QWI superlattice can be as high as 3×10^7cm/s and can supply reasonably large current or power if a large number of such structures are connected in parallel. Note that scheme of reducing the electron-optical phonon interaction is effective in small quantum boxes, in general, and may be quite important for optical device applications as well (Sect.6). A novel scheme to prepare such QWI superlattices will be discussed in Sect.7..

5. DIFFRACTION AND INTERFERENCE PHENOMENA AND VARIOUS QUANTUM INTERFERENCE TRANSISTORS

As described in Sect.3, class 1 quantum wires may serve as a system with small scattering probability and long coherence length. This feature is advantageous or prerequisite in realizing various kinds of quantum interference transistors, where the coherent interaction of multiple electron waves is to be controlled by external signals so as to reduce or increase the electron transmission through the structure. The planar superlattice FETs of Fig. 5 first proposed by Sakaki (1975-6) are the first of such examples. A similar proposal was later made for the double-barrier case (Luryi and Capasso, 1985). This FET makes use of Bragg interaction (diffraction) of electron waves with an artificial periodic potential V(x). Note that the gate can be used to vary the electron wavelength (through the electron concentration) as well as the miniband structure itself (through the potential). This Bragg interference transistor (BRINT) is predicted to exhibit both gate-controlled negative resistance and negative transconductances, which has indeed been demonstrated recently by Ismail et al. (1989) at low temperatures. If the period of planar superlattices is set around a few times 100Å, then the Bragg-reflection operation may persist even at room temperature. Note here that this BRINT concept holds valid for a large number of parallel quantum wire superlattices and therefore can achieve desirable current levels.

For the same reason, the use of high-mobility single mode (HM-SM) quantum wires is highly desirable in constructing the electrostatic Aharonov-Bohm FETs proposed by Datta et al. (1986), because electron wave interference in a ring geometry can be effective only when the electrons maintain their phases and modes. The FET of this scheme, however, has an inherent drawback that both current and operation voltage tends to be extremely low. Hence, it appears quite difficult for this device to be practical.

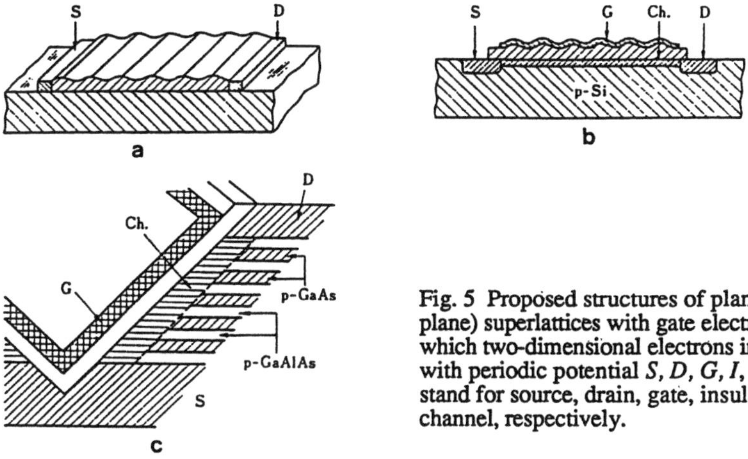

Fig. 5 Proposed structures of planar (in-plane) superlattices with gate electrode, in which two-dimensional electrons interact with periodic potential *S, D, G, I,* and *Ch* stand for source, drain, gate, insulator and channel, respectively.

6. PHYSICAL ORIGINS FOR THE BENEFIT OF USING QUANTUM-WIRE AND QUANTUM-BOX SYSTEMS AS LASER AND OTHER OPTOELECTRONIC MATERIALS

In 1982, the use of quantum wires (QWIs) and/or boxes (QBs) in the active region of laser diodes (LD) was first proposed and analyzed (Arakawa and Sakaki 1982). Further analysis of such lasers were made subsequently by Asada et al.(1986) and Arakawa and Yariv (1986). It has been theoretically found that both the threshold current J_{th} and its temperature variation (dJ_{th}/dT) should dramatically decrease, while the high-frequency response (the maximum modulation frequency) should improve by a factor of 3-6. These predicted improvements are due mainly to the sharpening of the gain spectrum $g(h\nu)$ or the density-of-states functions of QWI/QB systems (from the quasi-continuum one to a δ-function-like one) as shown in Fig. 6, though some advantages result from the smallness of the active

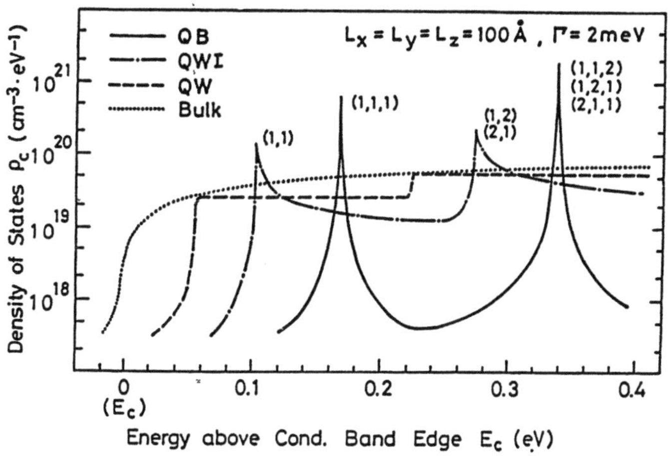

Fig. 6 Density of states of electrons in GaAs for bulk, quantum well (2D), quantum wires (1D) and quantum box (0D) structures. All the feature sizes are set to be 100Å and a Lorentzian level broadening of 2meV.

layer volume. One should note that the substantial sharpening of the gain spectrum can be achieved only when the two conditions (C-2) and (C-3) discussed in Sect. 1 are fulfilled and all the injected carriers are accommodated in the sharply-defined ground level. The use of quantum wires and quantum boxes as efficient electro-optic materials has been considered by Miller et al. (1988) for the case of the Stark effect and by Sakaki et al. (1990) for the case of the carrier-induced bleaching effect. In either case, the real advantages of these LASS systems result from the peaked feature of the joint density of states as in the case of lasers.

7. REQUIRED BREAKTHROUGHS FOR THE FABRICATION OF CLASS-1 LATERAL SUPERSTRUCTURES AND CURRENT STATUS

As described earlier, most of drastic physical phenomena and their device applications are to be realized only in class-1 LASS systems, where the ground quantum level is sharply defined ($\Delta E << E_f, kT_e$) and accommodates the major portion of carriers ($E(2) - E(1) > E_f, kT_e$). Since E_f and/or kT_e of typical devices are of the order of 10meV under normal operating conditions, it is important to set a large energy level spacing ($E(2) - E(1) >> 10meV$) and a small level broadening ($\Delta E < 5meV$) at the same time. This means that one must establish fabrication technology by which one can create various LASSs with characteristic length < 250Å with tolerable geometrical inhomogeneities. While remarkable progress in lithography continues to be made, these dimensions are quite difficult to handle. Hence, it appears highly desirable to look for alternative methods to prepare such LASS systems without resorting to lithography.

It has been proposed by Sakaki (1980) that a QWI structure can be formed by using inversion layers on the edge of multilayered quantum wells as shown in Fig. 2 (b). While the preparation of such structures was difficult because of the Fermi level pinning phenomenon at processed GaAs surfaces, the recent progress in the selective facet growth of quantum well structures has nearly allowed the fabrication of such systems (Fukui and Ando 1989). As the second alternative approach, Petroff et al. proposed a brilliant method of depositing half-monolayers of GaAs and AlAs alternately on vicinal GaAs substrates, where quasi regular steps are expected to be formed (1984). Recent experiments by Fukui and Saito (1987), Petroff et al. (1989), Tanaka and Sakaki (1988,1989), and Motohisa et al. (1989) have given positive evidence that the formation of laterally modulated structures with short periods (80 ~ 160Å) is indeed possible by the second method, although the composition modulation along the x-axis is not ideally stepwise. Figure 7 shows the structure of a novel planar superlatice, where AlAs grids are inserted in GaAs QWs with typical periods of 80 Å (Tanaka and Sakaki 1989). Figure 8 shows the polarization-dependent optical absorption of such structures, which indicates clearly the formation of PSL states.

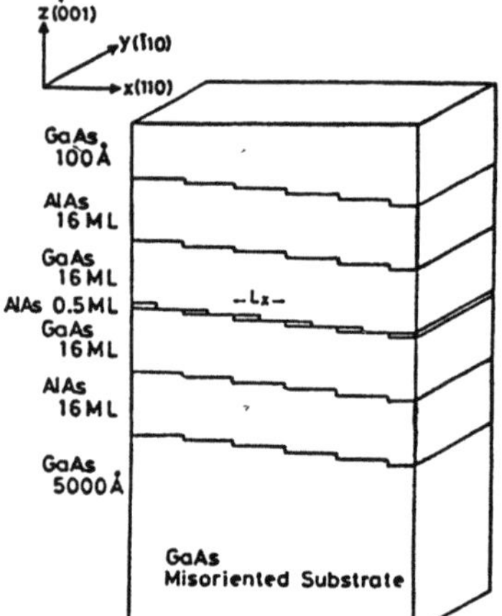

Fig. 7 Schematic illustration of the grid-inserted quantum well (GI-QW) structure, where a periodic array of AlAs bars is inserted in the center of GaAs quantum wells.

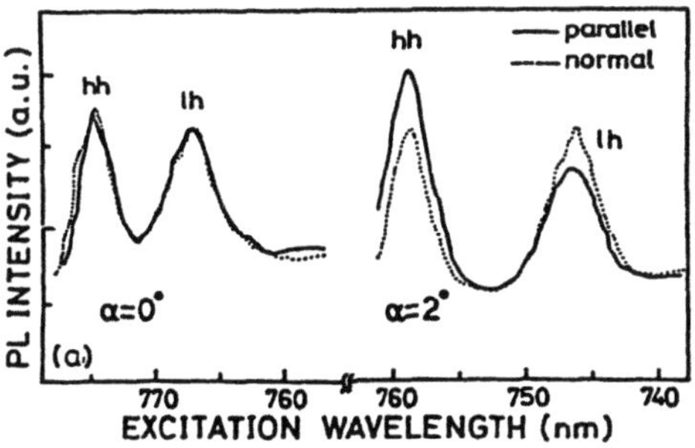

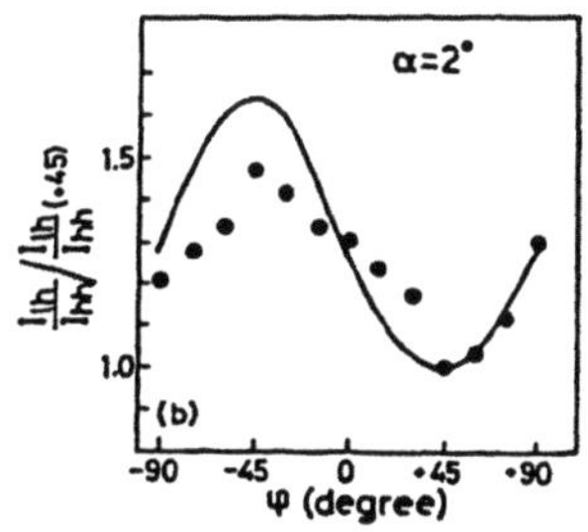

Fig. 8.(a) The photoluminescence excitation (PLE) spectra of grid-inserted quantum wells grown on a flat (100) substrate ($\alpha = 0°$) and a misoriented substrate ($\alpha = 2°$) which is tilted by 2 degrees from the (100) orientation. Note that the polarization angle φ of excitation laser light is set either parallel ($\varphi = -45°$) or normal ($\varphi = +45°$) to the grid. The presence of polarization dependence in the PLE peaks of electron-heavy hole (*e*-hh) excitons and those of electron-light hole (*e*-lh) excitons indicates the grid-induced anisotropy in the planar superlattices. (b) Ratio I_{lh}/I_{hh} of the peak height of *e*-hh and *e*-lh exciton in the PLE spectra of $\alpha = 0°$ and $\alpha = 2°$ is plotted as a function of φ after normalizing it by the ratio at $\varphi = +45°$. The theoretical predictions are also shown.

Future developments in material science, electron physics, and devices in LASS systems are likely to open exciting fields for new physics and electronics that span the conceptual gap between conventional solid state electronics and atomic electronics.

REFERENCES

For recent reviews and important references that are not listed here, see, for example, "Physics and Technology of Submicron Structures" ed. by H.Heinrich, G.Bauer, F.Kuchar, Springer Ser. Solid-State Sci., Vol.83 (Springer, Berlin, Heidelberg 1988)
Arakawa Y and Sakaki H 1982 Appl. Phys. Lett. **40** 893
Arakawa Y and Yariv A 1986 IEEE J. **QE-22** 1887

Electron Confinement in Quantum Dots

T.P. Smith, III

IBM T.J. Watson Research Center, P.O. Box 218,
Yorktown Heights, NY 10598, USA

Quantum dots represent a new class of electronic structures where the electronic configuration and degree of confinement or coupling can be tuned at will. These artificial mesoscopic structures mimic many atomic effects with a tremendous degree of flexibility. We have studied the transition from isolated to coupled quantum dots in quantum-dot and lateral-surface-superlattice structures. Our magnetocapacitance measurements probe three distinct regimes: a superlattice regime; a tight-binding regime; and an isolated-quantum-dots regime. In the presence of an external magnetic field we find that the coupling of the magnetic flux with the periodic potential depends on the strength of the potential modulation. In strongly-coupled quantum wires and quantum dots, the commensurability between the superlattice period and the cyclotron orbit is reflected in a modification of the envelope of the magnetocapacitance oscillations. In weakly-coupled quantum dots, a new structure is observed. Aperiodic oscillations in the density of states are related to fractal-like behavior in the energy dispersion relation first predicted by Hofstadter. In completely isolated quantum wires or quantum dots, the canonical magnetocapacitance oscillations are recovered with a Zeeman bifurcation of the states produced by the interplay of magnetic and spatial quantization. We will present our experimental work in light of theoretical models and discuss the fundamental physical phenomena that dominate the thermodynamic properties of electrons in quantum dots.

1. Introduction

The study of low-dimensional or quantum-confined systems has gained a great deal of attention in the past few years. A number of important discoveries and breakthroughs have heightened scientific interest in this area. The Aharonov-Bohm effect[1] and quantization of the conductance through ballistic point contacts[2, 3] are premier examples. There is also a great deal of interest from the technological point of view. This interest arises on both the primary (new device concepts) and the secondary (impact on existing devices) levels. The tools for nanostructure fabrication are the staples of the semiconductor industry and thus the connection is fairly direct. The study of quantum dots[4] is the result of tremendous advances in molecular-beam epitaxy, dry processing, and advanced lithography.

Springer Series in Solid-State Sciences, Vol. 97
Localization and Confinement of Electrons in Semiconductors
Editors: F. Kuchar · H. Heinrich · G. Bauer

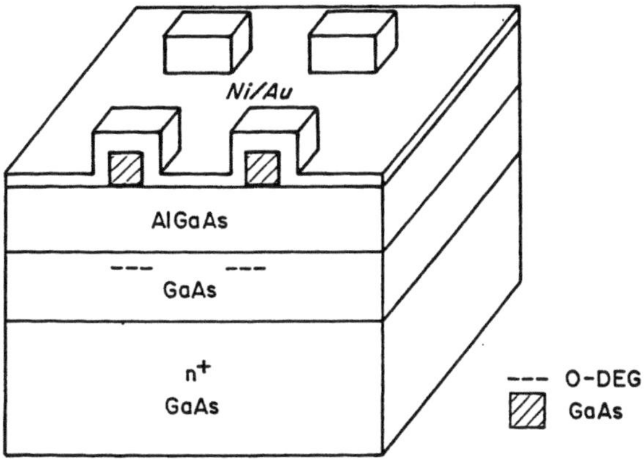

Fig. 1: Schematic diagram of a quantum dot capacitor: The 30 nm GaAs cap layer is etched away to deplete the underlying AlGaAs and confine the electrons at the heterointerface in a quasi-zero-dimensional system. The n+ GaAs is one electrode of the quantum capacitor and the Ni/Au metal on the top of the structure is the other.

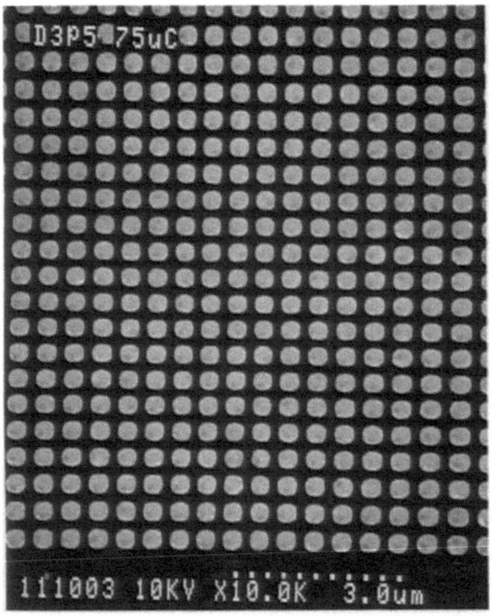

Fig. 2: Scanning-electron micrograph of an array of 300 nm quantum dots.

Figure 1 is a schematic view of a semiconductor heterostructure used in our experiments. Lateral confinement relies on depletion of a high-mobility two-dimensional electron gas. The lithographic features are about twice as large as the confining potential. Therefore, well defined quasi-zero-dimensional systems can have widths and diameters between 0.1 and 0.5 microns. Features of this size can be produced reproducibly and accurately with current electron-beam lithography systems. Figure 2 shows an array of quantum dots etched into a GaAs/AlGaAs heterostructure. In our experiments we use selective etching of a 300nm thick GaAs cap layer to deplete the free carriers. Selective etching produces a smooth etched surface and a reproducible etch profile; one key to obtaining good quantum confinement.

2. Spatial Quantization in Quantum Dots

The central problem in fabricating zero-dimensional systems is scattering by boundaries. Since the electrons are completely confined the boundary requirements are extremely stringent. Designing structures with tight lateral confinement that are not completely depleted of carriers is also a challenge. Nonetheless, there are a number of ways to probe quantum dots. Interesting results have been obtained using optical spectroscopy, [5, 6] tunneling,[7] and capacitance spectroscopy.[8, 9]

Figure 3 shows the results of capacitance measurements on three samples with different size quantum dots. We observe well-resolved oscillations in the derivative of the capacitance, reflecting zero-dimensional quantization. The period of the oscillations increases with decreasing dot size, and we have also found good qualitative agreement between our experimental results and preliminary model calculations.[10]

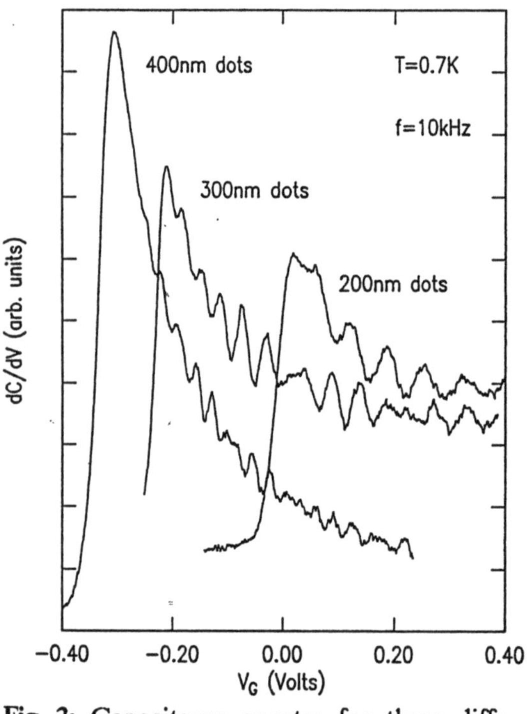

Fig. 3: Capacitance spectra for three different quantum dot samples: The spectra clearly show the size dependence of the spatial quantization.

Modelling these systems is extremely difficult. Electrons confined in an extremely small area no longer behave as part of a continuum. The behavior of individual electrons becomes dominant; changes in free energy, magnetization, or charge state become discrete. If the number of electrons and their spatial extent are small, then the Coulomb energy of the system can be dominant. Recent studies have examined this problem using tunneling through small metal[11] or superconducting junctions[12 - 14] and this was also pointed out by Silsbee and Ashoori.[15] The addition of a single electron can significantly chan the Coulomb energy of a quantum dot via discrete changes in the charging energy, $e^2/2C$. These effects may place a limit on the resolution of spatial quantization. If the Coulomb energy is larger than the energy level spacing then this will be the separation observed experimentally. If the two energies are comparable both effects can be important. The questions of charge quantization and spatial quantization are inextricably linked in electronic systems and the problem is even more complicated when the confining potential is electrostatic.

3. Magnetic Field Effects

One way to separate these effects may be the application of an external magnetic field. The effect of an external magnetic field on the Coulomb blockade should be small. In contrast, application of a magnetic field to a confined electronic system has a pronounced effect on spatial quantization. C. G. Darwin[16] was probably the first to address the problem theoretically. In essence, he showed that at low magnetic fields, such that the magnetic length is larger than or comparable to the size of the confining potential, there is a hybridization of the Landau levels with the levels resulting from the spatial confinement. As the magnetic length becomes much smaller than the confinement, spatial quantization effects become negligible and the free-electron result (standard Landau behavior) is recovered. Thus there is a smooth (albeit complicated) transition from spatial to magnetic quantization. This transition depends on the relative size of the electron systems with respect to the magnetic length. For example, the formation of Landau levels requires more than 40T if the electrons are confined in a 100Å box.

Figure 4 shows capacitance spectra for samples with 300nm quantum dots in the presence of an external magnetic field. Even at very low magnetic fields ($B < 0.2T$) there is a complicated rearrangement of the peaks in the spectra at positive gate biases. By about 1T all the peaks have shifted dramatically and there is a bifurcation of many of the peaks. Finally, at higher fields some of the peaks increase in size with other, weaker peaks moving between them. These stronger peaks are the precursors to Landau levels while the smaller peaks are the surface states or "whispering gallery" states which feed into the Landau levels and increase their degeneracy. We can trace the transition from spatial to magnetic quantization and see the influence of the orbital angular momenta of the confined electrons.

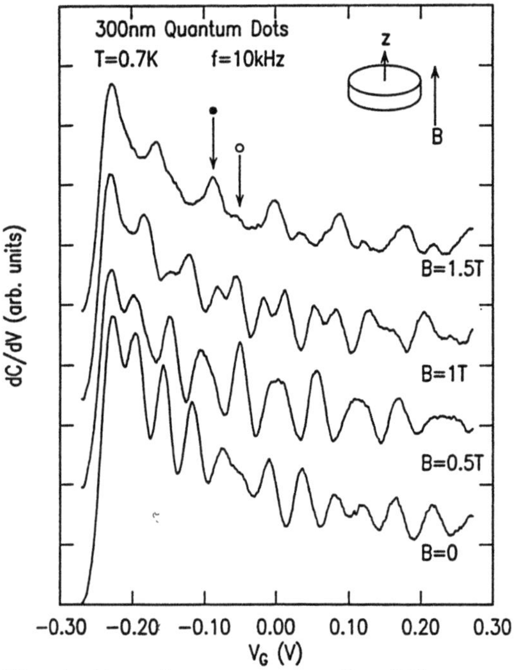

Fig. 4: Capacitance spectra for 300 nm quantum dots at various magnetic fields: The magnetic field is applied perpendicular to the original two-dimensional plane. (After [9])

4. Fractional Quantization of Quantum Dots

While external magnetic fields are an incisive probe of single-electron states, they can also give rise to many-body effects. This is particularly true at very high magnetic fields, where the fractional Hall effect[17] is seen in two-dimensional systems. In this case, the presence of an external magnetic field gives rise to a new quantum fluid. The electron-electron interactions combine with the magnetic field to produce a hierarchy of incompressible states. An important question is what effect boundaries have on the formation of the incompressible electron fluid states.

Figure 5 shows capacitance spectra for a sample with 300nm dots in magnetic fields up to 30T. Formation of fractional states at filling factors of 1/3 and 2/3 are seen between about 20 and 30T. The structure in the capacitance spectra becomes weaker with decreasing magnetic field and increasing temperature. At high magnetic fields the electrons respond as a bulk two-dimensional system. We also observe structure at fractional filling factors in samples with 400nm dots but not in samples with 200nm dots. This may be due to boundary scattering or the small number of electrons in the 200nm quantum dots. In any case, the fact that we see fractional states in these quantum-dot samples attests to their quality and the robustness of many-body interactions.

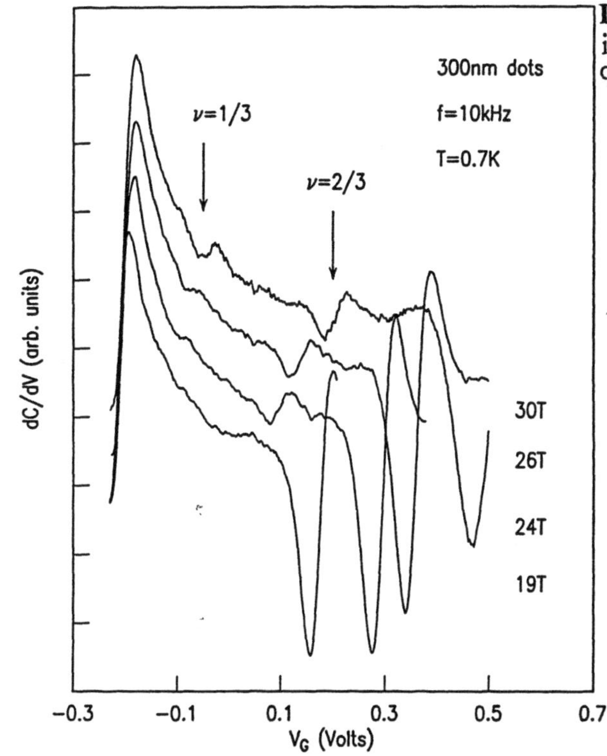

Fig. 5: Capacitance spectra in the extreme magnetic quantum limit. (After [23])

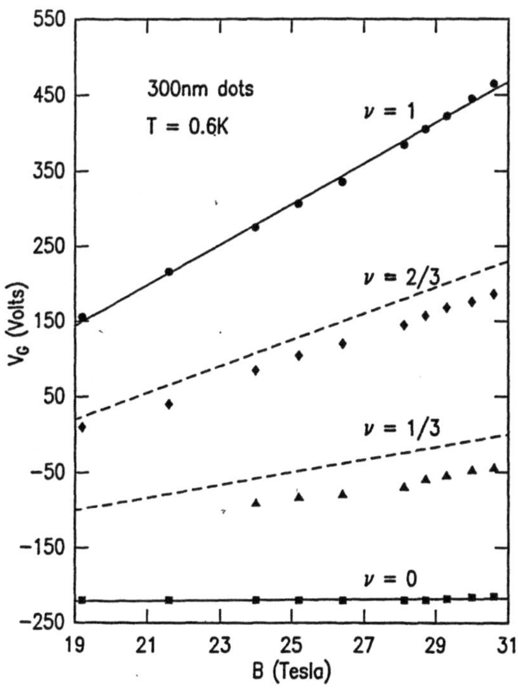

Fig. 6: The positions of the structure at fractional filling factors as a function of magnetic field. The dashed lines are calculated for v = 1/3 and v = 2/3 based on the positions of the lowest Landau level. (After [23])

5. Superlattice Effects

Thus far we have focussed on the properties of isolated quantum dots. However, if the barriers between dots are not too high quantum dots can couple with one another to form super-molecules or superlattices. In our experiments we have formed large-area quantum-dot lateral-surface superlattices (LSSL). The relevant length scales in such structures are the electron de Broglie wavelength, λ, and the superlattice periodicity, a (provided a is shorter than the phase coherence length). If a magnetic field is applied perpendicular to the LSSL plane, a third length scale, the magnetic length, l, becomes important. Weiss et al.[18] observed an oscillatory magnetoresistance effect in samples with a weak lD-periodic (a = 382 nm) potential modulation at magnetic fields, B, below 0.8T. They explained their results in terms of a commensurability criterion between the cyclotron orbit diameter, $2R_C$, and the superlattice period, a:

$$2R_C = (m + \phi)a \qquad m = 1, 2,...$$

where $R_C = 2\pi l^2/\lambda$ and ϕ is a phase factor. Winkler et al.[19] made a similar observation in a grating-gate structure (a = 500 nm) and explained the effect in terms of oscillations in the width of the Landau bands caused by the periodic potential modulation.

We have studied two types of structures: In the first, the periodic Schottky gate is used directly to change the periodic potential; in the second, a periodic metal gate is used as an etch mask, and the ungated areas are recessed about 10 nm using wet chemical etching, and a continuous metal gate is evaporated on top of the structure. The main difference between the two types is that the potential modulation strength and the Fermi energy, E_F, both vary with gate bias, V_G, in the former structure, while in the latter only E_F is changed and the modulation strength is set by the etch depth. The shape of the potential modulation produced by a 200 nm-period grid gate is nearly sinusoidal with a modulation strength, V_0 (peak-to-peak), ranging from 3 meV to 0 meV to 4 meV, as V_G is changed from +0.3 V to 0 V to -0.5 V[20]. In the relief-grid/grating LSSLs, V_0 is almost constant and equal to 6 meV.

Figure 7 shows the measured capacitance as a function of magnetic field for three different gate biases. A periodic potential produces clear modulation of the envelope of the Landau levels. At $V_B = 0.2V$, the left two arrows indicate fields where the commensurability condition of equation (1) is satisfied. However, the structure at B = 2.8T, does not fit this theory. We believe this structure is related to the strong 2D periodic potential in our samples. Our experiments make use of the LSSLs to create a strong 2D potential modulation, so that $V_0 \sim E_F \sim \hbar\omega_c$ and the electrons are confined in weakly coupled quantum dots.

In the planar grid-gate LSSL, both E_F and V_0 change with V_G. At $V_G = 0V$, there is no structure in the envelope since there is no potential modulation. At slightly positive or negative gate bias, V_0 is small, and oscil-

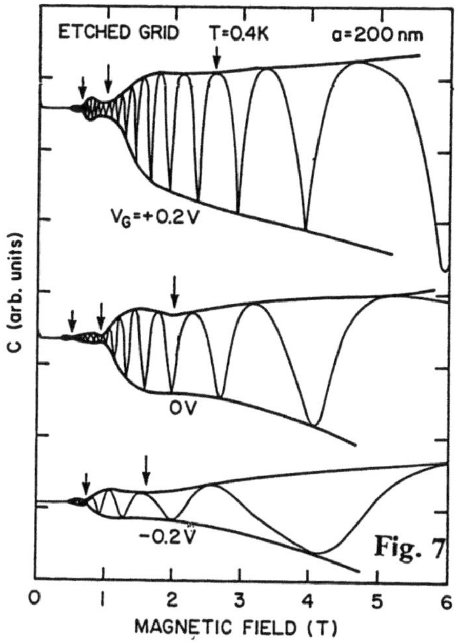

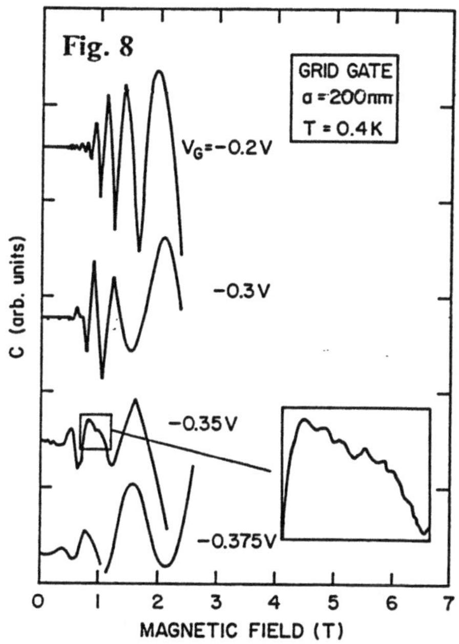

Fig. 7: Magnetocapacitance of the relief-grid-gate structure at three different gate bias conditions. The arrows indicate where a drop in the envelope is observed.

Fig. 8: Magnetocapacitance of the planar grid-gate structure at four different gate biases. The traces from top to bottom correspond to strong, weak and no coupling, respectively. The inset shows the fine structure at $V_G = -0.35V$, which is reproducible.

lations in the envelope of the Landau oscillations satisfy the commensurability criterion of equation (1). Similar to the etched grid-gate LSSL, a new structure in the envelope develops at B = 1.5T. In addition, smaller aperiodic oscillations with finer features superimposed develop at V_G below -0.31V. Figure 8 shows a spectrum for $V_G = 0.35V$. Finally, at V_G equal to -0.375 V or below, Landau oscillations are observed, showing only a slight deviation from the 1/B periodicity at low B. This corresponds to the case of isolated quantum dots, and the deviation from the 1/B periodicity is due to spatial confinement.[21]

As V_G is reduced from -0.2 to -0.3 V, the coupling between the quantum dots becomes weaker and a tight-bonding model rather than perturbation theory must be used to describe the problem. The most elegant and intriguing study of magnetic flux effects in this regime is due to Hofstadter.[22] He found that solutions to Harper's equation placed rationality conditions on the flux per lattice cell in the tight-binding regime. The allowed energy eigenstates for electrons in the presence of a magnetic field and a tight-binding, 2D periodic potential were summarized in a graph, where he plotted the energy versus number of flux quanta per unit cell, α, of the periodic structure. In a crystal lattice a value of $\alpha = 1$ implies an enormous B, but in a LSSL (a = 200 nm), B is of the order of 0.1 T. Since the pattern or spectrum repeats, modulo α, we are in the regime where we can probe the energy bands he predicted. The aperiodic behavior in the magnetocapacitance and the additional fine structure observed might reflect a part of the 'fractal-like' behavior of the energy bonds.

6. Conclusion

The study of quantum dots has revealed a number of interesting new physical phenomena but much remains to be done. The discoveries made using capacitance spectroscopy, resonant tunneling, and lateral transport are indicative of the wealth of physical effects in quantum dots and the large number of approaches possible.

Acknowledgement

This work is the result of collaboration with a number of people and it is a pleasure to acknowledge the contributions of J. A. Brum, L. L. Chang, L. Esaki, W. Hansen, R. Haug, J. M. Hong, K. Ismail, D. P. Kern, C. M. Knoedler, S. E. Laux, K. Y. Lee, and F. Stern. This work was supported in part by the Office of Naval Research.

References

1. R. A. Webb, S. Washburn, C. P. Umbach, and R. B. Laibowitz, *Phys. Rev. Lett.* 54 (1985) 2696.
2. B. J. van Wees, H. van Houten, C. W. J. Beenakker, J. G. Williamson, and C. T. Foxon, *Phys. Rev. Lett.* 60 (1988) 848.
3. D. A. Wharam, T. J. Thornton, R. Newbury, M. Pepper, H. Ahmed, J. E. F. Frost, D. G. Hasko, D. C. Peacock, D. A. Richie, and G. A. C. Jones, *J. Phys. C.* 21 (1988) L209.
4. T. P. Smith, III, J. A. Brum, J. M. Hong, C. M. Knoedler, H. Arnot, and L. Esaki, *Phys. Rev. Lett.* 61 (1988) 585.
5. H. Temkin, G. J. Dolan, M. V. Panish, and S. N. G. Chu, *Appl. Phys. Lett.* 50 (1987) 413.
6. C. Sikorski and U. Merkt, *Phys. Rev. Lett.* 62 (1989) 2164.
7. M. A. Reed, J. N. Randall, R. J. Aggarwal, R. J. Matyi, T. M. Moore, and A. E. Wetsel, *Phys. Rev. Lett.* 60 (1988) 535.
8. T. P. Smith, III, K. Y. Lee, C. M. Knoedler, J. M. Hong, and D. P. Kern, *Phys. Rev. B.* 38 (1988) 2172.

9. W. Hansen, T. P. Smith, III, K. Y. Lee, J. A. Brum, C. M. Knoedler, J. M. Hong, and D. P. Kern, *Phys. Rev. Lett.* 62 (1989) 2168

10. F. Stern, S. E. Laux, and A. Kumar, private communication.

11. R. E. Cavicchi and R. H. Silsbee, *Phys. Rev. Lett.* 52 (1984) 1453.

12. T. A. Fulton and G. J. Dolan, *Phys. Rev. Lett.* 59 (1987) 109.

13. P. J. M. van Bentum, H. van Kempen, L. E. C. van de Leemput, and P. A. A. Teunissen, *Phys. Rev. Lett.* 60 (1988) 369.

14. P. J. M. van Bentum, R. T. M. Smokers, and H. van Kempen, *Phys. Rev. Lett.* 60 (1988) 2543.

15. R. H. Silsbee and R. Ashoori, private communication.

16. C. G. Darwin, *Proc. Cambridge Phil. Soc.* 27 (1930) 86.

17. D. C. Tsui, H. L. Störmer, and A. C. Gossard, *Phys. Rev. Lett.* 48 (1982) 1559.

18. D. Weiss, K. V. Klitzing, K. Ploog, and G. Weinmann, *Europhys. Lett.* 8 (1989) 179.

19. R. W. Winkler, J. P. Kotthaus, and K. Ploog, *Phys. Rev. Lett.* 62 (1989) 1177.

20. K. Ismail, T. P. Smith, III, W. T. Masselink, and H. I. Smith, *Appl. Phys. Lett.*, to be published.

21. K. F. Berggren, T. J. Thornton, D. J. Newson, and M. Pepper, *Phys. Rev. Lett.* 57 (1986) 1768.

22. D. R. Hofstadter, *Phys. Rev. B* 14 (1976) 2239.

23. W. Hansen, T. P. Smith, III, K. Y. Lee, J. M. Hong, and C. M. Knoedler, *Appl. Phys. Lett.* 56 (1990) 168.

Quantum Dot Resonant Tunneling Spectroscopy

M.A. Reed, J.N. Randall, and J.H. Luscombe

Central Research Laboratories, Texas Instruments Incorporated,
P.O. Box 655936, MS 154, Dallas, TX75265, USA

Abstract. The fabrication and electronic transport through quantum dots (3-dimensionally confined semiconductor quantum wells) will be discussed. The spectrum corresponds to resonant tunneling from laterally-confined emitter contact subbands through the discrete quantum dot states. Transition selection rules for tunneling in these structures is derived, and momentum non-conservation is observed in these structures. An especially intriguing structure is two quantum dots separated by a tunnel barrier (a "quantum molecule"). Experimental results of a molecule structure are discussed.

1. Introduction

Carrier confinement to reduced dimensions has led to numerous important developments in basic semiconductor physics and device technology. Advances in fabrication technology and the understanding of electronic transport in low dimensional structures have led to fascinating discoveries [1]. Structures such as electron waveguides [2] and quantum point contacts [3] are excellent laboratories to study fundamentals of electronic transport in the quantum ballistic regime.

The creation of 3-dimensionally confined systems ("quantum dots") [4-6] is especially intriguing since these structures are analogous to semiconductor atoms, with energy levels tunable by the confining potentials. One method for creating these systems is the electrostatic confinement of an existing 2DEG [4,5]. Unfortunately, electronic transport that reveals quantum size effects in these systems only occurs for low temperatures ($\leq$ 1K) and voltages near equilibrium (typically $\sim \mu$V) [7].

A more interesting (and thus, more complex) situation would be the non-equilibrium quantum transport regime. However, quantum dot structures for electronic transport pose an experimental paradox distinct from 2D and 1D structures; establishing transport through the single electronic states of a quantum dot implies the states cannot be totally isolated. Additionally, contact to higher-dimensional carrier reservoirs are non-trivial from an experimental and analysis viewpoint.

Springer Series in Solid-State Sciences, Vol. 97
Localization and Confinement of Electrons in Semiconductors
Editors: F. Kuchar · H. Heinrich · G. Bauer

Recently, non-equilibrium electronic transport through quantum dots has been realized [6]. A configuration is adopted where the quasi-bound momentum component (and thus the resultant transport direction) is epitaxially defined in the form of a resonant tunneling structure, and additional confinement is fabrication-imposed. This configuration is distinct in that the current flow is through, not along, the interface. Because of this constraint, the behavior of a system operated far from equilibrium can be examined.

We present here a study of resonant tunneling through various quantum dot systems, and the bandstructure modeling necessary to understand the experimental electronic transport spectra.

2. Quantum Dot Fabrication and Tunneling

Our approach used to produce quantum dot nanostructures suitable for electronic transport studies is to laterally confine resonant tunneling heterostructures. The fabrication of these structures has been previously reported [6]. The approach essentially embeds a quasi-bound quantum dot between two quantum wire contacts. Creation of dots less than 500Å is possible, though we will show that the appropriate range for the typical epitaxial structure and process used is in the range 1000Å – 2500Å in diameter. A SEM of a collection of these etched structures is seen in Figure 1.

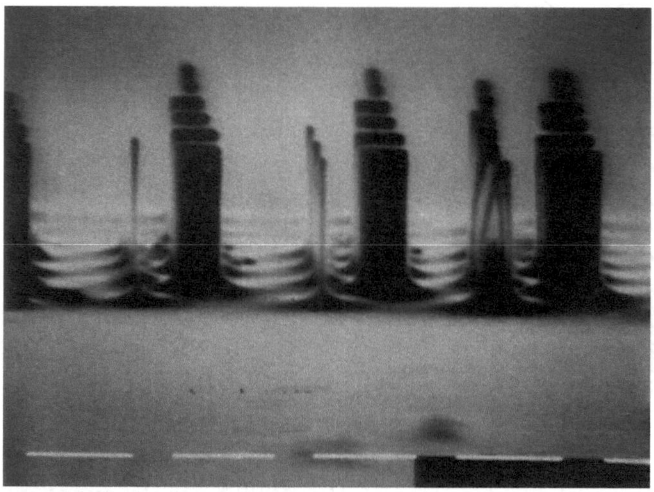

Fig. 1 : Scanning electron micrograph of an array of anisotropically etched columns containing a quantum dot. The horizontal white marker is 0.5 micrometer.

A spectrum of discrete states would be expected to give rise to a series of resonances in transmitted current as each state is biased through the emitter distribution. To observe lateral quantization of quantum well state(s), the physical size of the structure must be sufficiently small that quantization of the lateral momenta produces energy splittings > $3k_BT$. Concurrently, the lateral dimensions of the structure must be large enough such that pinch-off of the column by the depletion layers formed on the side walls of the GaAs column does not occur. Due to the Fermi level pinning of the exposed GaAs surface, the conduction band bends upward (with respect to the Fermi level in the contacts), and where it intersects the Fermi level determines in real space the edge of the central conduction path core. When the lateral dimension is reduced to twice the depletion depth or less, the lateral potential becomes parabolic though conduction through the central conduction path core is pinched off. Observation of tunneling through the discrete states of a quantum dot necessitates fabrication within this design criterion.

Figure 2 shows the current-voltage-temperature characteristics of a quantum dot resonant tunneling structure successfully fabricated within these constraints [6]. The structure lithographically is 1000Å in diameter and epitaxially is a n^+ GaAs contact/AlGaAs barrier/InGaAs quantum well structure. As the temperature is lowered, a series of peaks appears above and below the main negative differential resistance (NDR) peak. In

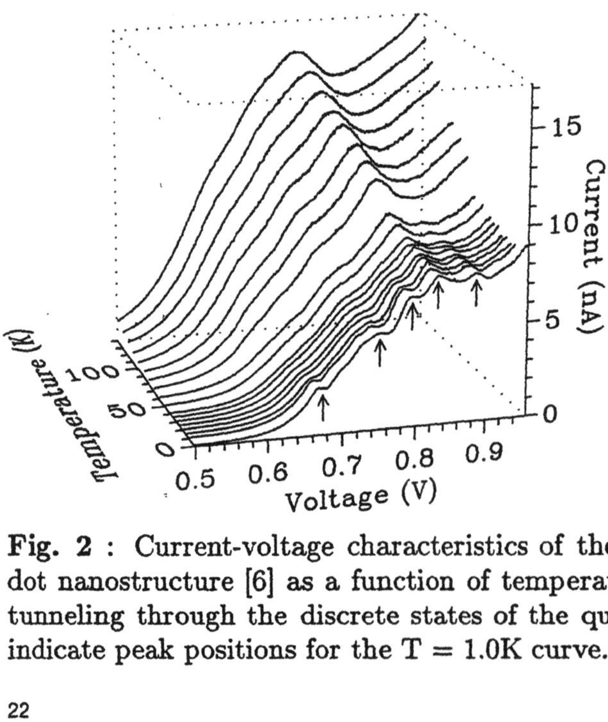

Fig. 2 : Current-voltage characteristics of the 1000Å single quantum dot nanostructure [6] as a function of temperature, indicating resonant tunneling through the discrete states of the quantum dot. The arrows indicate peak positions for the T = 1.0K curve.

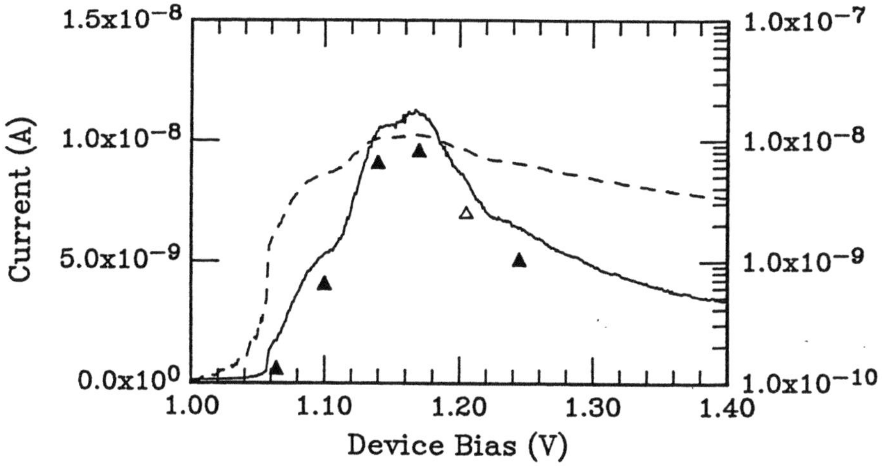

Fig. 3 : Current-voltage characteristics of a 900Å single quantum dot nanostructure at 4.2K. The arrows indicate peak positions.

the range of device bias 0.75V – 0.90V, the peaks appear equally spaced with a splitting of approximately 50 mV. Another peak occurs 80 mV below the equally spaced series.

Figure 3 shows the current-voltage characteristics of another quantum dot structure (in this case the quantum well contains no indium). The epitaxial structure consists of a (nominal) 65Å barrier, 44Å quantum well structure, with 150Å spacers. The characterization of the epitaxial structure is detailed elsewhere [8]. The radius of the column is 900Å. Peak splittings on the order of 35 – 40 meV are observed.

The existence of the fine structure in the tunneling characteristics of these laterally confined resonant tunneling structures indicates the formation of laterally confined electronic states. However, a full indexing of the spectrum is needed. To do this, a full 3D screening model of the quantum dot system is necessary.

3. 3D Resonant Tunneling Spectroscopy

We have modeled the full screening potential of the quantum dot system taking into account the effects of lateral confinement [9]. The model self-consistently obtains the electrostatic potential in a zero-current theory from Poisson's equation utilizing a Thomas-Fermi approximation for the electron density. The solution of the electrostatic problem then provides the potential responsible for lateral quantization of electron states. The radial bound states in the contacts provide the minima of the emitter and collector subbands. Likewise the discrete quantum well levels are obtained from a solution of the radial Schrödinger equation.

The boundary conditions necessary for a solution to the quantum dot screening potential are considerably more complicated than for the 1D problem. To obtain a boundary condition in the contact regions for the full quantum dot system, we first do a 1D self consistent calculation for the radial direction. The boundary conditions for this calculation are zero field at the origin and a Neumann condition at the external surface set by the Fermi level pinning surface charge. It is assumed that there is negligible variation of the potential in the vertical direction in the vicinity of where the contact boundary conditions are to be imposed. We assume a constant density of surface states per unit area, occupied according to a Fermi-Dirac distribution, with the value of the Fermi level pinning acting as a local chemical potential. The calculation adjusts the occupation of surface states to self-consistently achieve a constant vertical, zero bias, surface potential.

The equilibrium solution to the 3D screening problem of the experimental situation of Figure 2 is displayed in Figure 4. The electron potential energy surface is plotted as a function of radius (R) and epitaxial

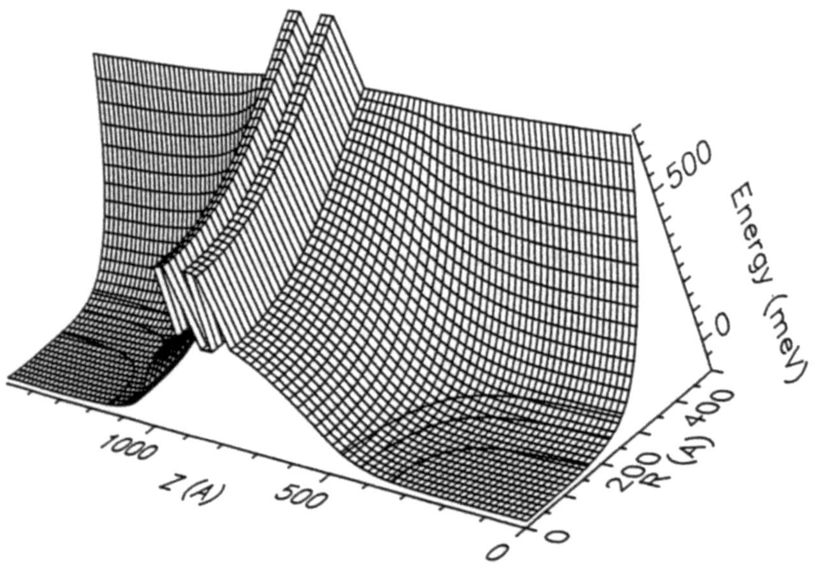

Fig. 4 : Self-consistent 3D band diagram of the single quantum dot structure (Figure 2), at equilibrium. The electron potential energy surface is plotted as a function of radius (R) and epitaxial (z) dimensions. The contours in the contact regions are the occupied laterally-defined subbands.

(z) dimensions. The energy scale is defined relative to the Fermi energy, thus the potential at the external radius equals 0.7V. The contours in the contact regions are the three occupied laterally-defined subbands that lie below the Fermi level. For clarity, the quantum dot energy levels are not drawn in Figure 4.

It is clear that the lateral depletion has a dominant effect on lifting the double barrier structure significantly above the level previously determined only by the z-doping profile. The relevant quantum dot states, determined by solving the radial Schroedinger equation, arise from the previous quantum well ground state ($n_z = 1$). The excited state ($n_z = 2$) quantum dot states are virtual.

It has been suggested [9] that the observed quantum dot spectrum can be explained as resonances when the quantum dot states are biased through the emitter subband states with increasing device bias. To determine if this mechanism quantitatively explains the spectrum, we solve the 3D self-consistent screening quantum dot model at applied bias, to determine the variation of the emitter and quantum dot energy levels with applied voltage.

Figure 5 shows the crossings of the emitter subband levels (n') with the quantum dot levels (n) as a function of applied bias, transposed onto the 1.0K slice of Figure 2. There is general agreement between the experimental and predicted peak voltage positions, especially the anomalously large splitting of the first resonance. It should also be noted that the experimental measurement is current, which implies that an integration over the density of emitter states should be done for a strict

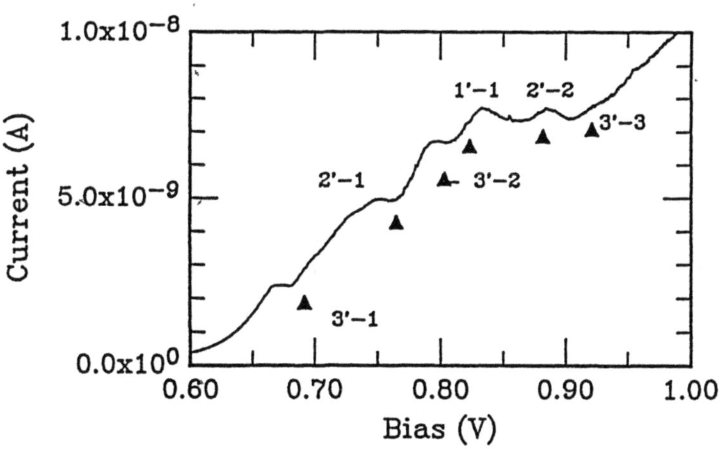

Fig. 5 : Current-voltage characteristic at T=1.0K of the single quantum dot (Fig. 2) with predicted resonant peak positions due to crossing of the laterally confined emitter (n') and dot (n) states.

comparison. The predicted $3' - 3$ transition appears to be absent in the spectrum, except for a very weak structure at 0.92V − 0.93V. However, this is expected since the collector barrier becomes sufficiently low such that the state becomes virtual. This has an important implication - verification that the observed resonances are due to states localized in the quantum dot.

An additional corroboration of this spectroscopy and the peak indexing is found in the temperature dependence of the quantum dot peaks. When the temperature is raised to ≥ 50K, the three lowest voltage peaks disappear and the spectrum is dominated by the single $1' - 1$ transition. This is expected when the subband spacing is less than $3k_BT$. In the high temperature limit, the structure emulates an unconfined 1D resonant tunneling diode, with the resonance determined by the $1' - 1$ crossing.

The observation of the momentum-nonconserving transitions ($n' \neq n$) show that $\hat{n}$ is <u>not</u> a conserved quantity in this quantum dot system. This is due to the hourglass topography of the electron energy surface determined partly by the z-dependent doping profile. This absence of a n, n' (radial) selection rule is natural from the radially changing, cylindrically symmetric geometry that breaks translational symmetry.

4. Quantum Molecules

The low temperature limit of these structures has been shown to be due to the thermal distribution of the laterally confined emitter states. An intriguing situation is to filter this distribution with another quantum dot, to achieve an input distribution as (theoretically) narrow as the probed states. The fabrication of such structures is relatively straightforward, in that the coupling of the two dots (i. e., a "molecule") is controlled epitaxially.

The starting epitaxial material is a double quantum well, triple barrier structure designed to have approximately coincident resonances in one bias direction. The structure is superficially similar to that of Nakagawa *et al.* [10] except that the coupling barrier is made thinner than the quantum well-to-contact tunnel barriers; 60Å outside $Al_{.4}Ga_{.6}As$ barriers, and 50Å and 65Å GaAs quantum wells separated by a 35Å $Al_{.4}Ga_{.6}As$ barrier. Lateral fabrication is the same as before.

Figure 6 shows the current-voltage-temperature characteristics one this molecule structure (radius of 650Å). NDR is clearly evident at low temperatures (peak-to-valley current ratios greater than 1.2:1 at T≤20K), which persists up to 200K. The splittings of the 6 obvious peaks are 75mV : 100mV : 75mV : 65mV : 125mV (lowest to highest).

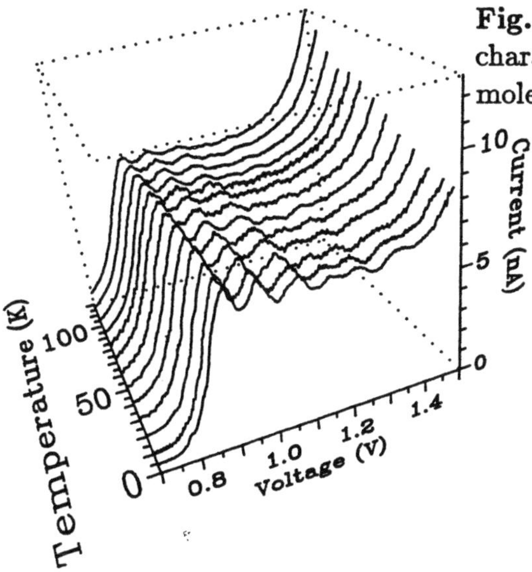

Fig. 6 : Current-voltage-temperature characteristics of the quantum molecule.

The situation is more complex than for the single dot case. For this experimental configuration, there are 5 lateral subbands occupied. If one naïvely treats the laterally quantized states in the two quantum dots as uncoupled states, and considers the crossings of the states with the emitter subbands, the resulting spectra is very dense with an average splitting of 5mV. Neglecting one of the wells only increases this spectral splitting by a factor of 2.

Instead, the spectrum appears to be explained as the resonant crossings of the (emitter) dot states with the downstream dot states. Since the confining potential of both dots is the same, the lower lying states in the two dots are equally spaced. Thus, there are j accidental degeneracies (j being the number of emitter dot states below the Fermi level).

5. Summary

Quantum dot structures provide a unique laboratory for the exploration of quantum transport through nanostructured semiconductors. The states can be varied with structural variables, allowing for the exploration of the fundamentals of quantum-confined electronic states, transitions, and selection rules involving those states. The ability to introduce an epitaxially-defined coupling parameter provides an intriguing tool into the investigation of quantum-localized electronic states.

6. Acknowledgement

We are indebted to our collaborators W. R. Frensley, R. J. Aggarwal, Y.-C. Kao, R. J. Matyi, T. M. Moore, and A. E. Wetsel. We thank R. T. Bate for constant encouragement and support, and R. K. Aldert, E. D. Pijan, D. A. Schultz, P. F. Stickney, and J. R. Thomason for technical assistance. This work was sponsored in part by ONR, ARO, and WRDC.

References

1. Entry to the literature can be found in: H. Heinrich, G. Bauer, and F. Kuchar, Physics and Technology of Submicron Structures (Springer-Verlag, New York, 1988); and M. A. Reed and W. P. Kirk, eds., Nanostructure Physics and Fabrication (Academic Press, San Diego, 1989).
2. G. Timp, A. M. Chang, P. Mankiewich, R. Behringer, J. E. Cunningham, T. Y. Chang, and R. E. Howard, Phys. Rev. Lett. **59**, 732 (1987).
3. B. J. van Wees, H. van Houten, C. W. J. Beenakker, J. G. Williamson, L. P. Kouwenhoven, D. van der Marel, and C. T. Foxon, Phys. Rev. Lett. **60**, 848 (1988); and D. A. Wharam, T. J. Thornton, R. Newbury, M. Pepper, H. Ahmed, J. E. F. Frost, D. G. Hasko, D. C. Peacock, D. A. Ritchie, and G. A. C. Jones, J. Phys. **C 21**, L209 (1988).
4. W. Hansen, T. P. Smith III, K. Y. Lee, J. M. Hong, and C. M. Knoedler, Appl. Phys. Lett. **56**, 168 (1990); and T. P. Smith III, K. Y. Lee, C. M. Knoedler, J. M. Hong, and D. P. Kern, Phys. Rev. **B38**, 2172 (1988)
5. B. J. Van Wees, L. P. Kouwenhoven, C. J. P. M. Harmans, J. G. Williamson, C. E. Timmering M. E. I. Broekaart, C. T. Foxon, and J. J. Harris, Phys. Rev. Lett. **62**, 2523 (1989).
6. M. A. Reed, J. N. Randall, R. J. Aggarwal, R. J. Matyi, T. M. Moore, and A. E. Wetsel, Phys. Rev. Lett. **60**, 535 (1988).
7. The maximum is $\sim E_F$; see J. G. Williamson, H. van Houten, C. W. J. Beenakker, M. E. I. Broekaart, L. I. A. Spendeler, B. J. van Wees, and C. T. Foxon, Phys. Rev. **41**, 1207 (1990).
8. M. A. Reed, J. N. Randall, J. H. Luscombe, W. R. Frensley, R. J. Aggarwal, R. J. Matyi, T. M. Moore, and A. E. Wetsel, Festkörperprobleme : Advances in Solid State Physics Vol. **29**, ed. U. Rößler (Vieweg, Braunschweig/Wiesbaden 1989).
9. G. W. Bryant, Phys. Rev. **B39**, 3145 (1989).
10. T. Nakagawa, T. Fujita, Y. Matsumoto, T. Kojima, and K. Ohta, Appl. Phys. Lett. **51**, 445 (1987).

Field-Effect Confined Electron Dots

J.P. Kotthaus[1], *A. Lorke*[1], *J. Alsmeier*[2], *and U. Merkt*[2]

[1]Sektion Physik, Universität München, Geschwister-Scholl-Platz 1, D-8000 München 22, Fed. Rep. of Germany

[2]Institut für Angewandte Physik, Universität Hamburg, Jungiusstr. 11, D-2000 Hamburg 36, Fed. Rep. of Germany

Abstract. Periodic arrays of electron dots with typical diameters of 100nm and below are prepared by lateral confinement of two-dimensional electron systems on InSb, Si, and GaAs via field effect. The characteristic excitations of the confined electrons are investigated with far-infrared spectroscopy at low temperatures and in high magnetic fields as a function of electron number. Field effect tuning also allows to observe the transition of isolated dots to an electron mesh.

1. Introduction

The artificial creation of electron dots in semiconductors in which few electrons are confined in all spatial dimensions to lengths in the range of 100nm and below has recently become possible by advances in submicron fabrication technologies [1-3]. Confinement to dots usually starts with heterostructures in which electrons can be confined to a two-dimensional plane. One route to achieve lateral confinement in the two-dimensional plane is by etching away the material around the dots [4-6]. However, the formation of surface states at the exposed edges results in depletion regions of varying size and thus makes it difficult to control the electronic dimensions of the dot, even with perfect geometrical control. In our studies electrons are confined in voltage-tunable potentials employing laterally periodic field electrodes with periods in the submicron range and suitable bias conditions [2,7-8]. The basic concept is that of field-effect devices traditionally used to create and tune two-dimensional electron systems in metal-oxide-semiconductor (MOS) or heterojunction transistors.

The investigation of the characteristic electronic properties of electron dot arrays is possible via tunneling [5] or capacitance [1,3] experiments. Here we examine characteristic electronic excitations of electron dots with far-infrared frequencies at low temperatures [2,7-10]. Application of strong magnetic fields perpendicular to the semiconductor interface allows us to use the cyclotron diameter as a size probe for lateral confinement. In sufficiently strong magnetic fields the electron system again behaves quasi-two-dimensional and exhibits bulk cyclotron-resonance-like excitations as well as edge modes connected with skipping orbit trajectories at the dot perimeter. Thus experimental investigation of the magnetic field dependence of the electronic excitations in dots allows us to infer on the characteristic size and shape of the confining potential as well as on the average electron number N_0 per dot.

Springer Series in Solid-State Sciences, Vol. 97
Localization and Confinement of Electrons in Semiconductors
Editors: F. Kuchar · H. Heinrich · G. Bauer

The study of infrared excitations also makes it possible to observe the transition from arrays of isolated dots to coupled dot arrays, i. e., electron meshes in which dots are connected by narrow constrictions through which electrons may travel ballistically from dot to dot [8]. In the strong coupling limit such meshes transform into lateral superlattices that have long been predicted to exhibit a fascinating spectrum of electronic states [11]. Field-effect controlled electron dot arrays now should make it possible by voltage variation alone to study the whole transition regime from artificial atom-like objects with controllable electron number to two-dimensional crystalline solids consisting of these quasi-atoms. The enormous tunability of these devices promises a rich variety of novel electronic properties that now become experimentally accessible.

2. Experimental notes

Three typical device structures used in our experiments are displayed in Fig. 1. They are fabricated starting with a mask consisting of a periodic array of typical 10^8 photoresist dots which is generated by holographic lithography with an argon laser using double exposure [9,10]. The InSb MOS-device shown in Fig. 1(a) relies on the pinning of the Fermi level in the band gap of InSb by a mesh-like Schottky electrode of NiCr. A semi-transparent uniform gate electrode deposited on top of a SiO_2 oxide fabricated by plasma-enhanced chemical vapor deposition (PECVD) then serves to induce electrons at the InSb-SiO_2 interface below the openings of the NiCr meshes

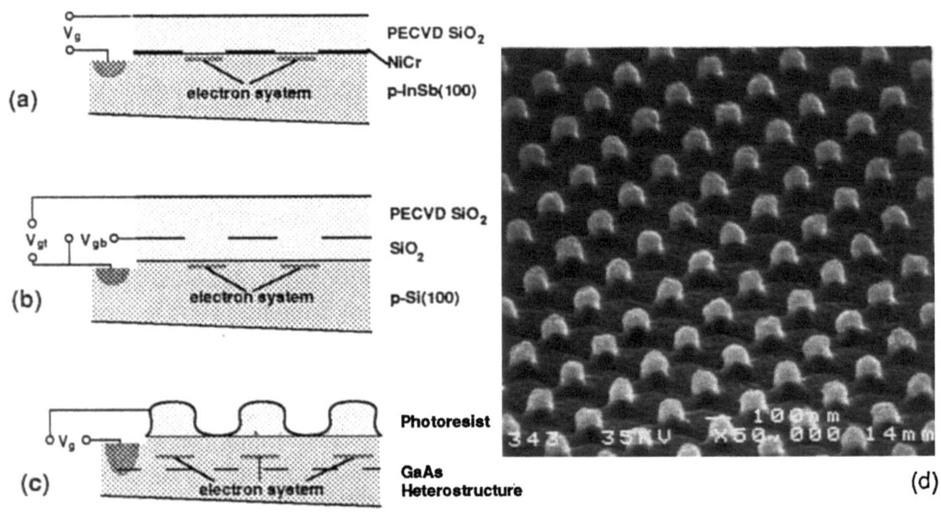

Fig. 1(a-c): Schematic cross sections of field-effect devices used to create electron dots on InSb (a), Si (b), and GaAs (c). Fig. 1(d): SEM picture of an InSb monitor sample showing photoresist dots covered with gold for SEM contrast enhancement. The grating constant is a=250nm, the size of the resist dots about 100nm in diameter.

under suitable bias V_g applied with respect to a substrate contact. A scanning electron micrograph of an InSb device is shown in Fig. 1(d).

Quite similar principles are used in the stacked gate MOS-device on p-Si illustrated in Fig. 1(b) [7]. Here the bottom gate is a NiCr mesh embedded in SiO_2 layers between the Si substrate and the semi-transparent top gate and is negatively biased by V_{gb} to cause electron depletion or hole accumulation below the mesh. With V_{gt} the top gate is positively biased to again induce electrons below the openings of the bottom gate. This versatile device allows nearly independent tuning of both the walls of the confining potential via V_{gb} and the number of confined electrons via V_{gt}. With the bias applied with respect to a substrate contact electrons can only be induced into the dots by generating electron-hole pairs using band-gap radiation. In the dark additional tunability then becomes possible by further raising V_{gt} by a so-called substrate bias voltage V_{SB} which increases the depletion layer and lowers the surface potential well below the band gap of Si without inducing additional minority carriers. Obviously there are other combinations of V_{gt} and V_{gb} that cause different modes of operation such as the generation of electron meshes and hole meshes and give this device enormous flexibility.

The GaAs field-effect transistor shown in Fig. 1(c) is based on a AlGaAs-GaAs heterostructure in which a two-dimensional electron layer is generated close to a AlGaAs-GaAs interface by modulation doping. Here a gate electrode that is periodically displaced via photoresist dots is used with negative bias to deplete electrons from the two-dimensional layer in a mesh-like pattern at those parts of the device where the gate is closest to the heterostructure. Since the GaAs substrate of such heterostructures needs to be semi-insulating for infrared transmission experiments one has to define a suitable ground plane with respect to which the gate bias can be applied, even when electrons are depleted from the inversion layer. Here we use a small bypass conductance in the AlGaAs layer below the electron system to provide such a ground plane as indicated by the heavy dashes in Fig. 1(c).

In our experiments we measure the transmission T of far-infrared radiation at low temperatures in a transmission waveguide in which the dot array completely covers the cross section of the waveguide of typically 3mm in diameter. Using square-wave modulation of the gate voltage with respect to an appropriate threshold voltage V_t at which there are no electrons in the dots we measure the relative change in transmission $-\Delta T/T=\{T(V_t)-T(V_g)\}/T(V_t)$. In the small signal approximation $-\Delta T/T \ll 1$ appropriate here, this quantity directly reflects the real part of the effective dynamic conductivity $\sigma(\omega)$ of the dot array which is proportional to the average areal density of the electrons per unit cell, i.e., $\bar{n}_s=N_o/a^2$. Here a is the period of the array and N_o the number of electrons per dot. The thus obtained transmission signal directly reflects the dynamic conductivity in the dots and also enables us to determine N_o.

3. Theoretical remarks

The dynamic conductivity of an electron dot is expected to exhibit resonances that are related to the energy level structure in the dot. In the following we will discuss the relation between the resonance frequency observed in infrared experiments and the energy level structure assuming for the sake of simplicity a parabolic confining potential in the two-dimensional plane. The energy levels in such a parabolic potential under the influence of a magnetic field B may be written as [12]

$$E_{n,m}=(2n+|m|+1)\hbar\sqrt{\Omega_0^2+\left(\frac{\omega_c}{2}\right)^2}+\frac{\hbar\omega_c}{2}m\ . \tag{1}$$

Here are n = 0,1,... the radial and m = 0, ±1, ... the azimuthal quantum numbers, Ω_0 the eigenfrequency of the harmonic confinement potential at B=0, and $\omega_c=eB/m^*$ the cyclotron frequency. In the absence of collective effects the allowed dipole transitions are

$$\omega_\pm=\sqrt{\Omega_0^2+\left(\frac{\omega_c}{2}\right)^2}\pm\frac{\omega_c}{2}\ . \tag{2}$$

The magnetic field dependence of the energy spectrum and allowed transitions are displayed in Fig. 2 using parameters that are appropriate for typical quantum dots on InSb.

In the above discussion we have neglected many electron effects that shift the B=0 infrared transition energy $\hbar\omega_0$ from the level spacing $\hbar\Omega_0$: In the simplest approximation these may be thought of arising from depolarisation effects [4] . If we characterize the restoring force provided by depolarisation by a frequency ω_{dep} we expect

$$\omega_0^2=\Omega_0^2+\omega_{dep}^2\ . \tag{3}$$

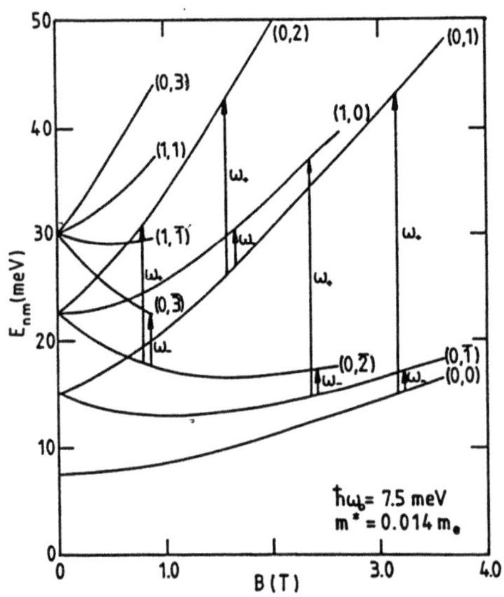

Fig. 2: Magnetic field dependance of the quantum levels in a parabolic confining potential according to Eq. (1). Arrows indicate allowed transitions ω_+ and ω_-. (From Ref.2)

Thus collective effects cause the infrared transition frequency to lie above the characteristic frequency of the confining potential. Depending on the relative importance of confinement vs. depolarisation the infrared response of an electron dot can be dominated either by quantization as is the case for recent experiments on InSb [2] or by depolarisation as is the case for large dots on GaAs [4]. In either case we expect the infrared frequency to be shifted to values above Ω_0 but to extrapolate to Ω_0 for vanishing electron density. A quite different and illuminating interpretation of the infrared frequency has recently been given by Brey et al. [13]. They point out that in parabolically confined systems the infrared frequency ω_0 measures the characteristic frequency of the bare confining potential whereas the level spacing $\hbar\Omega_0$ reflects the characteristic frequency of the screened potential. The surprising result is that the infrared frequency ω_0 thus is a direct measure of the bare potential which by the properties of the parabolic confinement coincides with the characteristic frequency of the screened potential properly corrected by depolarisation as given in Eq. (3). Though strictly true only for parabolic confinement this seems the most adequate interpretation of the experimental results obtained on electron systems that are electrostatically confined to two, one, or zero dimensions.

Since the infrared resonance energy does not immediately allow to decide whether classical or quantum confinement is dominant in a given electron dot one may use the magnetic field dependance of the resonance lineshape as an indicator for which type of confinement is dominant. This is illustrated in Fig. 3, which compares

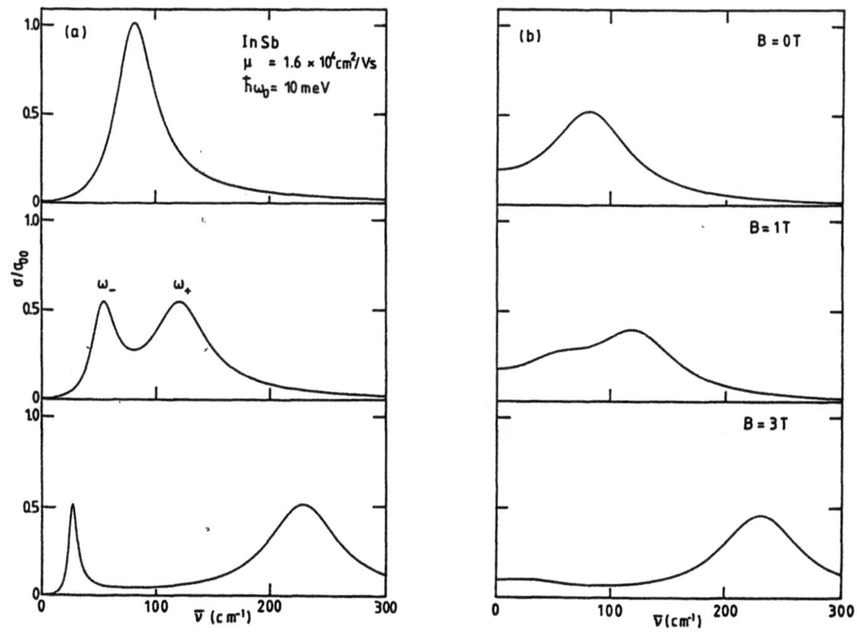

Fig. 3: Dynamic conductivity of electron dots on InSb as calculated with a classical (a) and a quantum mechanical (b) model, respectively. (From Ref. 10)

the dynamical conductivity in an electron dot on InSb calculated classically [14] with one obtained from the Kubo formula [15]. In both cases the magnetic field dependance of the oscillator strengths is $S_+/S_- = \omega_+/\omega_-$. For the classical case one expects the resonance amplitude of the ω_- and the ω_+ mode to be the same and independent of magnetic field. In contrast, the quantum mechanical calculation, assuming a constant phenomenological relaxation time, gives essentially the same halfwidths but different amplitudes for the ω_- and ω_+ mode. As will be shown below such a prediction is at least in qualitative agreement with experimental results representative for both regions of confinement.

4. Quantum dots versus classical discs

In the following we briefly summarize representative experimental results that fully cover the regime between quantum dots and classical quasi-two-dimensional electron discs. For a given confinement diameter quantization is strongest for electrons with a low effective mass. Thus it has been no surprise that large quantum effects have first been observed on InSb [2]. Since for a given electron number a low effective mass also yields a high dynamic conductivity $\sigma(\omega)$ in a dot, InSb is also best suited to observe infrared resonances of dot arrays with very low electron numbers down to $N_0=1$. The magnetic field dependance of the resonance energy, extracted from spectra taken on a dot array with period a=250nm in magnetic field sweeps employing fixed frequency far-infrared lasers, is displayed in Fig. 4. The experimental data can be well described by Eq. (2) if the non-parabolic band structure is introduced via a simple **k·p** estimate. As displayed in the inset the resonance energy was found to stay essentially constant for electron numbers varying between 1 and 20. This result, surprising at first, was originally explained as reflecting that many body corrections to the infrared transition energies are very small in parabolically confined dots [16]. It can be even better explained by that the infrared energy is a measure for the bare confining potential which is expected to hardly change as the electron number N_0 increases.

Dot arrays that can be tuned from quasi-classical behavior to quantum confinement can well be realized in the stacked gate Si-MOS device described above [7]. Figure 5(a) displays spectra from dot arrays on Si having a period of a=400nm and containing about $N_0 \approx 140$ electrons per dot. At low magnetic fields the spectral line shape deviates from the classical expectation indicated by the dashed line towards the quantum case (Fig. 3(b)). At high magnetic fields where the cyclotron radius becomes significantly smaller than the dot radius more classical behavior is found. The fact that quantum confinement is important in such dots under the bias conditions used in Fig. 5(a) can also be seen if one plots the square of the B=0 resonance frequency ν^2 vs. top gate voltage V_{gt} which controls the electron number N_0. Extrapolating towards vanishing N_0 we find a finite intercept. This is highest for strongest confinement, i.e., large depletion charge as controlled by positive V_{SB} and sufficiently negative V_{gb}. This is safely interpreted as reflecting quantization caused by confinement and yields spacings of neighboring quantum level as large as 9meV.

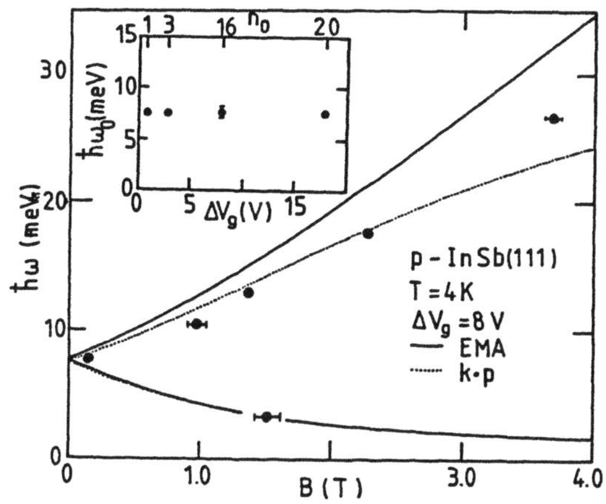

Fig. 4: Magnetic field dependance of the resonance energies in an array of quantum dots on InSb. The solid line is calculated according to Eq. 2, the dotted line takes non-parabolicity effects into account. The inset displays the dependence of the B=0 resonance energy on electron number. (From Ref. 10)

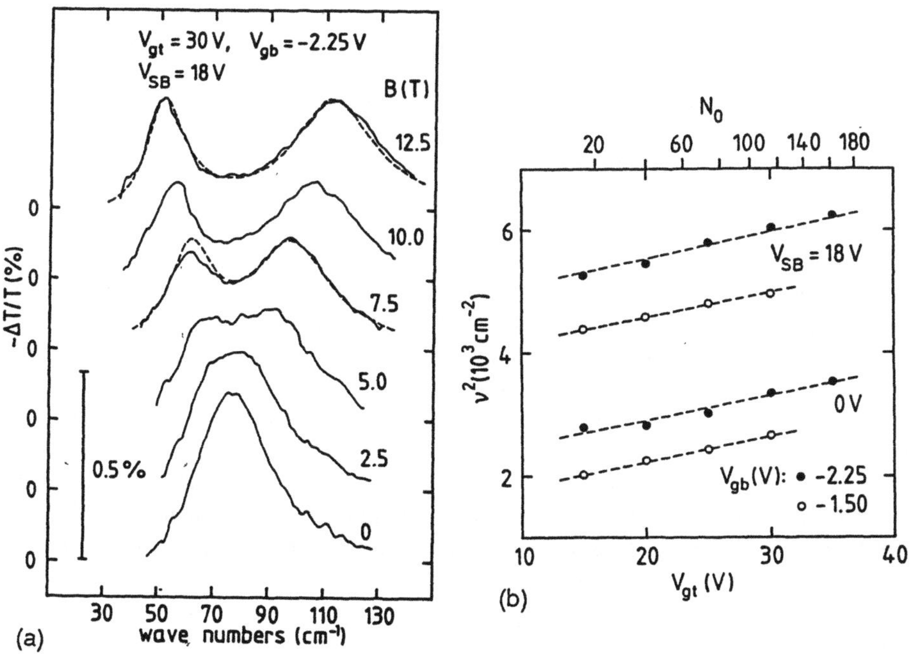

Fig. 5: Infrared excitations in an electron dot array on Si with period a=400nm. (a) Transmission spectra at various magnetic fields B. (b) Squared resonance frequencies at B=0 vs top gate voltage V_{gt}. The electron numbers N_0 displayed on top are only valid for V_{gb}=-2.25V. (From Ref. 7)

Most recently we have also been able to realize voltage-tunable electron dots with diameters around 200nm on GaAs heterostructures. As mentioned above the major problem is to imbed a ground plane between the semi-insulating substrate and the two-dimensional electron system that one wants to electrostatically confine to dots. We solved this problem by using a wide GaAs quantum well of 50nm width imbedded on both sides by a delta-doped AlAs-GaAs short period superlattice. In these structures a high resistance bypass in the lower superlattice serves as ground plane with respect to which a gate voltage can be applied. In such voltage tunable devices we were not only able to study the characteristic infrared modes as a function of N_0 down to $N_0 \approx 50$ but could also observe the transition from isolated dots to coupled dot arrays as discussed in the following.

5. Dot networks

To investigate the electronic excitations in dot networks in which electrons can travel ballistically from dot to dot, GaAs heterostructures are suited best because they can contain electron systems with elastic mean free paths well exceeding 1μm. On gated heterostructures as pictured in Fig. 1(c) and discussed above we were recently able to study the transition from isolated to coupled dots by infrared spectroscopy [8]. In Fig. 6 infrared spectra are displayed at finite magnetic fields and around the gate potential of V_g=-3V where the transition occurs. At V_g=-3.1V where we still have isolated dots we observe the two mode spectrum characteristic for single electron dots. The amplitude ratio indicates importance of quantum effects. At lower gate voltage where we know from transport and capacitance studies that the dots become connected we observe new features in the infrared spectra, namely, a second edge mode and a shoulder on the high energy side of the bulk mode. A detailed analysis of the resonance frequencies, partially presented in Fig. 6(c), reveals that in the coupled dot regime the new edge mode with lower frequency occurs at sufficiently large magnetic fields. The shoulder of the bulk mode appears at even higher magnetic fields. Both features directly reflect the coupling of the dots. The lower frequency edge mode is identified as a perimeter excitation of two adjacent dots with trajectories schematically shown in the inset (mode B) of Fig. 6(c). This mode requires a finite magnetic field such that the perimeter wave in the dot can be transmitted through the constriction between adjacent dots. Qualitatively, the cyclotron diameter needs to be sufficiently small to allow coupling of perimeter modes between adjacent dots. The high frequency mode appearing as a shoulder on the bulk-like dot excitation is shown to be a magnetoplasmon of the coupled dot crystal. For it to become possible complete cyclotron orbits have to fit into the constriction thus requiring an even higher magnetic field for observation.

The data in Fig. 6 demonstrate that infrared spectroscopy not only helps to understand electronic excitations in isolated dots but also can be used as a direct probe for studying extended excitations in coupled dot arrays. For safe identification of the observed modes it seems necessary to have tunable devices such that the whole regime between isolated dots and strongly coupled electron meshes can be investigated simultaneously. Only then a reliable identification of coupled dots modes seems possible.

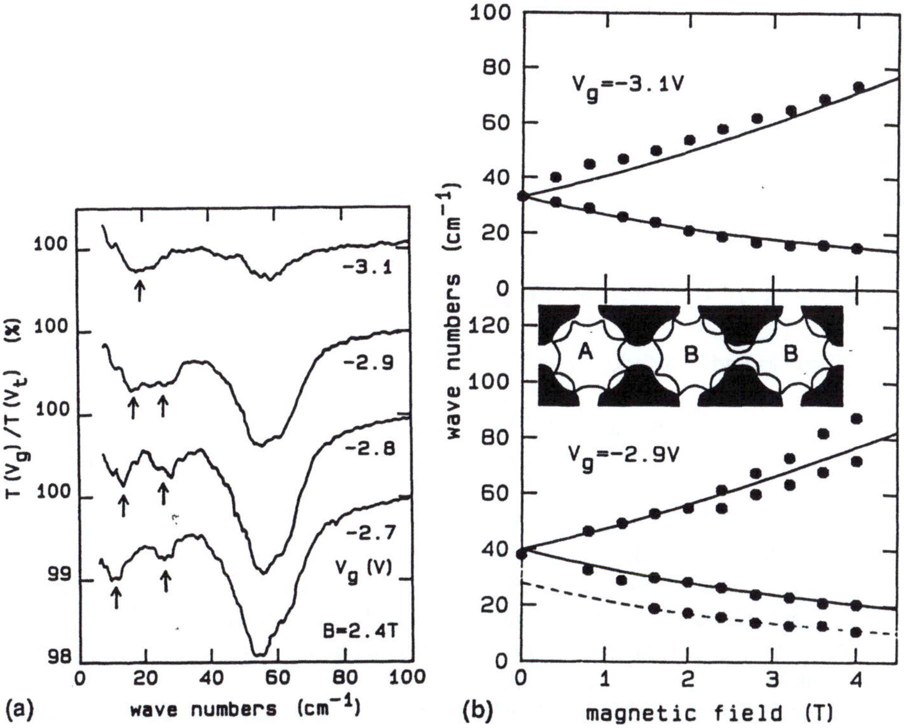

Fig. 6: Infrared resonances in an electron dot array on GaAs with period a=450nm at gate voltages where the coupling between adjacent dots develops. Shown are transmission spectra (a) and resonance positions (b) of isolated and connected dots as a function of magnetic field strength. The lines are calculated according to Eq. (2). The dashed line shows the expected dispersion of a type B trajectory as shown in the inset. (From Ref. 8)

6. Conclusions and perspectives

We have tried to show that field-effect tunable devices are best suited to investigate electronic excitations in electron dot arrays. Now that basic properties are reasonably understood there seem to be two exciting routes for dot physics. One is the field of dot atomic physics in which one wants to realize a periodic table of artificial electron dots and would like to observe fine structures in the spectra that allows the differentiation between, e.g., H-like and He-like dots. With present understanding such fine structures should only appear in dots with a non-parabolic confinement potential. Another exciting route is to study dot solids, i.e., arrays of coupled electron dots that finally become square lateral superlattices with all their intriguing properties [11,17]. For both routes one wants voltage tunable dot arrays that at present are best realized with field-effect devices as discussed above.

The work summarized here results from a large collaborative effort and we want to thank all our collaborators as identified in the original references for their contributions. We gratefully acknowledge financial support from the Volkswagen-Stiftung, the Deutsche Forschungsgemeinschaft and the Esprit Basic Research program.

References.

[1] T. P. Smith III, K. Y. Lee, C. M. Knoedler, J. M. Hong, and D. P. Kern, Phys. Rev. B **38**, 2172 (1988).
[2] Ch. Sikorski and U. Merkt, Phys. Rev. Lett. **62**, 2164 (1989).
[3] W. Hansen, T. P. Smith III, K. Y. Lee, J. A. Brum, C. M. Knoedler, J. M. Hong, and D. P. Kern, Phys. Rev. Lett. **62**, 2168 (1989).
[4] S. J. Allen, Jr., H. L. Störmer, and J. C. M. Hwang, Phys. Rev. B **28**, 4875 (1983).
[5] M. A. Reed, J. N. Randall, R. J. Aggarwal, R. J. Matyi, T. M. Moore, and A. E. Wetsel, Phys. Rev. Lett. **60**, 535 (1988).
[6] D. Heitmann, this volume
[7] J. Alsmeier, E. Batke, and J. P. Kotthaus, Phys. Rev. B **41**, 1699 (1990).
[8] A.Lorke, J. P. Kotthaus, and K.Ploog, to be published
[9] J. P. Kotthaus, Physica Scripta T**19**, 120 (1987).
[10] U. Merkt, Ch. Sikorski, and J. Alsmeier in "Spectroscopy of Semiconductor Microstructures" Eds.: G. Fasol, A. Fasolino, and F. Lugli, Plenum Press, in press.
[11] D. R. Hofstadter, Phys. Rev. B **14**, 2239 (1976).
[12] C. G. Darwin, Proc. Cambridge Phil. Soc. **27**, 86 (1930).
[13] L. Brey, N. F. Johnson, and B. I. Halperin, Phys. Rev. B **40**, 10647 (1989).
[14] B. A. Wilson, S. J. Allen , Jr., and D. C. Tsui, Phys. Rev. B **24**, 5887 (1981).
[15] M. Wagner, unpublished.
[16] A. V. Chaplik, , in Proc. of the 8th Int. Conf. on Electronic Properties of Two-Dimensional Systems, Grenoble, 1989, Surf. Science, to be published.
[17] L. P. Kouwenhoven, F. W. J. Hekking, B. J. van Wees, and C. J. P. M. Harmans, to be published, and L. P. Kowenhoven, this volume.

Quantum Confinement of Excitons by Strain Gradients

K. Kash, B.P. Van der Gaag, J.M. Worlock, A.S. Gozdz, D.D. Mahoney, J.P. Harbison, and L.T. Florez

Bellcore, 331 Newman Springs Road, Red Bank, NJ 07701, USA

Abstract. We describe recent efforts to achieve quantum wire confinement of excitons by strain-patterning GaAs quantum wells, and discuss considerations in the design of structures to achieve large potential wells for electrons. We examine the effect of shear strain on the modulation of the valence band edge and on the optical absorption of wire-confined excitons. Finally, we demonstrate recent success in the quest to observe quantum wire subbands.

1. Introduction

Almost since the first identification of a semiconductor band gap, it has been appreciated that the gap is sensitive to applied stress. Over twenty years ago, strain gradients were used to accelerate and move excitons [1], while shortly later it was noticed that inhomogeneous shear strain could be used to confine excitons to volumes of hundreds of microns in diameter in an indirect gap semiconductor [2]. This latter situation made possible many elegant experiments on excitonic interactions and on electron-hole liquids [2,3]. It was not until 1988, however, that it was proposed [4] that strain patterning could be used to achieve quantum confinement of carriers within a semiconductor.

One motivation for attempts to achieve lateral quantum confinement is anticipation of interesting modification of the optical properties of carriers and excitons in quantum wires and dots. Many interesting phenomena occurring in two-dimensional quantum wells should be transformed upon further change of the dimensionality of the semiconductor system [5]. Densities of states are modified (and become delta functions in the limit of strong three-dimensional confinement). Carrier-phonon coupling is changed as the density of states of carriers is modified, and thus energy relaxation processes are affected. Carrier-carrier interactions are modified both as a result of changes in densities of states and as a result of the modification of interparticle distances. Thus exciton binding energies and particle correlation energies can be strongly affected. In the extreme limit of highly quantum-confined, strongly coupled structures, lateral superlattices are created. Much of the focus of predictions has been on the case of extreme lateral confinement, where the zero-point energy of the carriers is much larger than the binding energy of the exciton, but many interesting possibilities such as large enhancements of third-order optical nonlinearities

Springer Series in Solid-State Sciences, Vol. 97
Localization and Confinement of Electrons in Semiconductors
Editors: F. Kuchar · H. Heinrich · G. Bauer

may actually occur for the regime of weak or moderate lateral confinement [5].

The class of III-V semiconductors, and especially the GaAs-AlGaAs system, is ideal for studying such effects not only because the exciton is relatively large (and thus lateral confinement effects are correspondingly easier to achieve), but also because it is one of the best-controlled systems with regard to growth and doping, and the best studied with respect to optical properties of planar quantum wells. A variety of methods for creating quantum wires or dots in GaAs-AlGaAs in particular have been under investigation in the last few years.

It is worthwhile to put strain confinement into the context of other patterning techniques [5]. Efforts to achieve lateral confinement of excitons in III-V semiconductors fall into two broad categories, that of patterned growth and that of post-growth patterning of planar quantum wells. The first includes growth of "tilted superlattices" and other structures on planar, vicinal substrates as well as growth of planar quantum wells on patterned substrates. The second category includes patterned intermixing, damaging (for example by ion implantation), or etching of planar layers, as well as strain patterning. As will be discussed in the conclusions, strain patterning offers some unique advantages over many of these other techniques, as well as its own peculiar limits.

2. Exciton Trapping: Modulation of the Conduction Band

Our technique for creating lateral confinement of excitons is shown in Fig. 1. A highly compressed layer of some material is deposited onto the semiconductor that we wish to pattern. (This material may be any of a wide variety, including a pseudomorphically strained semiconductor [6], a polycrystalline material, or an amorphous layer.) The compressed layer is typically on the order of 100 nm thick or less, and thus is three or four orders of magnitude thinner than the typical semiconductor substrate. The deformation is therefore entirely in the thin stressor layer, and the forces are directed in the plane of the layer, as shown in Fig. 1a.

To make a wire, we mask the segment shown in Fig. 1a and etch away the rest of the stressor layer. The resulting relaxation of the stressor is accompanied by a volume deformation of the underlying semiconductor, shown in the contour plot of Fig. 1b. The material under the center of the stressor is dilated; that near the edges is compressed. The stress integrated across this interface is zero, and thus the semiconductor is unstrained far from the stressor. For this example, the dilation decays by a factor of 2.5 in a distance of half the wire width below the center of the stressor. It should be emphasized that decay lengths for the strain modulation are comparable to the width of the stress modulation at the interface, regardless of the details of the structure that produces the strain modulation. This characteristic requires that the strain-modulated structure be located quite close to the stress-modulating interface, on the scale of the lateral extent of the stress modulation.

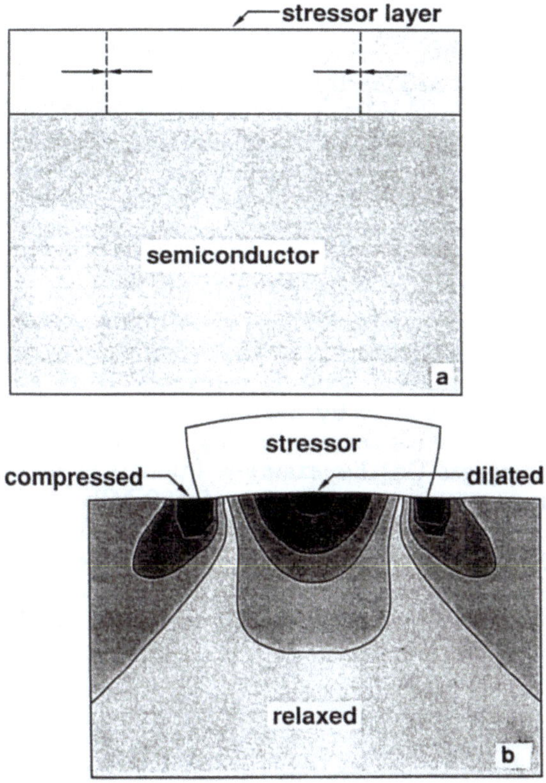

Fig. 1. (a) Stressor layer before patterning and etching. Forces are directed parallel to the interface, as shown by arrows, and the substrate is undeformed. (b) Contour plot of volume dilation of substrate resulting when a rectangular cross section wire is patterned and etched from the stressor layer. Contours are at equal intervals of the dilation. Portions of the semiconductor under the wire edges are under compression; the portion under the center is dilated. The deformation is exaggerated here. From calculations similar to those discussed in [7].

The pattern shown was generated by numerical integration of the equations of elasticity, using the anisotropic elastic coefficients of GaAs and a relatively incompressible, isotropic stressor layer [7]. The results apply equally well to the example described or to a structure created from two materials with dissimilar thermal expansion coefficients, which are put together unstressed at some elevated temperature, then cooled.

The modulation of the spherically symmetric conduction band edge states is given by $D_c\Delta$ where D_c is the hydrostatic deformation potential of the conduction band and $\Delta=\epsilon_{yy}+\epsilon_{zz}$ is the volume dilation (the wire axis being oriented along the x axis). The maximum dilation in the semiconductor occurs at the stressor-semiconductor interface at the center of the wire. Thus a potential well for excitons is created under the stressor by virtue of the conduction band modulation. As we will see later, the stress pattern generated in this way does not necessarily tend to confine holes to the center of the wire. (This situation is related to the earlier work on confinement of excitons and electron-hole liquids in Si and Ge. There, a potential well for excitons was also created by conduction band modulation, but in that case it was not by hydrostatic strain, but rather by shear strain, which lifted the degeneracy of the lowest energy Γ or X conduction band minima [2,3].)

Evidence for the creation of a lateral potential well for excitons in our structure is shown in Fig. 2 in a comparison of photoluminescence spectra from patterned and unpatterned regions of a GaAs quantum well [7,8]. Here a single, 9 nm GaAs quantum well is separated from an amorphous carbon layer of 160 nm thickness by a 20 nm thick AlGaAs barrier of composition 0.3 Al. The carbon layer was created by RF plasma deposition from a carbon-containing gas (in this example, butane) at high substrate/target voltage to produce a stress on the order of 500 MPa, or 5 Kbar. Wires were defined by electron beam lithography and formed by oxygen reactive ion etching. The photoluminescence spectrum of the well patterned by an array of 350 nm wide wires reveals a band gap that is shifted from that of an adjacent, unpatterned region by 31 meV—the depth of the exciton potential well.

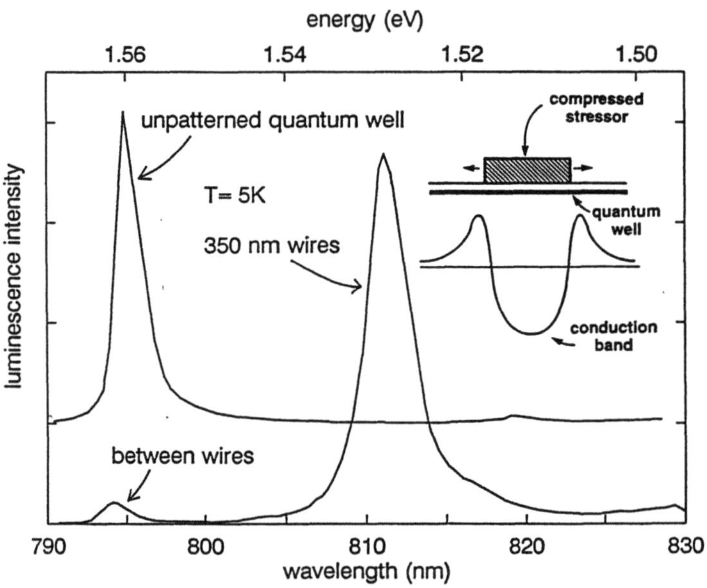

Fig. 2. Comparison of the luminescence spectra of an unpatterned, 9 nm wide GaAs quantum well with an upper barrier of 20 nm $Al_{0.3}Ga_{0.7}As$, and the same well patterned with an array of 350 nm wide, 160 nm tall amorphous carbon wires, showing the red-shifted luminescence of excitons trapped under the wires [8].

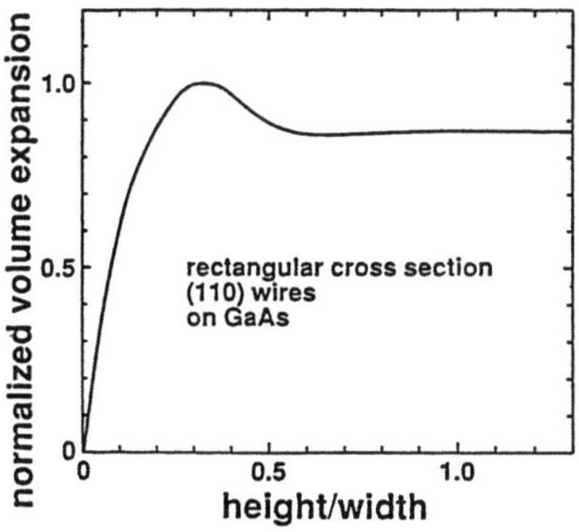

Fig. 3. Normalized volume dilation in GaAs at the center of the stressor-semiconductor interface versus the ratio of wire height to width. The normal to the interface is oriented along [100], and the wire axis along [110]. From calculations similar to those discussed in [7].

How large are the subband splittings for electrons in such a finite, near-parabolic potential well, shown in the inset to Fig. 2? From the curvature we estimate a subband spacing of approximately 1 meV. We note that this splitting should scale inversely with the structure size and as the square root of the potential well depth.

The shape of the strain-induced potential well depends, of course, on the elastic properties of the materials, the shape of the stressor cross section, and the crystallographic orientation of the wire axis. If we fix these parameters, the pattern will scale linearly with the magnitude of the initial biaxial stress and the size of the stressor. In Fig. 3, we plot the calculated dilation under the center of a wire of rectangular cross section as a function of its aspect ratio. (This calculation was made using the elastic properties described for Fig. 1.) In general, there is an optimal ratio; for smaller ratios of height to width the maximum dilation occurs not at the center but near the edges of the wire, creating a double potential well for electrons. For taller-than-optimal wires the deformations at opposite edges interfere destructively, and the upward bowing of the stressor is suppressed. This example is shown for wire axes oriented along the [110] crystallographic axis of GaAs; the normal to the interface is the [100] axis. For wires oriented instead along a [100] axis, we predict a maximum dilation of a factor of 1.35 greater. This result is directly related to the orientation dependence of Poisson's ratio for the crystal [9].

3. Properties of The Exciton: The Valence Band

Because the valence band edge of the III-V semiconductors has p symmetry, it is modulated by shear as well as by hydrostatic strain. The shear deformation potentials, which describe distortions along either the [110] or the [111] crystalline directions, are in general larger than the valence band

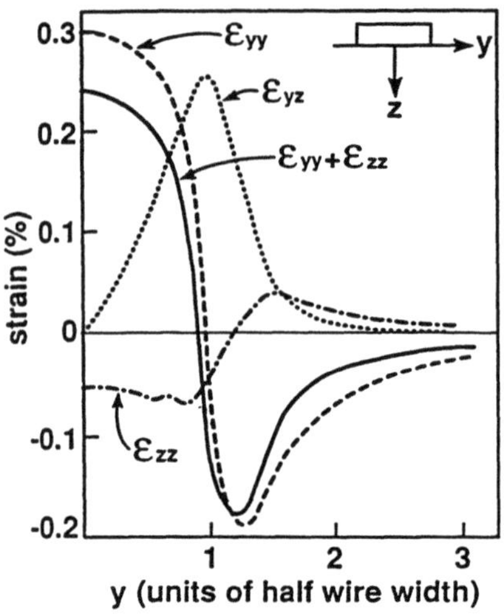

Fig. 4. Components of the strain tensor at the stressor-semiconductor interface of the wire shown in Fig. 1 [7]. There are large shear terms $\epsilon_{yy} - \epsilon_{zz}$ under the center of the stressor (y=0) as well as large shear terms ϵ_{yz} at its edges (y=1).

hydrostatic deformation potential [10]. For the structure of Fig. 1, the shear terms can be as large as the peak hydrostatic strain at the interface (see Fig. 4.). Thus, it is primarily the shear strain terms, rather than the hydrostatic strain, that will control the modulation of the valence band in our structures. This situation allows the modulation of the conduction and valence bands to be separated to some extent. Furthermore, as will be seen later, it leads to the prediction of some very complicated effects on the exciton wave function due to the spatial variation of the mixing of heavy and light holes induced by the shear strain.

The chosen orientation of both the wire axis and the normal to the interface along [100] minimizes piezoelectric effects, directing the strain-induced polarization along the axis of the wire. We will neglect piezoelectric effects, of great potential interest for other structures, in what follows.

It is straightforward (but tedious) to obtain the valence band edge as a function of position. We define the wire axis along x, the growth direction, or normal to the quantum well, along z, and orient both of these axes along [100] crystallographic directions. The symmetries of the problem require that ϵ_{xx}, ϵ_{xy}, and ϵ_{xz} be zero.

The strain Hamiltonian is [11,12]

$$D_v \epsilon_{ii} + b\left(J_i^2 - \frac{J^2}{3}\right)\epsilon_{ii} + d\left(J_iJ_j + J_jJ_i\right)\epsilon_{ij}$$

where cyclic permutations of the subscripts are assumed. D_v, b, and d are the valence band hydrostatic deformation potential and the two shear deformation potentials, respectively. We find that the band edges vary as

$$\delta E_{\pm} = D_v \Delta \pm \sqrt{\alpha^2+\beta^2+\gamma^2}$$

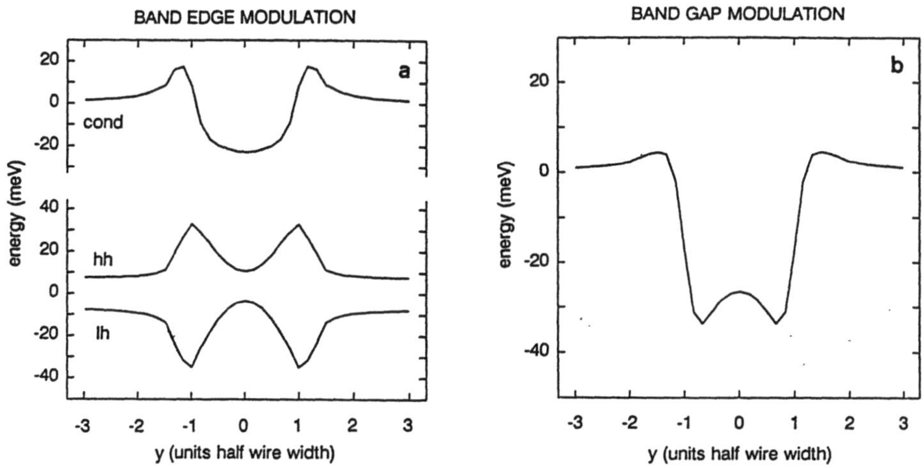

Fig. 5 (a) Calculation of the band edge modulation of the structure of Fig. 1 at the interface, showing "heavy hole" maxima at the edges of the wire, and (b) the band gap modulation (the exciton potential well) of this structure. Note the double minimum of the lateral potential well, and the small barriers to lateral trapping of the excitons generated outside the wire.

where

$$\alpha = \frac{A}{2} + b\left(\frac{\epsilon_{yy}}{2} - \epsilon_{zz}\right)$$

$$\beta = \frac{\sqrt{3}}{2} b\epsilon_{yy}\left(1 + 2\left(\frac{\epsilon_{yz}d}{\epsilon_{yy}b}\right)^2\right)^{\frac{1}{2}}$$

and

$$\gamma = \frac{\sqrt{3}}{2} d\epsilon_{yz}.$$

Here A is the splitting of the light and heavy hole.

We use values of 1.1, -2 and -5 eV for the deformation potentials D_v, b and d. The band edges are shown in Fig. 5a as a function of lateral position within the quantum well. The strain in the plane of the well, ϵ_{xx}, is chosen to be 0.4 % at the center of the wire, and the splitting of the heavy and light hole in the unpatterned well to be 14 meV. The conduction band shows a potential well at the center of the wire and potential barriers at the compressively stressed edges. The minimum of the higher-lying hole band occurs at the edges, rather than at the center, of the wire. This band is labeled "heavy hole" in the figure because of its origins, but it must be kept in mind that the strain strongly mixes heavy and light holes.

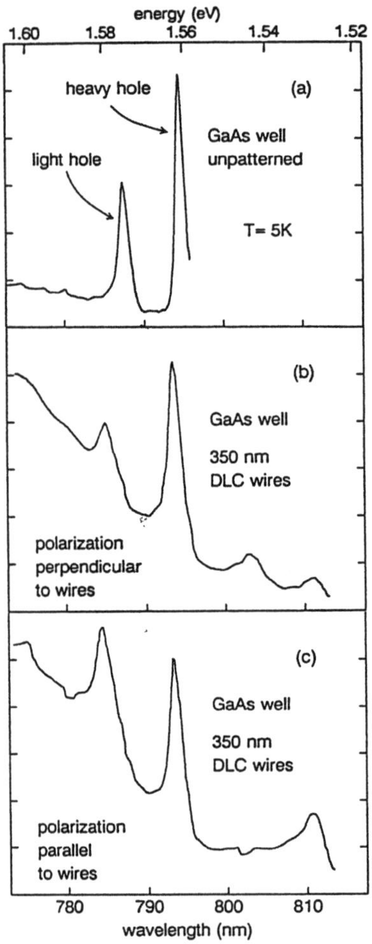

Fig. 6. Excitation spectra of the sample of Fig. 2, showing (a) the light and heavy hole n=1 excitons of the unpatterned quantum well, (b) the two lower energy, wire-confined exciton peaks, as well as the light and heavy hole exciton peaks from regions between the wires, and (b) and (c) the polarization selection rules of the excitons trapped in the wire that result from the strong mixing of the light and heavy hole by the anisotropic strain. [8].

For structures much larger than the diameter of the exciton in the unpatterned well, the potential well for the exciton is given by the band gap variation. The well for the "heavy hole" exciton, shown in Fig. 5b, is almost square, with a shallow double minimum. The behavior of the heavy hole minima near the eges of the wire almost completely offsets the lateral barrier in the conduction band. Hence, the lateral barriers to exciton trapping are small.

The effect of the mixing of heavy and light holes by strain is seen in excitation spectra, Fig. 6, of the emission shown in Fig. 2 [8]. The n=1 light and heavy hole peaks of the unpatterned well are resolved in Fig. 6a. The excitation spectrum of the patterned region, Fig. 6b, shows four exciton peaks. The two higher energy exciton peaks arise from absorption of light by excitons created in the slightly compressively stressed regions between the wires. The two lower energy peaks are the two n=1 quantum well excitons trapped in the wires. The large polarization anisotropy of the "wire

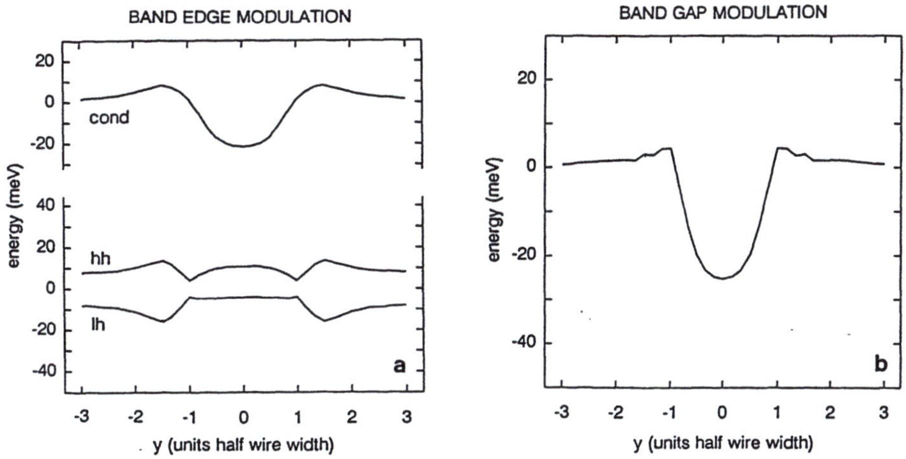

Fig. 7. Calculations of the modulation of (a) the band edges and (b) the band gap of the structure of Fig. 1 at a depth of 1/4 of the wire width from the interface. The magnitude of the strain has been adjusted to give the same band gap shift at the center of the wire as in Fig. 5. Note the almost parabolic potential well created for the exciton at this depth.

excitons", Fig. 6b and 6c, is well explained by calculations of the strong mixing of the light and heavy hole by the anisotropic strain at the center of the wire. Although the exciton absorption is thus strongly modified by the strain pattern, the 5 meV exciton linewidth is much larger than the 1 meV estimated subband splitting due to the lateral confinement, and quantum wire subbands are not resolved.

It is intriguing to note the way in which one may modify the exciton wave function as one scales down the size of the strain pattern of Fig. 5a. When the structure is small enough that the effective field can polarize the exciton, the heavy hole exciton becomes spatially indirect, with the hole confined at the edge of the wire. This system is thus the quasi-one-dimensional analogue to the two-dimensional indirect exciton system that is found in type II quantum wells and in type I coupled wells under an applied electric field [13,14].

If one removes the quantum well from just under the stressor-semiconductor interface to a distance equal to one-quarter of the wire width, the valence band edge modulation and band gap variation look quite different from the first example (Fig. 7). The exciton potential well is almost harmonic. Since the hole band edge modulation is small, even for highly confining structures the hole trapping into the wire will occur via the Coulomb attraction to the electron, analogous to hole confinement in small CdSe crystallites [15]. The envelope wave functions of the excitons should be very different for the two examples of Figs. 6 and 7. Their calculation, which is non-trivial, must incorporate the spatial variation of the hole mixing.

4. Observation of Quantum Wire Subbands

The first observation of quantum wire subbands produced by strain confinement is shown in the excitation spectrum of the wire-confined excitons shown in Fig. 8. The structures are similar to those that yielded the excitation spectra of Fig. 5, but the linewidth of the heavy hole exciton in the unpatterned 12 nm wide GaAs well is sharper (1.5 meV), and the wires are narrower (180 nm). Three, possibly four subbands are resolved in the splitting of the lower energy wire-confined exciton. These splittings are approximately 2.3 meV, which is the magnitude expected for this particular structure. This is an impressively large subband splitting when one considers that it is created by a stressor of width 180 nm and a potential well depth of only 22 meV compared to the band gap of the unpatterned well of 1.545 eV.

This result encourages one to think that with smaller width wires and larger stresses one may hope to attain the regime of large, as well as moderate, quantum confinement by strain patterning. Amorphous carbon has been deposited with stresses of 5 GPa, or 50 kbar [16]—ten times that used in the examples here, and of the same order as that which requires the use of a diamond anvil cell. The wire width may also be made much smaller than 180 nm. However, the requirement of a sufficiently thick barrier layer between the quantum well and the carbon layer puts a lower limit on how close to the surface one may place the quantum well, and thus puts a lower limit on the size of stressors that will be effective in straining the well.

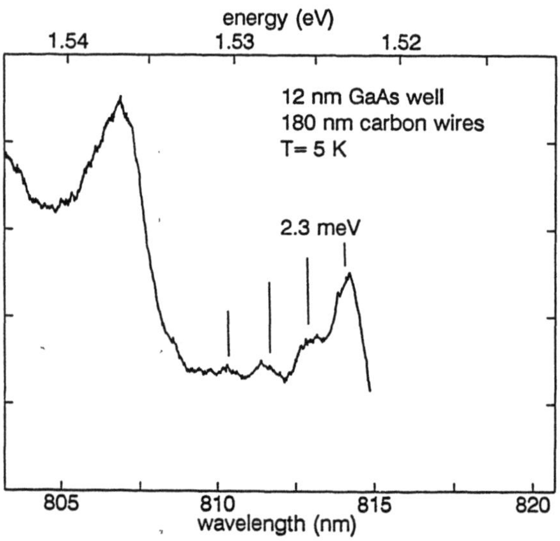

Fig. 8. Observation of quantum wire subbands in the excitation spectrum of a 12 nm wide GaAs well, patterned by 180 nm wide, 100 nm thick wires oriented along [100]. The well is capped by a 20 nm thick AlGaAs barrier, with 0.3 Al composition, and a 30 nm GaAs cap layer.

5. Conclusions.

As pointed out in the introduction, strain confinement is only one of many techniques for making quantum wires and dots. Patterned etching to produce lateral quantum confinement of excitons has met with great success recently [17], as has growth of quantum wells on patterned substrates [18]. Furthermore, in light of recent advances in patterned growth on planar, vicinal substrates [19-21], it seems that control of growth to the point of being able to create uniform arrays of highly laterally confined or strongly laterally coupled structures is likely. It may be difficult to do the same with strain gradients.

However, there are a few qualities that are, at least in combination, unique to strain confinement. The type of strain patterning shown here allows a great deal of freedom in the design of the shape of potential wells. In particular, it is as straightforward to produce quantum dots as it is to make quantum wires, and this is unlikely to be the case for patterned growth. It is also possible to design electron and hole potential wells that are very different, perhaps to the extent of forming spatially indirect excitons. Furthermore, strain confinement leaves the quantum well continuous and undamaged, unlike most of the post-growth patterning techniques mentioned earlier. In addition, it can be equally well applied to a number of semiconductor systems (not just the III-V materials).

We have demonstrated lateral quantum confinement, including resolved quantum wire subbands, by patterned strain. This result makes it possible to look for a large number of interesting phenomena in the regime of moderate lateral quantum confinement of excitons. Furthermore, with the use of higher strains and narrower wires, it should be possible to use strain confinement to achieve strong lateral quantum confinement of excitons.

References

[1] V.S. Bagaev, T.I. Galkina, O.V. Gogolin and L.V. Keldysh, Pis'ma Zh. Eksp. Teor. Fiz. **10**,309 (1969) [JETP Lett. **10**,195 (1969)].

[2] J.P. Wolfe, R.S. Markiewicz, C. Kittel and C.D. Jeffries, Phys. Rev. Lett. **34**,1292 (1975).

[3] See, for example, J.P. Wolfe and C.D. Jeffries, in *Electron-hole Droplets in Semiconductors* (North-Holland Press, Amsterdam, 1983), p.431, and R. S. Markiewicz and T. Timusk, *ibid.*, p. 543.

[4] K. Kash, J.M. Worlock, M.D. Sturge, P. Grabbe, J.P. Harbison, A. Scherer and P.S.D. Lin, Appl. Phys. Lett.**53**,782 (1988).

[5] Many (but not all) important references to experimental and theoretical work are listed in the review paper by K. Kash on optical properties of III-V semiconductor quantum wires and dots, to be published in J. Luminesc.

[6] K. Kash, R. Bhat, Derek D. Mahoney, P.S.D. Lin, A. Scherer, J.M. Worlock, B.P. Van der Gaag, M. Koza and P. Grabbe, Appl. Phy. Lett. **55** ,681 (1989).

[7] K. Kash, J.M. Worlock, Derek D. Mahoney, A.S. Gozdz, B.P. Van der Gaag, J.P. Harbison, P.S.D. Lin and L.T. Florez, Surf. Sci. **228**, 415 (1990).

[8] K. Kash, J.M. Worlock,A.S. Gozdz, B.P. Van der Gaag, J.P. Harbison, P.S.D. Lin and L.T. Florez, to be published in Surf. Sci. **229**, April 1990.

[9] Robert M. Jones, *Mechanics of Composite Materials* (Hemisphere Publishing Corporation, New York,1975), p. 37.

[10] *Landolt-Boernstein Numerical Data and Functional Relationships in Science and Technology*, Volume III/17a of *Physics of Group IV Elements and III-V Compounds*, O. Madelung, ed. (Springer-Verlag, Berlin-Heidelberg, 1982).

[11] G.E. Pikus and G.L. Bir, Fiz. Tverd. Tela **1**, 1642 (1959). [Sov. Phys. Solid State B1, 136 (1959).]

[12] W.H. Kleiner and L.M. Roth, Phys. Rev. Lett. **2**, 334 (1959).

[13] T. Fukuzawa and T.K. Gustafson, to be published in IEEE J. Quantum Electron.

[14] J.E. Golub, K. Kash, J.P. Harbison and L.T. Florez, to be published in Phys. Rev. Rapid Commun.

[15] A.I. Ekimov, Al.L. Efros, M.G. Ivanov, A.A. Onuschenko and S.K. Shumilov, Solid State Commun. **62**, 465 (1987).

[16] G. Gille and B. Rau, Thin Solid Films **120**, 109 (1984).

[17] M. Kohl, D. Heitmann, P. Grambow and K. Ploog, Phys. Rev. Lett. **63**, 2124 (1989), and D. Heitmann, this volume.

[18] E. Kapon, D.M. Hwang and R. Bhat, Phys. Rev. Lett. **63**, 430 (1989).

[19] M. Tsuchiya, J.M. Gaines, R.H. Yan, R.J. Simes, P.O. Holtz, L.A. Coldren and P.M. Petroff, Phys. Rev. Lett. **62**, 466 (1989).

[20] Masaaki Tanaka and Hiroyuki Sakaki, Appl. Phys. Lett. **54**, 1326 (1989), and H. Sakaki, this volume.

[21] S.J. Allen, Jr., P. Bastos, H.M. Cox, F. DeRosa, D.M Hwang and L. Nazar, J. Appl. Phys. **66**, 1223 (1989), and H.M. Cox, private commun.

Dynamic Excitations of Quantum Dots in AlGaAs-GaAs

T. Demel, D. Heitmann, P. Grambow, and K. Ploog

Max-Planck-Institut für Festkörperforschung, Heisenbergstr. 1,
D-7000 Stuttgart 80, Fed. Rep. of Germany

Abstract: We review far infrared (FIR) experiments on very small quantum dot structures containing down to only 25 electrons per dot on discrete quantized energy levels. These structures have been prepared by deep-mesa-etching of modulation-doped AlGaAs-GaAs heterostructures. The FIR response exhibits a set of resonances which split in a magnetic field into branches with positive and negative B-dispersion. We will explain this FIR response starting from two different approaches: (*i*) from a classical 2D electron plasma of finite size and (*ii*) from an atom-like system with discrete energy levels where, if several levels are occupied, transitions are strongly influenced by collective effects.

In modern semiconductor heterostructures, quantum well systems and metal-insulator semiconductor devices it is possible to realize two-dimensional electronic systems (2DESs) in the sense that a strong confinement potential for electrons or holes in the direction perpendicular to the layered structure, which will be called the z-direction in the following, leads to quantized energy levels E_z^k. Such 2DESs have been fabricated with extremely high qualities and have initiated an enormous amount of fundamental research and many interesting applications. Currently there is an increasing interest in the preparation and investigation of systems with further reduced dimensionality where, due to an ultrafine lateral confinement, the original free dispersion of the electrons in the lateral direction is also quantized. A potential acting in the x-direction gives rise to a 1DES with an energy spectrum $E^{ik}(k_y) = \hbar^2 k_y^2/2m^* + E_x^i + E_z^k$. Ultimately, for a confining potential in both the x- and y-directions, quantum dots, artifical "atoms" with totally discrete energy spectra, $E^{ijk} = E_x^i + E_y^j + E_z^k$, are formed (e.g., Refs. 1-7).

We will review here far infrared (FIR) experiments on quantum dot structures which have been prepared by deep-mesa-etching [8] of modulation doped $AlGaAs - GaAs$ heterostructures [7]. Various samples have been prepared where the numbers of electrons per dot, N, was varied, down to $N = 25$. The dynamic FIR response shows a set of resonances which split in an external magnetic field B into two sets of branches with positive and negative B-dispersion. We will explain this excitation spectrum starting from two different approaches, (*i*) from an edge magnetoplasmon model of a classical 2DES of finite size, (*ii*) from an "atom"-like model. We will discuss that the optical excitations in such an atom, occupied by several electrons, are strongly influenced by collective effects, which tend to hide the internal energy structure of the atom. In our structures we are able to resolve additional excitation modes and resonant anti-crossing behaviour, which gives insight into the internal energy structures of the atom.

Quantum dot samples were prepared by starting from modulation doped $AlGaAs - GaAs$ heterostructures. An array of photoresist dots (with a period of $1000nm$ in both the x- and y-directions) was prepared by a holographic double exposure. Using an anisotropic plasma etching process, rectangular $200nm$ deep grooves were etched all the way through

Editors: F. Kuchar · H. Heinrich · G. Bauer

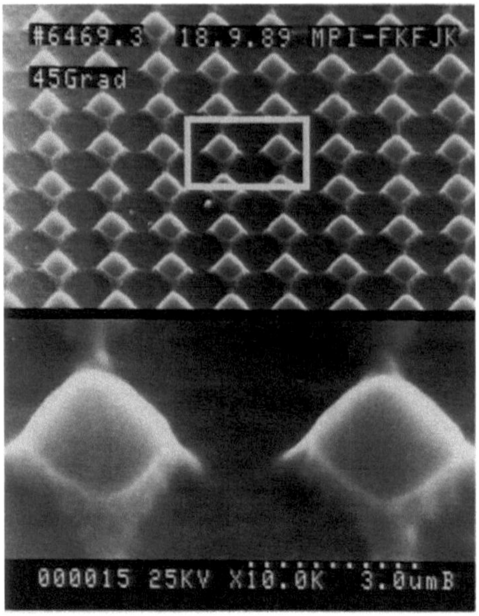

Fig. 1 Electron micrograph of deep mesa-etched quantum dot structures in $AlGaAs - GaAs$. The geometrical size of the dots is $600nm \cdot 600nm$ with rounded corners of radius $R = 100nm$. Electrons are confined in the center of the box by the positive donors in the $AlGaAs$ and by negatively charged surface states at the edges.

the $10nm$ thick $GaAs$ cap layer, the $53nm$ thick Si-doped $AlGaAs$ layer, and the $23nm$ thick undoped $AlGaAs$ spacer layer into the active $GaAs$, leaving quadratic dots with rounded corners and geometrical dimensions of about $600nm$ by $600nm$ (see Fig. 1 and inset of Fig. 2). With this technique of 'deep mesa etching' it was possible to realize 1DESs in linear stripe systems with typical energy separation for the 1D subbands of about $2meV$.[8] It was found that the actual width of the electron channels was smaller than the geometrical width, indicating a lateral edge depletion of 100 to $120nm$. For the dot structures here with increased etched surface area, this depletion is even more pronounced. Actually, we have prepared samples which have, in the dark, no mobile electrons. Via the persistent photoeffect we could then increase the number of electrons per dot, N, in steps up to $N = 210$. We determine N from the strength of the FIR absorption at high magnetic field using a Drude-type model for bound electrons in Ref. 9, formula 7. The potential, that confines the electrons and thus determines the radius of the 2D disc, depends on the remote ionized donors and, in a self-consistent way, on N. We have estimated the radius from the observed resonance frequency and formula (1), which will be explained below. DC transport is inherently not possible in our quantum dot structures, thus we cannot, as in the case of 1DES, determine the level spacing directly from magnetic depopulation. However, we can assume that the energy levels in the dots have a similar separation to those in the 1D wires [8] which were prepared with exactly the same methods. Thus the dots have a discrete energy spectrum with typically 1 to 2 meV energy separation.

The FIR experiments were performed in a superconducting magnet cryostat, which was connected via a waveguide system to a Fourier transform spectrometer. The transmission $T(B)$ of unpolarized FIR radiation through the sample was measured at fixed magnetic fields, B, oriented normally to the surface of the sample. The spectra were normalized to a spectrum $T(B_0)$ with a flat response. The resolution of the spectrometer

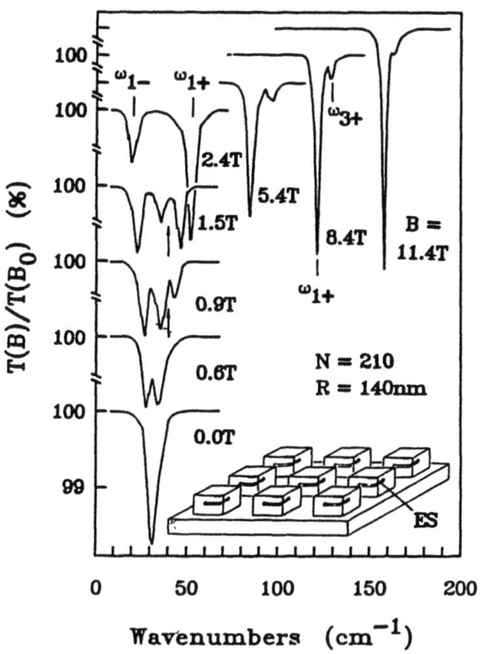

Fig. 2 Relative transmission of unpolarized FIR radiation for a dot structure with radius $R = 140nm$ and $N = 210$ electrons per dot. In a magnetic field $B > 0$ the resonance splits into two branches, ω_{1+} and ω_{1-} with positive and negative B-dispersion, respectively. For $B > 4T$ an additional resonance, ω_{3+} is observed. At $\omega \approx 40cm^{-1}(\uparrow)$ there is a pronounced anti-crossing intersection. The inset shows schematically the dot structures. The actual dots have rounded corners. ES denotes the confined electron system.

was set to $0.5cm^{-1}$. The temperature was $2.2K$. The measured sample area was 3×3 mm^2, thus we measure 10^7 dots.

Experimental spectra for a sample with a dot radius $R = 140nm$ and $N = 210$ electrons per dot are shown in Fig. 2. For $B = 0$ one resonance is observed at $\omega_{01} = 32cm^{-1}$. With increasing B the resonance splits into two resonances, one, ω_{1-}, decreases in frequency, the other, ω_{1+}, increases. For $B > 4T$ a second resonance, ω_{3+}, can be resolved which also increases with B. The most interesting observation is the obvious resonant coupling which occurs at a frequency of about $40cm^{-1}$. Experimental resonance positions for three samples are depicted in Fig. 3. Figure 3b shows the resonance positions from a sample containing 80 electrons per dot, Fig. 3c a sample with only 25 electrons per dot. For this latter sample the dip in $T(B)/T(B_0)$ is only 0.4% which is the experimental limit to determine reliably position and signal strengths for the current spacings of the dots. Here we cannot clearly resolve the ω_{1-} branch because of the limited sensitivity of our spectrometer at small frequencies. Dispersions similar to those in Fig. 3 were measured on a series of samples with slightly different values of R, N and corresponding ω_{01}. The interesting observation is the significant resonant anti-crossing at $\omega \approx 1.4\omega_{01}$, which was found for all our samples.

The FIR resonances observed here are, except for the resonant splitting and the higher modes very similar to earlier observations on larger finite-sized 2DES in GaAs-heterostructures (dots with $R \approx 1.5\mu$m) [9]. Even for electron-systems on LHe with large radii $R \approx 1cm$, a similar dispersion and higher modes but no anti-crossings were observed.[10,11] The experiments in Refs. 9 - 11 have been interpreted as depolarisation, or, equivalently, as edge magnetoplasmon modes [12 - 15]. However, recently also for quantum dots in the $InSb$ systems containing only a very small number of electrons per dot ($N < 50$) resonances with the same dispersion as our ground modes $\omega_{1\pm}$ have been

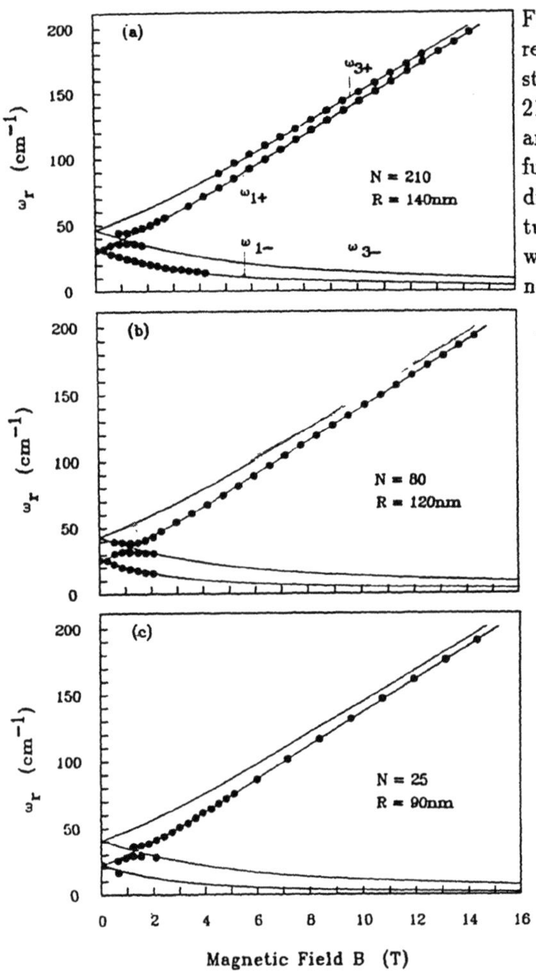

Fig. 3 Experimental B-dispersion of resonant absorption in quantum dot structures with $R = 140nm$ and $N = 210$ (a), $R = 120nm$ and $N = 80$ (b), and $R = 90nm$ and $N = 25$ (c). The full lines are fits with the theoretical dispersion (equation (2)). All structures show an anti-crossing of the ω_{3-} with the ω_{1+} mode, which is caused by nonlocal interaction.

observed, but without anti-crossing.[6] For our quantum-confined systems here and the experiments in Ref. 6 one would expect, at a first glance, that the adequate description of the FIR response are transitions between the discrete energy levels of the "atoms". However, it is known from FIR experiments (e.g. Ref. 16) on 1DES with quantum confined states and from theory [17,18], that the optical response of such low dimensional systems is, with increasing number of electrons, strongly influenced by collective effects which gives the excitations increasingly the character of local plasmon resonances. To elucidate this behaviour for quantum dot structures and to explain the experimentally observed excitation spectrum we will approach this problem from two limits: (*i*) plasmon type of excitation in a 2DES of finite size and (*ii*) transition in an "atom" including collective effects.

A simple way to describe the FIR response of a disc-like 2DES is to start from linear edge plasmons (e.g. Refs. 12,14), i.e., excitations which exist at an edge of a semi-infinite 2DES. These excitations have the dispersion $\omega_{ep}^2 = 0.81\omega_p^2(q)$ where $\omega_p^2(q) =$

$N_s e^2 q/2m^* \epsilon_0 \epsilon_{eff}$ is the 2D plasmon frequency. For a disc the circumference quantises the q-vectors in values $q = i/R \quad (i = 1, 2, ...)$ thus that at $B = 0$

$$\omega_{0i}^2 = 0.81 N_s e^2 i / 2m^* \epsilon_0 \epsilon_{eff} R \quad . \tag{1}$$

In a magnetic field one calculates a set of double branches (e.g. Refs. 12-15)

$$\omega_{i\pm} = \sqrt{a_{i1}\omega_{0i}^2 + a_{i2}(\omega_c/2)^2} \pm a_{i3}\omega_c/2 \tag{2}$$

with $a_{i1} = a_{i2} = a_{i3} = 1$ (ω_c is the cyclotron frequency). This model already agrees very well with more sophisticated theories which determine more accurately the coefficients a_{ik} $(k = 1, 2, 3)$ of ω_{0i} and ω_c. The exact values of the coefficients and the spacings of higher modes depend slightly on the exact modelling of the 3D density profile for the electron distribution. For example, the squared frequency ratios at $B = 0, \omega_{02}^2/\omega_{01}^2$ $(\omega_{03}^2/\omega_{01}^2)$, are 2 (3) for a step-like density profile [15] and 3/2 (15/8) for a density profile $n(r) = n_0\sqrt{1 - r^2/R_0^2}$ $(r \leq R_0)$, which is the classical self-consistent solution of a parabolic confinement potential (n is the density, r the radial distance) [19].

We have fitted in Fig. 3 the dispersion with formula (2), using a_{21} as a fitting parameter and all other $a_{ik} = 1$. Taking into account the nonparabolicity of m^* in *GaAs*, we find a very good agreement with the experimental dispersion. From arguments given below we believe that the confining potential for our structures has a nearly parabolic form. Hence coupling for higher modes occurs dominantly with the $\omega_{3\pm}$ modes.[19] We thus label in the following the higher mode $\omega_{3\pm}$. From the fit we find $\omega_{03} \approx 1.5\omega_{01}$, which agrees well with the expected frequency ratio $\sqrt{15/8}$ for a parabolic confinement. (However, it is also close to the step-like density profile value of $\omega_{02} = \sqrt{2} \cdot \omega_{01}$.) From the fit it becomes clear that the splitting is caused by an anti-crossing of the ω_{3-} and the ω_{1+} mode. The interesting question is which interaction causes this splitting and what determines its strength. This anti-crossing is not expected in the models discussed above (Eq. (2)) since the modes ω_{i+} and ω_{i-} represent, respectively, left and right circularly polarized eigenmodes, which are as such decoupled.

A first possible reason for the splitting is a geometrical effect in the sense that our dots do not have a perfectly circular shape. This would, in principle, enable a mode coupling. However, we do not believe that this is a significant effect, since (i) such effects are independent of the dimensions and should thus be also present in the experiments of Ref. 9 where samples of a very similar shape show no splitting; (ii) for our samples one would expect that a splitting due to the geometrical shape effect should be especially pronounced at large N and R where the electronic system extends closer to the geometrical edge. The surprising observation is that the experimental interaction strengths, expressed by the relative splitting $\Delta\omega/\omega$, increase with decreasing radius.

These observations, and the fact that this splitting is not observed on larger *GaAs*-systems with a very similar shape [9], in particular not on the LHe-systems of Ref. 10 where beautifully sharp intersecting edge magnetoplasmons were found with, however, no interaction at all, leads to the conclusion that the smallness of our structures is the important parameter. This automatically draws the attention to nonlocal interaction. Nonlocal effects are well-known for homogeneous 3DES and 2DES. They arise from the inherent finite compressibility of the Fermi gas, the "Fermi pressure", and lead to corrections $f \cdot q^2 v_F^2$ (v_F = Fermi velocity) for the squared plasma frequency, where the prefactor f is 3/5 in 3DES and 3/4 in 2DES [20]. A microscopic description of the effects, which we here call "nonlocal", is that plasma oscillations are not only driven by Coulomb interaction, but also by the Fermi velocity of the individual electrons. These effects become important at small plasmon wavelengths and correspondingly high values of q. A famous manifestation of nonlocal interaction is observed if photons are incident on a boundary of a semi-infinite 3DES. Within a classical, local description the continuity conditions require an infinitely high, δ-shaped density profile at the boundary. The Fermi pressure counteracts this δ-shaped profile which in turn induces a coupling of the incident transverse photons with longitudinal plasmons [21]. Directly related with these effects are also the widely discussed "additional boundary conditions" (ABC) [22]. Thus if a plasma is modulated or confined on a very small scale, nonlocal effects become important and a new type of mode coupling is possible. In homogeneous 2DES nonlocal effects are small under usual experimental conditions because the relevant parameter, a^*q, is small as compared to 1 (a^* is the effective Bohr radius). However, nonlocal interaction in 2DES can be clearly observed in resonant magnetic field experiments [23] where it leads to an anti-crossing of the 2D magnetoplasmon dispersion with harmonics $n \cdot \omega_c$ of the cyclotron frequency.

For 0DES there is so far no theoretical treatment of the resonant nonlocal interaction. Only non-resonant nonlocal effects on the linear 1D edge magnetoplasmon dispersion have been considered until now [14,15]. Therefore, it is interesting to compare the 2D nonlocal interaction strength with the splitting that is observed here. The strength of the nonlocal resonant coupling in 2D, measured in terms of the frequency splitting, is $\Delta\omega/\omega_0 = 2.6\sqrt{q}$ (q in units of nm^{-1})[23]. If we use the same model for the interaction here and for $q = 1/R$ we find $\Delta\omega/\omega_{01} = 0.27$ for the sample of Fig. 3c with $R = 90nm$, which agrees surprisingly well with the experimental value of 0.30. This very close agreement might be accidental to a certain degree and should be compared with a so far not available rigorous theory. In particular we expect that the nonlocal interaction also depends on the exact three-dimensional density profile of the electron system. However, this close agreement demonstrates that the experimentally observed splitting is of the expected order for nonlocal interaction. Moreover, independent of the absolute value we expect within our simple analogy $\Delta\omega/\omega_{01} \propto \sqrt{1/R}$. If we scale the experimental splitting of 0.30 for the sample of Fig. 3c ($R_1 = 90nm$) with $\sqrt{R_1/R_2}$ we find 0.24, which agrees very well with the experimental splitting of 0.23 for the sample in Fig. 3a ($R_2 = 140nm$). Thus, the dimensionality dependence also confirms that the interaction is caused by nonlocality. (Note, for the $R = 1500nm$ $GaAs$ structures in Ref. 9 we expect $\Delta\omega/\omega_{01} = 0.07$ with $\Delta\omega/2\pi = 40$ GHz. This splitting is less than the linewidth and can thus not be observed.)

So far we have approached the dynamic response using the model of a classical 2DES of small size. However, for our quantum dots with only 25 electrons, which occupy discrete energy levels separated by $2meV$, one might well ask whether a classical description is still appropriate or whether a description within an atom-like model is not required. The simplest way to treat a laterally 2D confined system in a magnetic field is to assume a parabolic confining potential, which is actually not a bad assumption, as we will see later. We consider the one-electron Schrödinger equation:

$$H(\vec{r},\vec{p}) = \frac{1}{2m^*}\left[\vec{p} + \frac{e}{c}\vec{A}(\vec{r})\right]^2 + \frac{1}{2}m^*\Omega_0^2 r^2, \qquad (3)$$

where Ω_0 characterizes the confining potential. The energy eigenvalues have been calculated by Darwin [24]

$$E_{n,\ell} = (2n + |\ell| + 1)\cdot\sqrt{(\hbar\Omega_0)^2 + (\hbar\omega_c/2)^2} + \ell\cdot\hbar\omega_c/2 \qquad (4)$$

where n= 0,1,2 ... and $\ell = 0, \pm 1, \pm 2$, are, respectively, the radial and azimuthal quantum numbers. This energy spectrum is shown in Fig. 4a. From calculations of the matrix elements one finds that dipole allowed transitions have transition energies

$$\Delta E^{\pm} = \sqrt{\hbar^2\Omega_0^2 + (\hbar\omega_c/2)^2} \pm \hbar\omega_c/2, \tag{5}$$

which have thus exactly the same B dispersion as the classical lowest plasma modes $\omega_{1\pm}$ in Eq. (2).

What happens now if this external potential Ω_0 is "filled" with more than one electron? From self-consistent band structure calculations for 1DES, e.g. of Laux *et al.* [25], it is known that electron-electron (ee) interaction, i.e., the self-consistent screening, strongly decreases the separation of the subbands. E.g., for the parameters of Ref. 25 one electron per $20nm$ in the $40nm$ wide wire decreases the subband spacing from the unscreened "empty" separation $\hbar\Omega_0 = 5.6meV$ to $\hbar\Omega = 3meV$. This screened one-particle excitation can be determined in the Shubnikov-de Haas type of magnetic depopulation experiment [1,8]. However, is it possible to observe this screened energy difference $\hbar\Omega$ in an optical dipole excitation? The answer for a parabolic confinement is *No* ! It has been shown for 1D [26] and 2D [27,28] parabolic confinement that the Hamiltonian for N electrons, which is the sum of (3) over all N electrons including ee-interaction, separates in a center of mass (CM) motion and into relative, internal motions. This CM motion satisfies exactly the one-electron Hamiltonian (3) and is the only allowed optical dipole excitation. Thus, the optical response of quantum dots with parabolic confinement represents the collective CM excitation at the frequency of the unscreened external potential, i.e., for $B = 0$, at $\hbar\Omega_0$ and not at the screened one-particle energy $\hbar\Omega$. Thus, for a parabolic confinement, dipole excitations are insensitive to ee-interaction and reflect the energy spectrum of the "empty" atom. Shikin *et al.* [29] have drawn the same conclusion from a classical electron system in a parabolic confinement. This quantum mechanical CM motion corresponds thus exactly to the classical dipole plasma mode, the lowest plasmon mode $\omega_{1\pm}$, if we evaluate Eq. (2) for a parabolic confinement. We would like to note here that often (e.g. Refs. 17,18) the optical response is calculated in RPA, where one starts from the screened one-particle spectrum with energy separation $\hbar\Omega$. The optical resonance frequency $\hbar\omega_r = \sqrt{\hbar^2\Omega^2 + \hbar^2\omega_d^2}$ is shifted with respect to $\hbar\Omega$ by the so-called depolarisation shift, which is characterized by $\hbar\omega_d$. For the parabolic confinement, in this language, we thus have that with increasing number of electrons the decrease of the screened one-particle spacing $\hbar\Omega$ with respect to $\hbar\Omega_0$ is exactly cancelled by the increase of the depolarisation shift.

The self-consistent band structure calculations, in particular for split-gate devices, show that the parabolic confinement is actually a very good approximation.[25] This is reflected for our structures by the fact that we observe dominantly only the lowest mode, Eq. (5). From this we can conclude that the unscreened potential $\hbar\Omega_0$ in our samples is approximately given by the optical response frequency of about 2 to $4meV$. On the other hand, having demonstrated that it is possible to realize artificial atoms, one is of course in a next step interested to study the fine structure, internal degrees of freedom and ee-interaction in the atom. The fact that we observe weak higher order modes and resonant coupling, which is not allowed in a strictly parabolic confinement, shows that for our structures weak, but definite nonparabolic terms exist in the potential. This makes our experiments sensitive to ee-interaction and the internal fine structure. Moreover, since the nonparabolic effects are weak, we can treat them on a perturbative level. On this level

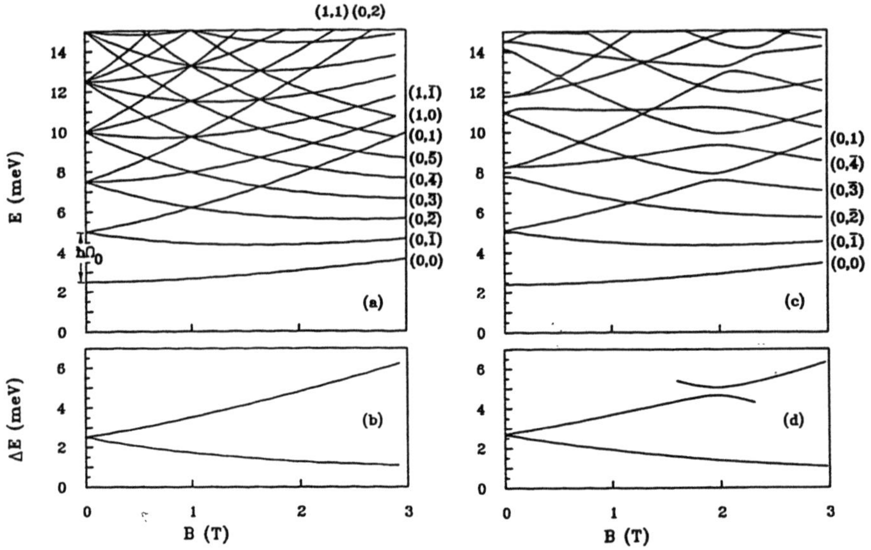

Fig. 4 Calculated one-electron energy spectrum for (a) a parabolic confinement potential and (c) for a potential that includes a noncircular nonparabolicity term, see text. The lower parts show transition energies.

we can still assume that the FIR response reflects the one-electron energy spectrum of the "empty" atom. This allows us to estimate the influence of geometrical effects. We have calculated the one-electron energy spectrum for a potential $V(x,y) = \frac{1}{2}m^*\Omega_0^2(x^2+y^2) + c(x^4+y^4)$, where c was chosen such that the equipotential line at the highest occupied energy level for a system with $N = 25$ extends 0.1 more in the [1,0]-direction then in the [1,1]-direction. The results are shown in Fig. 4c. The degeneracy of crossing energy levels, whose azimuthal quantum numbers differ in the parabolic confinement (Fig.4a) by 4, is now lifted. The experimentally observed anti-crossing corresponds to transitions between the ground state (0,0)and the split energy levels of the (0,1) state. The equipotential lines for the chosen value of c at the highest occupied energy level for a system looks very similar to the geometrical shape of the dots, which is quadratic with rounded corners. Although it is expected that this shape of the confining potential overestimates the nonparabolicity deep inside the dot, the calculated anti-crossing of the (0,1) and the $(0,\bar{3})$ state is not strong enough to reproduce the experimentally observed splitting of $1meV$. This indicates that geometrical effects are <u>not</u> the dominant contributions to our observed anti-crossing. A similar perturbative treatment of the nonlocal effects, the Fermi pressure, which now becomes a "quantum pressure", is a tough problem. Calculations are under way.[28]

In conclusion, we have realized quantum dot structures containing only 25 electrons per dot on discrete energy levels separated by 1 to $2meV$. The confining potential has a nearly parabolic form, which implies that optical dipole excitations are only sensitive to the collective center-of-mass motion. However, due to small nonparabolic terms we are able to observe internal excitations in the artifical atom. In particular, we resolve a resonant anti-crossing, which arises from the quantum-mechanical analogue of the classical Fermi pressure.

We acknowledge financial support by the BMFT.

References

1. K.-F. Berggren, T.J. Thornton, D.J. Newson, and M. Pepper , Phys. Rev. Lett. **57**, 1769 (1986)
2. B.J. van Wees, H. van Houten, C.W.J. Beenakker, J.G. Williamson, L.P. Kouwenhoven, D. van der Marel, and C.T. Foxon , Phys. Rev. Lett. **60**, 848 (1988)
3. M.A. Reed, J.N. Randall, R.J. Aggarwal, R.J. Matyi, T.M. Moore, and A.E. Wetsel, Phys. Rev. Lett. **60**, 535 (1988)
4. T.P. Smith III, K.Y. Lee, C.M. Knoedler, J.M. Hong, and D.P.Kern , Phys. Rev. **B38**, 2172 (1988)
5. W. Hansen, T.P. Smith III, K.Y. Lee, J.A. Brum, C.M. Knoedler, J.M. Hong, and D.P.Kern , Phys. Rev. Lett. **62**, 2168 (1989)
6. Ch. Sikorski and U. Merkt , Phys. Rev. Lett. **62**, 2164 (1989) and J.P. Kotthaus, A. Lorke, J. Alsmeier, and U. Merkt, this volume
7. T. Demel, D. Heitmann, P. Grambow, and K. Ploog , Phys. Rev. Lett. **64**, 788 (1990)
8. T. Demel, D. Heitmann, P. Grambow, and K. Ploog , Appl. Phys. Lett. **53**, 2176 (1988)
9. S.J. Allen, Jr., H.L. Störmer, and J.C. Hwang , Phys. Rev. **B28**, 4875 (1983)
10. D.C. Glattli, E.Y. Andrei, G. Deville, J. Poitrenaud, and F.I.B. Williams , Phys. Rev. Lett. **54**, 1710 (1985)
11. D.B. Mast, A.J. Dahm, and A.L. Fetter , Phys. Rev. Lett. **54**, 1706 (1985)
12. V.B. Sandomirskii, V.A. Volkov, G.R. Aizin, and S.A. Mikhailov, Electrochimica Acta **34**, 3 (1989)
13. J.-W. Wu, P. Hawrylak, and J.J. Quinn , Phys. Rev. Lett. **55**, 879 (1985)
14. A.L. Fetter , Phys. Rev. **B32**, 7676 (1985)
15. A. Fetter , Phys. Rev. **B33**, 5221 (1986)
16. T. Demel, D. Heitmann, P. Grambow, and K. Ploog , Phys. Rev. **B38**, 12732 (1988)
17. W. Que and G. Kirczenow , Phys. Rev. **B38**, 3614 (1988)
18. W. Que and G. Kirczenow , Phys. Rev. **B37**, 7153 (1988)
19. V. Shikin, S. Nazin, D. Heitmann, and T. Demel to be published
20. F. Stern , Phys. Rev. Lett. **18**, 546 (1967)
21. M. Anderegg, B. Feuerbach, and B. Filton , Phys. Rev. Lett. **27**, 1565 (1971)
22. F. Sauter, Z. Phys. 203, 488 (1967) , A.R. Melnyk and M.J. Harrison , Phys. Rev. Lett. **21**, 85 (1968)
23. E. Batke, D. Heitmann, J.P. Kotthaus, and K. Ploog , Phys. Rev. Lett. **54**, 2367 (1985)
24. C.G. Darwin, Proc. Cambridge Philos. Soc. **27**, 86 (1930)
25. S. Laux, D.J. Frank, and F. Stern Surf. Sci. **196** 101 (1988)
26. L. Brey, N. Johnson, and P. Halperin , Phys. Rev. **B40**, 10647 (1989)
27. P. Maksym and T. Chakraborty, to be published
28. R.R. Gerhardts, to be published
29. V. Shikin, T. Demel, and D. Heitmann, Workbook of EP2DS, p. 519, Grenoble (1989) and Surf. Sci. in press

Inelastic Light Scattering by Electrons in Microstructured Quantum Wells

G. Abstreiter and T. Egeler

Walter Schottky Institut, Technische Universität München, Am Coulombwall, D-8046 Garching, Fed. Rep. of Germany

Abstract. Experimental results of electronic excitations in microstructured GaAs quantum wells are discussed. Micro–Raman spectroscopy and grating coupler effects are used to extend the accessible wave vector range of in–plane single particle and plasmon excitations in multi–layered two–dimensional electron systems. In lateral quantum wire structures a strongly anisotropic dispersion of the plasmon resonances is observed, which can be understood in terms of plasmons confined normal to the wires.

1. Introduction

The first observation of inelastic light scattering by free carriers in semiconductors was reported by Mooradian and Wright [1] in doped n–type GaAs more than twenty years ago. In this pioneering work plasmons of a degenerate electron gas, coupled to LO–phonons were studied. Shortly afterwards Mooradian [2] also reported light scattering by single particle excitations in n–GaAs. The early work stimulated extensive light scattering research of electron and hole plasmas in semiconductors (for reviews see Refs. [3,4,5]). Especially resonant inelastic light scattering experiments [6] showed that the method is sensitive enough to observe electronic excitations with densities of the order of 10^{11} cm^{-2}. This was followed by the proposal [7] and the first observations of resonant light scattering by quasi–two–dimensional (2D) electron systems in selectively doped GaAs–(AlGa)As heterojunctions [8] and multiple quantum wells [9] about ten years ago. Numerous publications have appeared since then, which demonstrate the versatility of light scattering for the study of various properties of 2D carrier systems. These include single–particle and collective intersubband transitions in various potential wells which allow a separation of the depolarization shift, 2D plasmons in layered electron systems and in–plane single particle excitations. The light scattering work of 2D–systems of the past ten years has been reviewed for example in [10,11]. So far the dispersion of the different excitations could be studied only in a limited wave vector range ($q \lesssim 1.7 \times 10^5$ cm^{-1}) in normal backscattering geometry from the surfaces. In the present paper we first review shortly new developments in electronic light scattering by 2D systems with large

Springer Series in Solid-State Sciences, Vol. 97
Localization and Confinement of Electrons in Semiconductors
Editors: F. Kuchar · H. Heinrich · G. Bauer

scattering wave vectors which become accessible by micro–Raman scattering [12] and by using grating coupler effects [13,14]. In the last part we concentrate on first light scattering results in wire structured electron systems in GaAs quantum wells [15,16].

2. Electronic excitations in systems of different dimensionality

Elementary excitations of free carrier systems fall into two main categories: collective excitations and single particle excitations. Resonant electronic light scattering allows a separation of the two kinds of excitations by analyzing the polarization of incident and scattered light. Polarized spectra (incident and scattered light polarized parallel to each other) usually display collective excitations, i.e. plasmons which are coupled to LO–phonons in polar semiconductors. Single–particle excitations are observed in crossed polarizations. They are characterized by uncorrelated excitations of electrons below the Fermi energy to empty states above the Fermi energy. In backscattering geometry the scattering wave vector is given by $q \simeq 2 \cdot 2\pi\eta/\lambda$, which is about $7 \cdot 10^5$ cm^{-1} for GaAs in the red spectral range (η is the refractive index and λ the laser wavelength). It leads to the kind of excitations shown in Fig. 1a for an isotropic 3D–electron system. The cutoff frequency of the single particle excitations is at about $q \cdot v_f$, where v_f is the Fermi velocity. In 3D they have approximately triangular lineshape. The plasmon frequency is proportional to $\sqrt{n_{3D}}$, with n_{3D} being the three–dimensional carrier concentration. It exhibits a weak dispersion which in random phase approximation (RPA) is given by $\omega_p^2(q) = \omega_p^2(0) + 3q^2 v_f^2/5$.

In 2D–systems, which are realized for example in narrow quantum wells or heterostructures, the motion of the carriers is quantized normal to the layers and is free in parallel direction ($q_\parallel$). Typical excitations of such systems are shown in Fig. 1b. In usual backscattering geometry normal to the layers ($q_\parallel = 0$) collective and single particle intersubband excitations are observed, which reveal the subband splittings ω_{01} and the depolarization shift. Recent experiments on extremely high mobility samples seem to indicate that depolarized spectra (scattering via spin density fluctuation) do not give exactly the one particle energies ω_{01}, but are subject to measureable exciton–like shifts to lower energies [17], which were neglected in most of the earlier experiments in GaAs. In–plane single–particle and collective excitations are only observed for finite $q_\parallel$. Due to the high refractive index of GaAs the maximum achievable in–plane wave vector in backscattering from the surface is given by $q_\parallel = (4\pi/\lambda) \sin\alpha \lesssim 1.7 \times 10^5$ cm^{-1} for laser energies around the E_0 and $E_0 + \Delta_0$ energy gap. The angle α is defined between the incident (scattered) light and the normal to the sample surface. The dispersion of both plasmons and single particle excitations of layered electron systems have been studied in this limited wave vector range [10,11,18–21]. An extension of the accessible in–plane wave vector is discussed below (see also [12,13,14]).

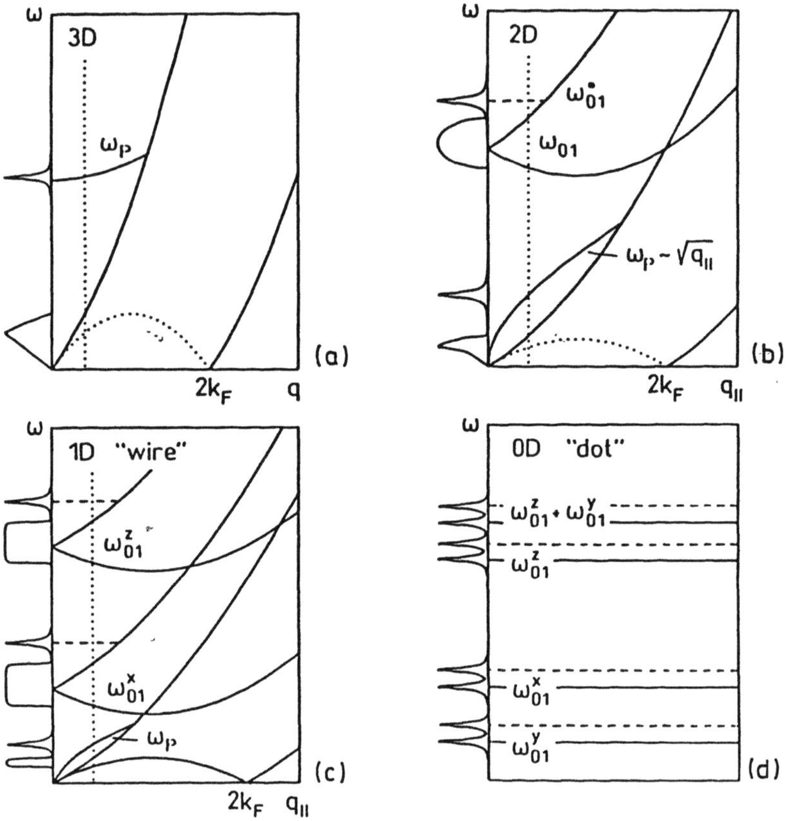

Fig. 1 Single particle and collective excitations in 3, 2, 1 and 0 dimensional electron systems (schematically)

In Fig. 1c and 1d the possible excitations in 1D and 0D systems are shown schematically. In quantum dots, confinement exists in all three space directions which finally results in an atomic–like situation with no dispersion in any direction. In 1D systems (quantum wires) free motion of carriers is still possible along one direction ($q_{\parallel}^{y}$). For the other two directions different confining potentials are assumed in Fig. 1c. This leads to different single particle and collective excitations in all three directions. The restriction of free motion in one dimension also changes the single particle excitation along the wire drastically. No low frequency excitations are possible for a finite wave vector $q_{\parallel}^{y}$ along the wire. First experimental results of electronic excitations in wire structured quantum wells or heterostructures have been published recently [15,16]. Our results, especially on confined plasmons, will be discussed in this paper. To the best of our knowledge, no results have been published so far for quantum dots.

3. In–plane excitations in layered electron systems

Until recently in–plane excitations in 2D systems were observed only in a relatively small wave vector range. In backscattering from the surface of a layered electron gas the total wave vector is given by $q = 2 \cdot 2\pi\eta/\lambda$, which divides into $q_{\parallel} = (4\pi/\lambda) \sin\alpha$ and $q_{\perp} = (4\pi\eta/\lambda) \sqrt{1-\sin^2\alpha/\eta^2}$, depending on the angle α. A maximum in–plane wave vector is achieved in back–scattering geometry from the edge of the layered structure (cleavage plane). Such geometries have been realized recently by using low temperature micro–Raman scattering [12]. Results of single–particle excitations and plasmon excitations in cleaved and wedged GaAs multilayer structures are shown in Fig. 2 and 3. The maximal achievable in–plane wave vector is $q_{\parallel} = 4\pi\eta/\lambda \simeq 7 \cdot 10^5$ cm^{-1} using a red laser line for excitation. The sample consists of 25 GaAs quantum wells with a thickness of 60 Å and a carrier concentration of about 9×10^{11} cm^{-2} per well. The in–plane single particle excitations shown in Fig. 2 exhibit the characteristic lineshapes of 2D systems. The solid lines are calculated lineshapes using the Lindhard–Mermin dielectric function of the 2D electron systems. All spectra can be fitted by assuming a temperature of the electron gas of $T_F = 13$ K. Scattering

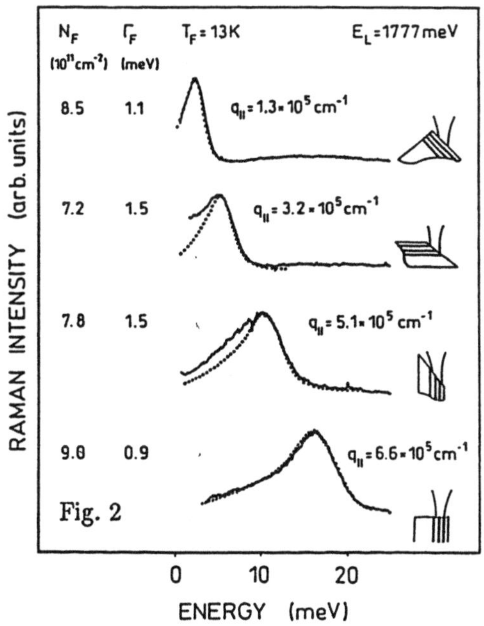

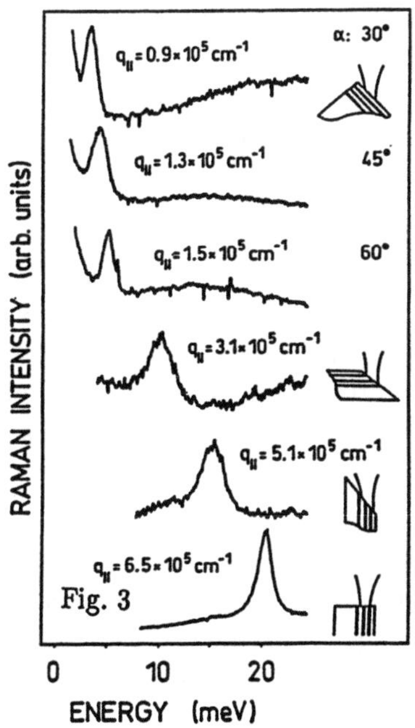

Fig. 2 Depolarized Raman spectra of layered electron systems. The dotted lines are theoretical fits to the observed single–particle excitations using the parameter given in the figure. The in–plane wave vector is determined by the scattering geometries which can be realized with micro–Raman spectroscopy as shown schematically.

Fig. 3 Plasmon excitations (polarized spectra) of layered electron systems measured with micro–Raman spectroscopy from the as-grown surface ($q_\parallel = 1.3 \cdot 10^5$ cm^{-1} for $\alpha = 45°$) and from the cleavage plane reveal a carrier concentration of about 9×10^{11} cm^{-2} and an electron damping of $\Gamma_F \simeq 1$meV. The polished surfaces show somewhat lower quality which results in a slightly reduced carrier concentration and larger damping.

Plasmons in layered electron systems are characterized by the Coulomb interaction between the layers. The dispersion of a purely 2D system is given by $\omega^2_{2D}(q_\parallel) = 2\pi n_{2D} e^2 q_\parallel / m^* \epsilon$, where n_{2D} is the 2D carrier concentration and ϵ the dielectric constant. The coupling between the layers leads to a plasmon band which is determined for an infinite system by $\omega(q_\parallel, q_\beta) = \omega_{2D}(q_\parallel)\, S(q_\parallel, q_\beta)$, where $S(q_\parallel, q_\beta) = \sqrt{\sinh q_\parallel d / (\cosh q_\parallel d - \cos q_\beta d)}$ is a structure factor. For a finite number N of electron gas layers N plasmon branches are obtained [22] which can be approximated by the above–mentioned formula with $q_\beta = (2\pi/Nd)\beta$ with $\beta = 0,1,\ldots N-1$ for a large number of layers. The branches of the plasmon band which are probed by inelastic light scattering depend on the ratio $q_\parallel / q_\perp$. In backscattering from the as–grown surface $q_\perp$ is large so the lower (acoustic) plasmon modes are observed. From the cleavage plane $q_\perp = 0$, therefore the more 3D upper branch of the plasmon ($\beta = 0$) is probed. We see again from the spectra of Fig. 3 that the edges polished under an angle have less quality compared to the grown or cleaved surfaces. The plasmon dispersion, however, can be measured up to $q_\parallel \simeq 7 \times 10^5$ cm^{-1}.

Another way to extend the accessible wave vector range is the so–called grating coupler effect [13,14]. A periodic structure on top of the layered electron system can couple larger in–plane wave vector into the underlying semiconductors due to diffraction. The total wavevector is then given by $q^m_\parallel = q_\parallel + m \cdot g$ where $m = m_I + m_S$ is the diffraction order of incident and scattered light and $g = 2\pi/a$ is the reciprocal lattice constant of the periodic grating. Examples of collective in–plane excitations in GaAs quantum wells are shown in Fig. 4 [14]. A periodic etch mask of photo resist stripes with a period of 8100 Å was produced by holographic lithography. The grating was etched into the sample in an optimized reactive ion etching process using $SiCl_4$ plasma [23]. Eighteen quantum wells remained underneath of this grating with a carrier concentration of about $5 \cdot 10^{11}$ cm^{-2}. Plasmons can be excited up to the seventh order now, even in backscattering with normal incidence on the surface. The peaks in Fig. 4 are labeled by the suitable diffraction order index m. For $\alpha = 0°$ the scattering wavevector is $q^m_\parallel = mg$. With decreasing α we have to add $q_\parallel = (4\pi/\lambda)\sin\alpha$ to obtain the total in–plane wavevector. With this grating coupler technique, the dispersion of layered plasmons could be measured up to about $q_\parallel = 5.5 \times 10^5$ cm^{-1}[14].

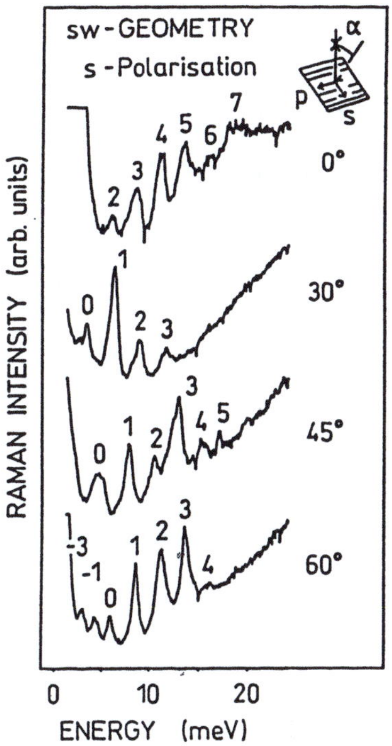

Fig. 4: Plasmon excitations measured with resonance Raman spectroscopy in layered quasi 2D electron systems. The different plasmon peaks correspond to different in–plane wave vectors $q_{\parallel}$ which are accessible via diffraction of light due to the grating coupler which is etched into the sample. The numbers label the grating order. The 0 order peaks are similar to the plasmons observed in the upper three spectra of Fig. 3. The seventh order corresponds to an in–plane wave vector of $q_{\parallel} = 5.5 \times 10^5$ cm^{-1} (from [14]).

4. Electronic excitations in wire structured quantum wells

In the last section we discuss very recent results on inelastic light scattering by free carriers in lateral quantum wire superlattices. The samples are prepared in a similar way as the etched grating coupler. The resulting periodic wires are shown schematically in Fig. 5. The geometrical width of the individual wires is about 4000 Å. Each wire contains five sheets of electrons in GaAs quantum wells with a thickness of 60 Å and a 2D carrier concentration of about $5 \cdot 10^{11}$ cm^{-2}. The actual lateral width of the electron systems is expected to be considerably smaller than the geometrical width of the wires due to surface depletion. This is shown schematically on the right–hand side of Fig. 5. The Fermi level pinning at the surface due to surface states and damage causes a depletion layer on each side, which leads to an approximately parabolic potential well toward the surface with a flat part in the middle [24]. A subband splitting due to this confinement in x–direction of about 1 to 2 meV is expected.

We have investigated both collective and single–particle excitations in such wire structured system by resonant inelastic light scattering. The dispersion of these excitations were studied in backscattering by tilting the laser parallel and perpendicular to the wires. In Fig. 6 polarized spectra are shown for three tilt angles α

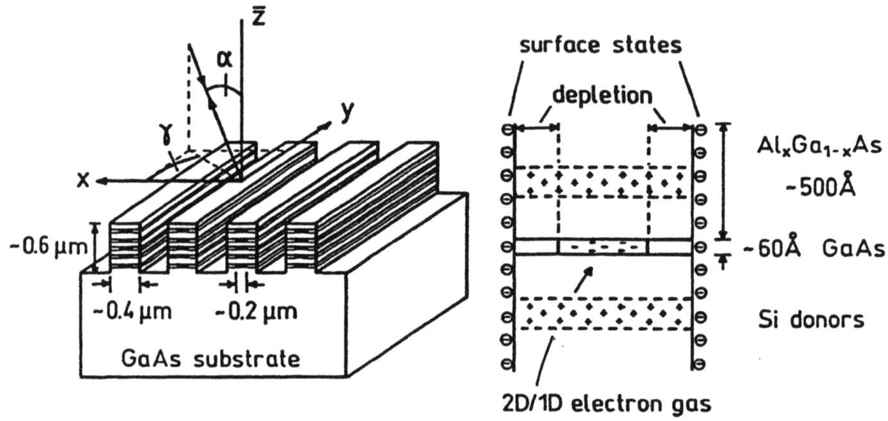

Fig. 5 Typical geometry of the wire structured samples achieved by holographic lithography and reactive ion etching. Each wire contains five modulation doped GaAs quantum wells. Details of a wire are shown schematically on the right–hand side of the figure.

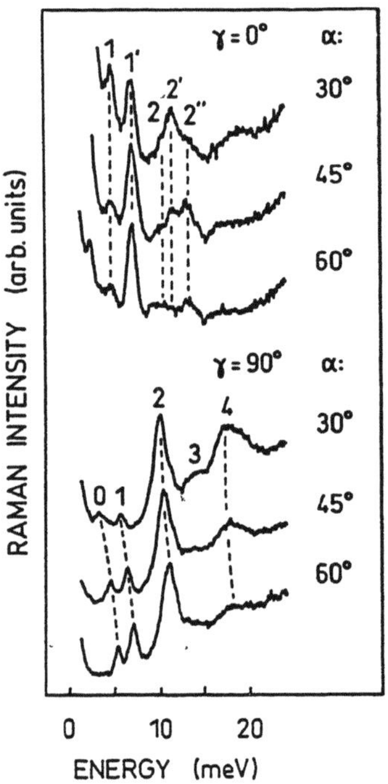

Fig. 6: Typical plasmon excitations in the wire structured sample for various tilt angles α normal to the wires ($\gamma = 0^\circ$) and along the wires ($\gamma = 90^\circ$). The peak positions remain unchanged for $\gamma = 0^\circ$ (no dispersion). With $\gamma = 90^\circ$ a considerable dispersion of the plasmon modes is observed, which is strongest for the lower order modes.

in each direction. Several plasmon excitations are observed in each spectrum. No shift of the peak position is observed with tilt angle α normal to the wires ($\gamma = 0^\circ$). On the other hand the peaks shift to higher energies when α is increased in the direction along the wires. The dispersion is stronger for the low energy excitations. The vanishing dispersion perpendicular to the wires shows that Coulomb coupling between

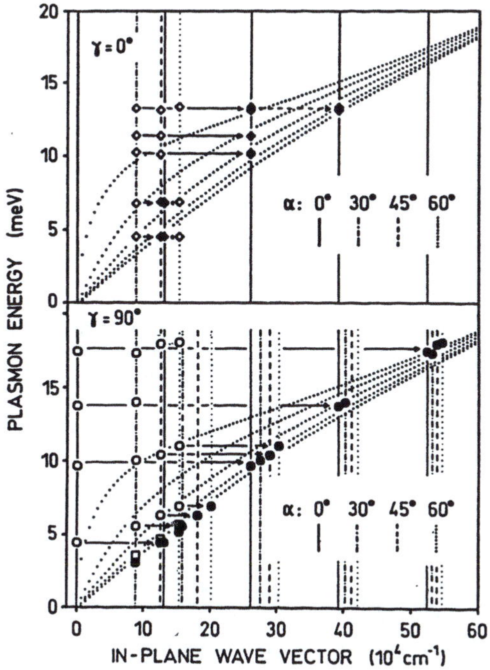

Fig. 7: Dispersion of plasmon excitations in the wire structured sample for $\gamma = 0°$ and $\gamma = 90°$. The allowed values for the in–plane momentum for the different angles α are represented by the vertical lines. The effective wire width determined from this analysis is $w_e \simeq 2400$ Å.

neighbouring wires is very weak and we observe resonances determined by the isolated wires. The spectrum of plasmon modes obtained parallel to the wires reflects qualitatively all the predictions made by Eliasson et al. [25] for a laterally bounded single layer 2D electron gas. To include the multilayer structure, we use the simplified model of confined plasmons. The interpretation of the plasmon resonances is based on the dispersion shown in Fig 7. The dotted lines are the five plasmon branches calculated for $n_{2D} = 5.1 \cdot 10^{11}$ cm^{-2}. We keep this plasmon dispersion of the layered 2D electron gas and impose standing wave conditions for the plasmons in direction perpendicular to the wires. Then the in–plane component of the plasmon perpendicular to the wires is restricted to values $q_m = m \cdot \pi / w_e$, where w_e is the effective wire width which takes the partial lateral depletion into account. The component parallel to the wires remains unrestricted: $q_\parallel = \sqrt{q_m^2 + q_\parallel^y(\alpha)^2}$. These values are represented as vertical lines in Fig. 7. The peaks are assigned to the quantized momenta, for which their energy coincides with one of the plasmon dispersion branches. The peaks in Fig. 6 are labeled according to the quantization index m. Different branches of the plasmon dispersion are distinguished by primes. From the lowest confined plasmon, labeled by 0, we obtain a perpendicular momentum of 1.3×10^5 cm^{-1} which results in an effective wire width of $w_e \approx 2400$ Å. These experiments represent the first experimental study of plasmon dispersion and plasmon confinement in lateral quantum wire superlattices by resonant electronic light scattering. A detailed analysis of the results is presented in [16].

In conclusion, we have shown that resonant electronic light scattering is a versatile tool to study the dispersion of single particle and plasmon excitations in electron gases of different dimensionality in a wide wavevector range. Especially the possibilities opened by micro–Raman spectroscopy and grating couplers, and the new properties of quantum wires and quantum dots, which become now accessible, will lead to stimulating and exciting developments in the near future.

Acknowledgements. The work reported here was performed in close collaboration with G. Weimann, W. Schlapp, T. Demel and D. Heitmann. We are grateful for the experimental and technological help and for many stimulating discussions. The project is supported financially by the Deutsche Forschungsgemeinschaft via Schwerpunkt "Physikalisch–technische Grundlagen von III–V Halbleiterstrukturen" and via the "Gottfried–Wilhelm–Leibniz Programm".

References

[1] A. Mooradian and G.B. Wright, Phys. Rev. Lett 16, 999 (1966)
[2] A. Mooradian, Phys. Rev. Lett. 20, 1102 (1968)
[3] A. Mooradian, in "Festkörperprobleme" IV, ed. by O. Madelung, (Pergamon Vieweg, Braunschweig 1969) p. 74
[4] M.V. Klein, in "Light Scattering in Solids" ed. by M. Cardona, Topics Appl. Phys., Vol. 8, 2nd ed. (Spinger, Berlin, Heidelberg 1983) p. 147
[5] G. Abstreiter, M. Cardona, and A. Pinczuk, in "Light Scattering in Solids" IV, ed. by M. Cardona, G. Güntherodt, Topics Appl. Phys., Vol. 54 (Springer, Berlin, Heidelberg 1984) p. 5
[6] A. Pinczuk, G. Abstreiter, R. Trommer, and M. Cardona, Solid State Commun. 30, 429 (1979)
[7] E. Burstein, A. Pinczuk, and S. Buchner, in "Physics of Semiconductors", ed. by B.L.H. Wilson (The Institute of Physics, London, 1979) p. 1231
[8] G. Abstreiter and K. Ploog, Phys. Rev. Lett. 42, 1308 (1979)
[9] A. Pinczuk, H.L. Störmer, R. Dingle, J.M. Worlock, W. Wiegmann, and A.C. Gossard, Solid State Commun. 32, 1001 (1979)
[10] G. Abstreiter, R. Merlin, and A. Pinczuk, IEEE J. Quantum Electron. QE 22, 1771 (1986)
[11] A. Pinczuk and G. Abstreiter, in "Light Scattering in Solids" V, ed. by M. Cardona and G. Güntherodt, Topics Appl. Phys. Vol. 66 (Springer, Berlin, Heidelberg 1989) p. 153
[12] T. Egeler, S. Beeck, G. Abstreiter, G. Weimann, and W. Schlapp, Superlattices and Microstructures, 5, 123 (1989)
[13] T. Zettler, C. Peters, J.P. Kotthaus and K. Ploog, Phys. Rev. B 39, 3931 (1989)
[14] T. Egeler, G. Abstreiter, G. Weimann, T. Demel, D. Heitmann, and W. Schlapp, Surface Science, in press
[15] J.S. Weiner, G. Danan, A. Pinczuk, J. Valladares, L.N. Pfeiffer, and K. West, Phys. Rev. Lett 63, 1641 (1989)
[16] T. Egeler, G. Abstreiter, G. Weimann, T. Demel, D. Heitmann, and W. Schlapp, submitted for publication
[17] A. Pinczuk, S. Schmitt–Rink, G. Danan, J.P. Valladares, L.N. Pfeiffer, K.W. West, Phys. Rev. Lett. 15, 1633 (1989)
[18] D. Olego, A. Pinczuk, A.C. Gossard, and W. Wiegmann, Phys. Rev. B 25, 7867 (1982)

[19] G. Fasol, N. Mestres, H.P. Hughes, A. Fischer, and K. Ploog, Phys. Rev. Lett. 56, 2517 (1986)
[20] A. Pinczuk, M.G. Lamont, A.C. Gossard, Phys. Rev. Lett. 56, 2092 (1985)
[21] G. Fasol, N. Mestres, M. Dobers, A. Fischer, and K. Ploog, Phys. Rev. B 36, 1536 (1987)
[22] J.K. Jain and P.B. Allen, Phys. Rev. Lett. 54, 2437 (1985)
[23] T. Demel, D. Heitmann, P. Grambow, and K. Ploog, Appl. Phys. Lett. 53, 2176 (1988)
[24] S.E. Laux, F. Stern, Appl. Phys. Lett. 49, 91 (1986)
[25] G. Eliasson, Ji–Wei Wu, P. Hawrylak, J.J. Quinn, Solid State Commun. 60, 41 (1986)

Localization and Confinement of Electrons in Heterostructures by Focused-Ion-Beam Insulation-Writing

A.D. Wieck

Max-Planck-Institut für Festkörperforschung, Heisenbergstr. 1, D-7000 Stuttgart 80, Fed. Rep. of Germany

A focused beam of Ga^+ ions, swept under normal incidence over the surface of AlGaAs-GaAs heterostructures, decreases dramatically the conductivity of the 2-Dimensional Electron Gas (2DEG) in laterally well-defined areas and lines. We present examples for this insulation-writing, leading to ballistic transport in quasi-one-dimensional channels, coupling of quantum Hall states, and In-Plane-Gated (IPG) transistors. The striking feature of all these systems is that the confining electric field is tunable and in the same plane as the 2DEG. This aspect provides an important flexibility for applications in research and in device fabrication.

The advent of lateral constrictions in the 2-Dimensional Electron Gas (2DEG) in AlGaAs-GaAs heterostructures to 1-Dimensional (1D) channels has led to the important discovery of the conductivity quantization in multiples of $2e^2/h$. [1,2] In these experiments, the tunability of the channel is provided by metallic surface gates, forming a field effect transistor. [3] The inhomogeneous electric field between two closely spaced gates has components parallel to the 2DEG, which confine the electrons to a 1D channel. This split-gate structure forms the 1D-channel at a certain negative bias voltage, which is needed to deplete the electrons directly under the gates (Fig.1a).

Very recently, we presented a totally different technique to create such 1D-systems by Focused Ion Beam (FIB) direct writing. [4] It is the aim of this contribution to review briefly the most striking features of this so-called In-Plane-Gated (IPG) system and to demonstrate its flexibility in different experimental fields.

If a modulation-doped $Al_xGa_{1-x}As-GaAs$ heterostructure is bombarded by accelerated (100keV) Ga^+ ions, the conductivity of its 2DEG decreases dramatically. Fig.2 shows the Current-Voltage (IV) characteristic

Springer Series in Solid-State Sciences, Vol. 97
Localization and Confinement of Electrons in Semiconductors
Editors: F. Kuchar · H. Heinrich · G. Bauer

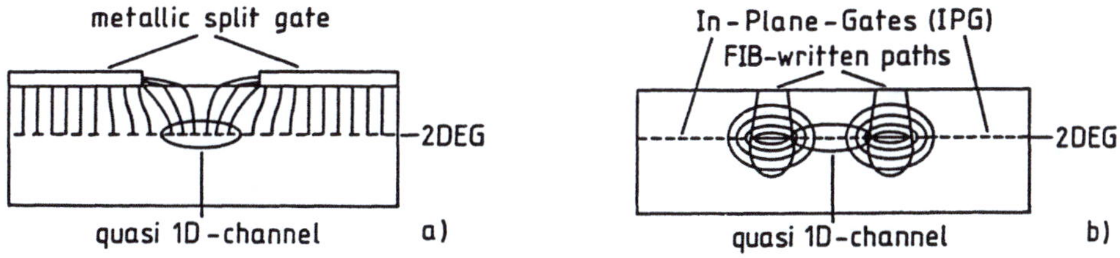

Fig.1: Schematic cross-section of *a)* split-gate and *b)* In-Plane-Gate configuration, respectively. The lines between the gates and the electron gas indicate the electric field distribution.

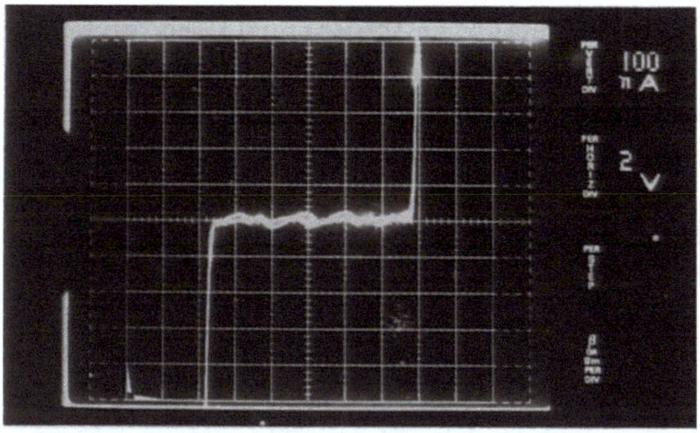

Fig.2: Current-Voltage (IV) characteristic at T=77K of an insulation-written FIB-line (100keV Ga^+, spot diameter 100nm, dose $2 \times 10^{13} cm^{-2}$). For voltages $|V_g| \leq 6V$ the interface is insulating, whereas reversible breakthrough occurs at higher voltages.

of such an insulating FIB-line between two ohmic contacts to the 2DEG. This insulation behavior persists up to several Volts even at room temperature. [5] Potential-contrast images obtained with a scanning electron microscope show that the lateral depletion length, inducing the observed insulation, is of the order of $1\mu m$. [6] It is therefore possible to create lateral electric field strengths of the order of several $10^4 V/cm$. These field strengths are already present in the growth direction of the heterostructure and confine the electrons to the 2DEG. Here, we use the additional lateral electric field to confine the 2DEG to a 1D channel. This field is tunable by the external bias, which we will call gate voltage V_g, and lies predominantly in the plane of the 2DEG (Fig.1b). In this way, we use the 2DEG adjacent to a channel defined by two FIB-paths as electrostatic gates (Fig.3). The striking feature of these In-Plane-Gates (IPG) is that

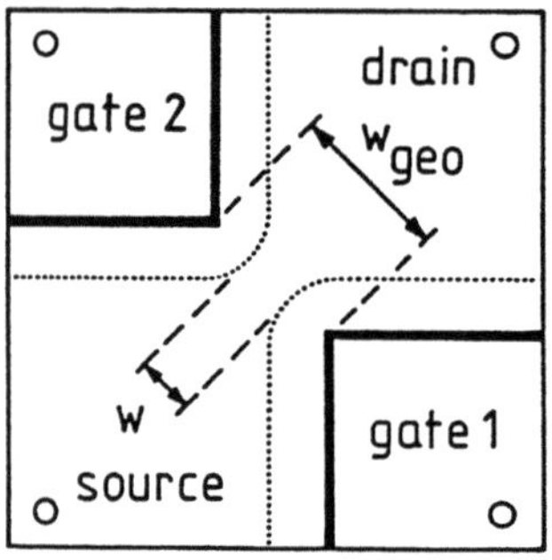

Fig.3: Schematic top view of an IPG-structure. The two heavy lines mark FIB-written paths and the circles symbolize ohmic contacts to the gates, source and drain, respectively.

they consist of the same 2DEG structure as the channel itself, giving an enormous flexibility to the design of the structure.

Channel characterization in the low temperature (T=1.1K) and low source-drain voltage ($V_{SD} \leq kT/e$) limit is performed with a lock-in technique: Driving an alternating current of constant amplitude with a frequency of 8Hz through the channel, we detect the voltage along the channel, record and convert it to resistance and conductance. Various currents ranging between 1 and 10nA led to the same results. The conductivity increases linearly with V_g, if a constant series resistance of the order of $1k\Omega$ is subtracted. The resistance and conductance of a channel with a geometrical width of $2.8\mu m$ are plotted in Fig.4. The structure is totally depleted at the threshold voltage of -4.1V. At $V_g = -3.4V$, the $m = 1$ filling factor (one 1D-subband occupied) is clearly visible. Structures at higher m=2,3,4 are evident after expanding the vertical scale for the conductivity. The initial high quality of the heterostructure (mobility $\mu = 6.7 \times 10^5 cm^2/Vs$) is thus conserved in a μm distance from the FIB path. The observation that the structures are less developed than in the split-gate experiments [1,2] is due to the fact that the channel length is approximately given by W_{geo} and thus relatively large.

A slightly different structure than that in Fig.3 is shown in Fig.5: Here, we use only one In-Plane-Gate to deplete the channel asymmetrically against the opposite FIB-path. The 4-point resistance R of this structure for different gate voltages is plotted versus the magnetic field B in Fig.6. At positive V_g, the channel conductivity is enhanced and Shubnikov-de Haas oscillations are clearly visible. When decreasing the gate voltage these oscillations shift to lower magnetic fields, indicating a decrease of the 2D carrier density in the channel region. Also the amplitude of the oscillations decreases. At gate voltages of about -2.7V, the amplitude nearly vanishes, raising again at decreasing V_g below -2.98V. This behaviour is

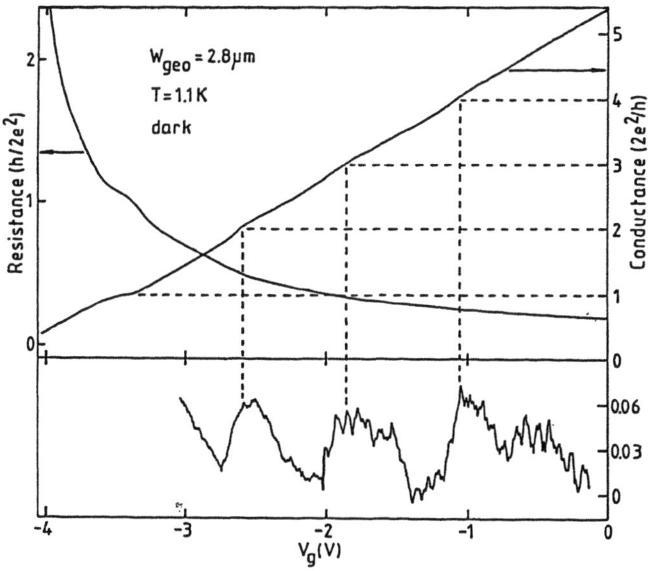

Fig.4: Resistance and conductance of an IPG-structure. Steplike structures are visible at integral multiples m of the fundamental conductance $2e^2/h$. For better resolution at $m \geq 2$, the difference between the measured conductivity and a theoretical straight line has been plotted on an expanded vertical scale at the bottom of the figure.

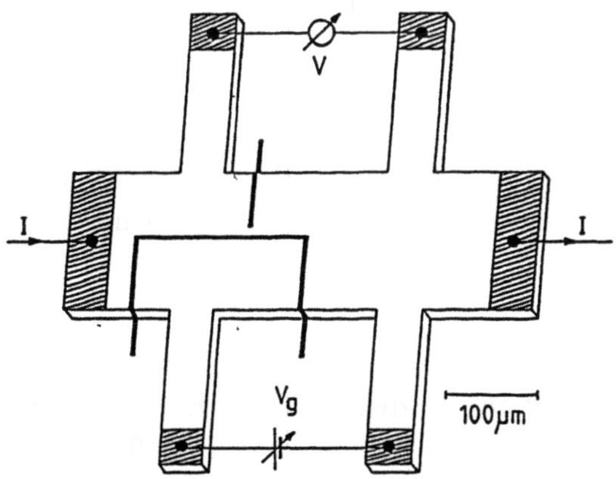

Fig.5: Sketch of the mesa-etched heterostructure for the measurement in Fig.6. Shaded areas mark contact regions, which are connected to the electronic setup as indicated in the figure. Heavy lines symbolize FIB-written paths, which define the In-Plane-Gate area on the left side at the bottom and the channel between the gate and the end of the straight FIB-line. The channel has a geometrical width of $2\mu m$.

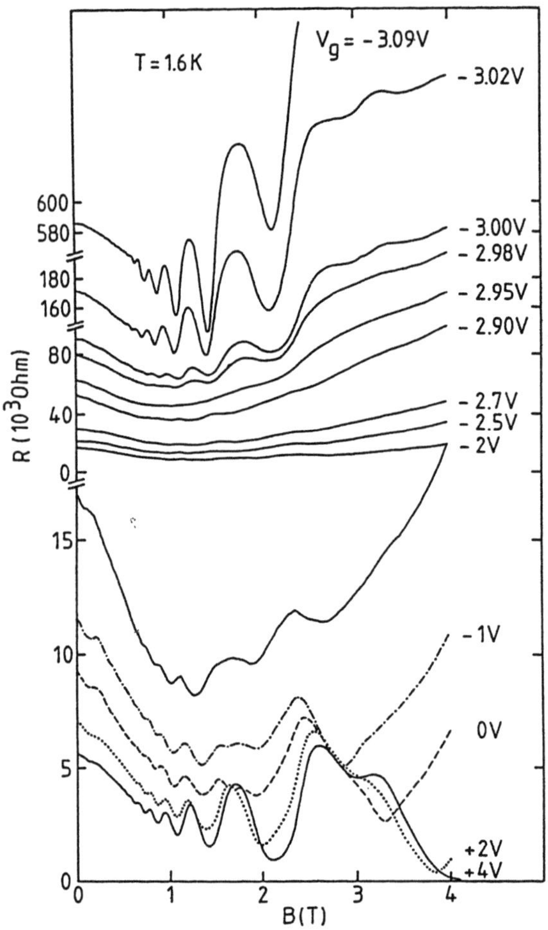

Fig.6: Four-point resistance R of the channel and the adjacent 2DEG measured as sketched in Fig.5 versus the magnetic field B perpendicular to the 2DEG.

tentatively explained by an interference between the Shubnikov-de Haas oscillations of the tuned channel and the undisturbed 2DEG, respectively. Interesting details in Fig.6 are also the precursor at $B \approx 0.1T$ and $-2V \leq V_g \leq +4V$ and the fact that the resistance R decreases with decreasing gate voltage for certain magnetic fields ($1.7T \leq B \leq 2T$ and $2.6T \leq B \leq 3.4T$).

At the higher temperature of T=77K, quantum effects are no longer visible. However, the tunability of the channel is maintained, as shown in Fig.7. Here, $V_{SD} \leq 1.2V$ runs to voltages considerably larger than kT/e, and we obtain the field effect transistor operation with reasonable drain

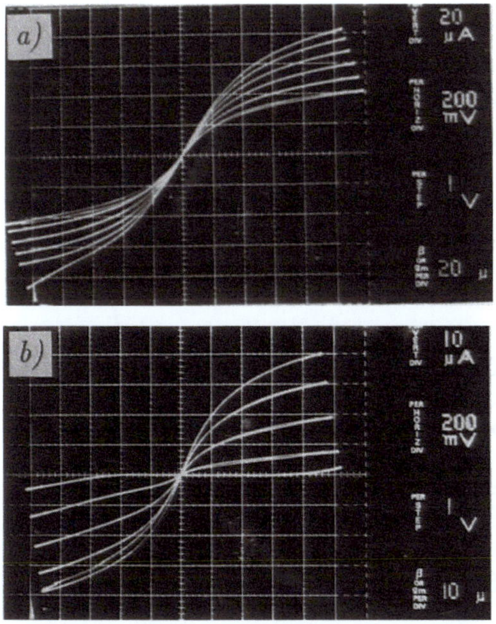

Fig.7: Current-Voltage (IV) characteristics of an IPG-channel with $W_{geo} = 4.2\mu m$ at $T = 77K$ and different gate voltages. The central point of the diagram is the coordinate origin. In *a)* the curve with the lowest absolute current values corresponds to $V_g = 0V$, the higher ones to positive V_g with $1V$ spacing. In *b)* the curve with the highest absolute current values corresponds to $V_g = 0V$, the lower ones to negative V_g with $-1V$ spacing.

currents of up to $80\mu A$. It should be noted that the leakage currents are in the pA range (see Fig.2), which means that the power gain of the IPG-transistor exceeds 10^7.

In conclusion, I have presented a new concept of electrostatic conductivity control in heterostructures by means of lateral electric fields induced by external bias voltages. The versatility of the concept is demonstrated by experiments related to ballistic transport, magneto-transport and in the classical limit at T=77K to field-effect transistor operation.

I would like to thank my colleagues Thomas Bever and Klaus Ploog for their collaboration and the Bundesministerium für Forschung und Technologie for financial support.

References

[1] B.J. van Wees, H. van Houten, C.W.J. Beenakker, J.G. Williamson, L.P. Kouwenhoven, D. van der Marel, and C.T. Foxon, Phys.Rev.Lett.**60** 848 (1988).

[2] D.A. Wharam, T.J. Thornton, R. Newbury, M. Pepper, H. Ahmed, J.E.F. Frost, D.G. Hasko, D.C. Peacock, D.A. Ritchie, and G.A. Jones, J.Phys.C**21** L209 (1988).

[3] W. Shockley, Proc.IRE **40** 1365 (1952).

[4] A.D. Wieck and K. Ploog, Appl.Phys.Lett.**56** 5 Mar 1990.

[5] A.D. Wieck and K. Ploog, German Patent Application P 3914007.5 (1989).

[6] A.D. Wieck and K. Ploog, Surf.Science, in press (1990).

Transport Through Single and Coupled Quantum Dots: The Formation of a Mini-Band Structure

L.P. Kouwenhoven[1], *B.J. van Wees*[1], *F.W.J. Hekking*[1], *K.J.P.M. Harmans*[1], *and C.E. Timmering*[2]

[1]Faculty of Applied Physics, Delft University of Technology, P.O. Box 5046, 2600 GA Delft, The Netherlands

[2]Philips Research Laboratories, 5600 JA Eindhoven, The Netherlands

Abstract. We describe the formation of a mini-band structure in a one dimensional chain of N barriers at periodic positions in terms of transmission probabilities. For a single quantum dot it is shown experimentally that resonant transmission occurs whenever the Fermi energy coincides with the energy of a magnetically induced zero dimensional (0D) state. In a sequence of 15 coupled quantum dots, these 0D-states group together and form a mini-band structure, which is reflected in conductance measurements.

1. Introduction.

Early in the seventies, Esaki and Tsu [1] proposed that the periodic potential in a structure of alternating layers of different semiconductor alloys can form a superlattice with a period much larger than the period of the host atomic lattice. Besides the band structure of the atomic crystal, it was shown that the superlattice can give rise to a second band structure in the same material. Due to the smaller Brillouin zones, bandwidths and gaps, the latter was called a mini-band structure. A few years later Esaki and Chang [2] found experimentally a negative differential conductance in the non-linear transport properties of a vertically grown superlattice, which was argued to be a direct consequence of the mini-band structure. Since these early works, much effort has been put in transport and optical studies on grown superlattices.

A disadvantage of grown superlattices is that the mini-band structure and the Fermi-energy are fixed after the growing process. This leaves only non-linear methods to probe the electron states in the mini-band structure, for instance by measuring non-linear transport properties [2] or by optical excitation of electrons into higher mini-bands and measuring the resulting luminescence spectra [3]. We have used a different technique to induce the periodic potential, which allows us to tune the Fermi energy through the mini-band structure. We have fabricated a lateral one dimensional (1D) crystal in a two dimensional electron gas (2DEG) by means of electrostatic depletion induced by metallic gates on top of a GaAs/AlGaAs hetero structure. The 1D crystal consists of a sequence of 15 coupled quantum dots, in which the spatial quantization is realized in all three directions. Note that in the so-called 1D superlattices, the periodic potential is in one direction and the electron

Springer Series in Solid-State Sciences, Vol. 97
Localization and Confinement of Electrons in Semiconductors
Editors: F. Kuchar · H. Heinrich · G. Bauer © Springer-Verlag Berlin, Heidelberg 1990

motion is still free in the other two directions. In a single quantum dot, discrete electron states can arise either due to the lateral confinement or by a combination of this with a magnetic field. These zero dimensional (0D) states develop into mini-bands separated by energy gaps, due to the coupling which exists between adjacent quantum dots. As will be discussed, the linear conductance reflects the formation of a mini-band structure in the 1D crystal, when the Fermi energy is tuned through the bands by varying the voltage on the gates.

The outline of this paper is as follows. We will illustrate the formation of a band structure from a transport point of view in section 2. Experiments are discussed in section 3 and 4. Section 3 concentrates on transport through magnetically induced 0D-states in a single quantum dot [4] and section 4 treats the transport properties of a series of 15 coupled quantum dots [5]. Some conclusions are summarized in section 5.

2. Formation of a band structure in a finite 1D crystal.

To calculate the band structure of an infinite crystal one can incorporate the translational invariance of the lattice in the electron wave function by means of Bloch's theorem. In a finite crystal, consisting of a small number of unit cells, the translational invariance is broken and the boundary conditions have to be included in a calculation of the electron states. We will illustrate the formation of a band structure in a finite 1D crystal with on both sides infinite 1D leads, by calculating the transmission probability T_N of an electron wave through a 1D chain of N barriers at periodic positions [6]. To illustrate the procedure, an explicit derivation for the two barrier system (or single quantum dot) in fig.1a will be described in detail, while for the N barrier system of fig.1b we refer to ref.7 for a full derivation. The starting point is the scattering matrix $S=\binom{r \; t'}{t \; r'}$ which relates the outgoing waves from a single barrier to the incoming waves. For an asymmetric barrier in a magnetic field (which breaks time reversal symmetry) the complex transmission and

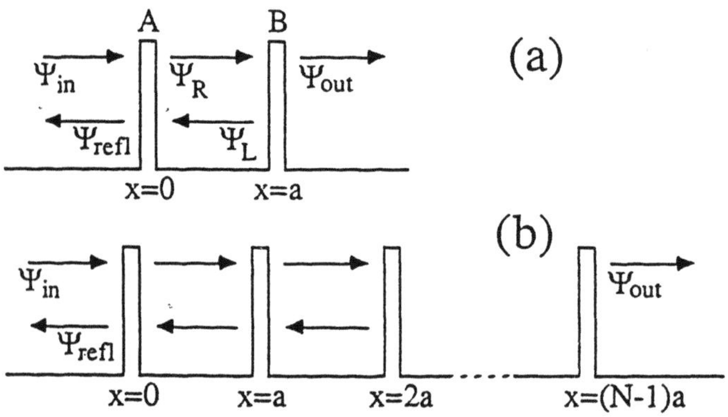

Figure 1. Schematic representation of an electron wave in a 1D 2-barrier (a) and N-barrier (b) system.

reflection amplitudes t and r for an incoming wave in one direction differ from the amplitudes t' and r' for a wave in the opposite direction. For simplicity we will assume symmetric barriers, which gives t=t' and r=r'. Conservation of current yields the condition that S is unitary, which leads to the relations: $tt^*+rr^*=1$ and $t/t^*=-r/r^*$. To allow unequal barriers we label them by A and B. The right- and left-moving waves in the cavity between the two barriers at x=0 in fig.1a are given by: $\Psi_R=t_A\Psi_{in}+r_A\Psi_L$ and $\Psi_L=r_B\Psi_R\cdot\exp(i\theta)$, in which θ is the phase acquired by an electron wave after travelling back and forth once within the cavity. With the outgoing wave $\Psi_{out}=t_B\Psi_R$ one finds for the transmission amplitude $t_2=\Psi_{out}/\Psi_{in}$ through the two barriers:

$$t_2=\frac{t_A\,t_B}{1-r_A\,r_B\,\exp(i\theta)} \tag{1a}$$

and with the definitions $T_A=t_At_A^*$, $R_A=r_Ar_A^*$, $T_B=t_Bt_B^*$, and $R_B=r_Br_B^*$ for the transmission probability $T_2=t_2t_2^*$:

$$T_2=\frac{T_AT_B}{1-2\sqrt{R_AR_B}\cdot\cos(\theta+\theta_r)+R_AR_B} \tag{1b}$$

θ_r is the total phase shift after reflection at the barriers A and B, and follows from the imaginary parts of r_A and r_B. The striking aspect of eq.1 is that even with the single barrier transmissions $T_A=T_B<1$, the two barriers in series have a transmission $T_2=1$ whenever the total phase $\theta+\theta_r$ is an integer times 2π. This aspect is well-known as resonant tunneling. From an energy point of view, the cavity forms a finite 1D system in which discrete 0D-energy-states arise (when $T_A=T_B\approx0$), similar to particle-in-a-box states. When the energy of an incoming electron exactly coincides with a 0D-state, the electron is transmitted with probability equal to 1.

For N barriers in series (see fig.1b), one can calculate the transmission $T_N=t_Nt_N^*$ from a numerical solution of the recursive formula for the transmission amplitude t_N [5,7]:

$$t_N=\frac{t\,t_{N-1}}{1-r\,r_{N-1}\,\exp(i\theta)} \tag{2}$$

in which t denotes the transmission amplitude of a single barrier. The current conservation relations also hold for t_N and r_N.

In fig.2 the transmission T_N versus phase θ is plotted for a different number N of equal barriers with transmissions T=0.90 (the constant phase shift θ_r is ignored). The lowest curve with N=2 oscillates with a period 2π, which corresponds with consecutive 0D-states. The oscillations in T_2 are broadened due to the large coupling (T is nearly 1) to the 0D-states. For small barrier transmission ($T\approx0$) these oscillations appear as sharp Lorentzian-shaped peaks. Increasing the number of barriers N, the number of oscillations increases accordingly. As can be seen in fig.2, a change in phase of 2π contains a number of oscillations equal to the number of

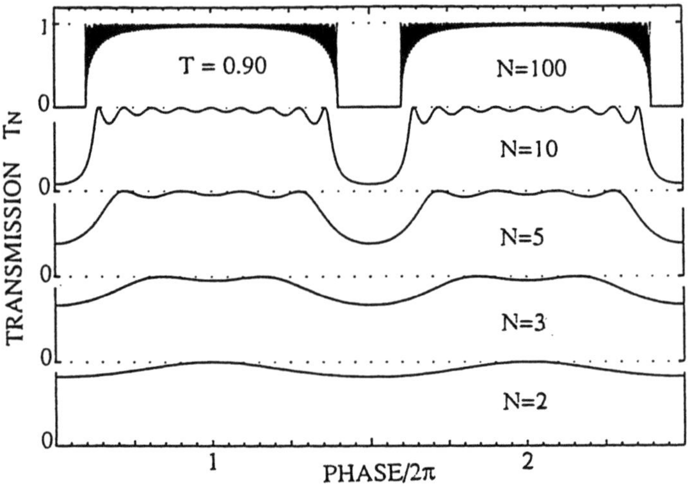

Figure 2. Transmission T_N versus phase $\theta/2\pi$ of a periodic 1D chain of N barriers, calculated from eq.2 with single barrier transmissions T=0.90.

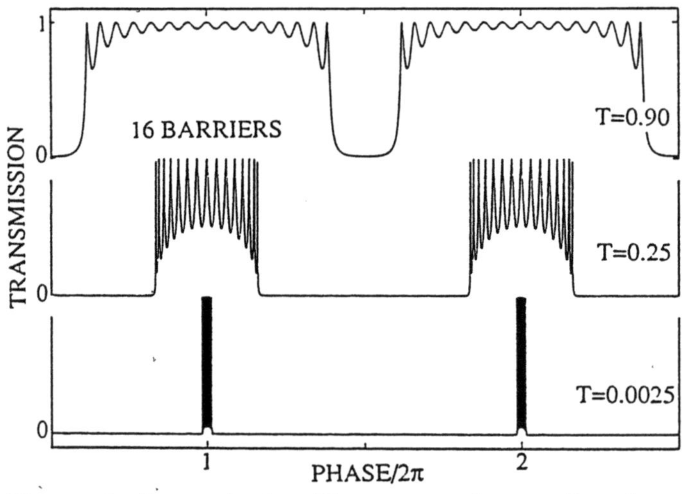

Figure 3. Transmission T_N versus phase $\theta/2\pi$ of a periodic 1D chain of 16 barriers, for different single barrier transmissions T.

cavities N-1. Furthermore, the oscillations which originate from the 0D-states formed in each cavity, group together in bands of high transmission T_N. These bands are separated by regions of low transmission T_N, which are associated with the gaps. In fig.3 the transmission T_N is shown for N=16 barriers, but now for three different values of the transmission T through the single barriers, illustrating the effect of the coupling between the adjacent cavities. Weak coupling (i.e. small T) results in a band structure with narrow bands and large gaps ("tight-binding regime"), while strong coupling (i.e. T nearly 1) yields wide bands and small gaps ("nearly-free-electron regime").

The transmission T_N is directly related to the conductance G by the Landauer formula: $G=T_N \cdot e^2/h$, which therefore also reflects the band structure if one is able to vary experimentally the phase θ. In zero magnetic field, the phase is given by: $\theta=2\pi(2a/\lambda_F)$, i.e. it is determined by the number of half Fermi wavelength's $\lambda_F/2$ matching the distance a between adjacent barriers. Smith et al. [8] have reported the observation of 0D-states for B=0 in a single quantum dot device in which they were able to vary the distance a. However, in a device with more than two barriers this is experimentally very difficult to realize. One possible alternative for varying the phase θ is by changing the Fermi energy and thereby λ_F. This has been done recently by means of a gate covering a single quantum dot [9].

We have chosen a different approach to vary the phase θ. We have used the fact that in a high magnetic field the current carrying electron states of a 2DEG are confined to 1D edge channels, which are located at the sample boundary [10]. It has been shown that on short distances of order μm, the scattering between edge channels is absent in a sufficiently high magnetic field [11]. Edge channels can therefore be treated as independent 1D current channels, which make them an ideal starting point for confinement to zero dimensions. By confining an edge channel in a quantum dot it forms a 1D loop (see fig.4). Due to the Aharonov-Bohm effect, an electron which makes one revolution around the dot acquires a change in phase of $2\pi(\phi/\phi_0)$, where $\phi=BA$ is the magnetic flux, A denotes the area enclosed by the edge channel loop and $\phi_0=h/e$ is the flux quantum [12]. In the case of a high magnetic field, the phase θ can be varied either by changing the magnetic field B or the enclosed area A. The correspondence between the phase $\theta=2\pi(2a/\lambda_F)$ at B=0 and $\theta=2\pi\phi/\phi_0$ at B≠0 has been pointed out by Büttiker et al. [13] for a 1D ring structure, which encloses an Aharonov-Bohm flux $\phi=BA$.

3. Transport through 0D-states in a single quantum dot.

In this section we describe the experimental realization of magnetically induced 0D-states in a single quantum dot, and how they appear in the conductance. Fig.4 shows schematically the layout of the split-gate geometry, which is fabricated on top of a GaAs/AlGaAs hetero structure. The ungated 2DEG has an electron density of $2.3 \cdot 10^{15} m^{-2}$ and a transport mean free path of 9μm. A negative voltage of -0.2V on both gate pairs A and B depletes the electron gas underneath the gates and creates a quantum dot with a diameter of 1.5μm in the 2DEG. The narrow channels separating the gate pairs are already pinched-off at this gate voltage. Two 300nm wide quantum point contacts (QPCs) allow for electronic transport through the quantum dot [14]. The transport properties of each individual QPC can be measured by applying a voltage to only one gate pair and zero voltage to the other.

Above B=1T the transport through the quantum dot takes place adiabatically in edge channels (i.e. absence of scattering between the edge channels) [15]. In this

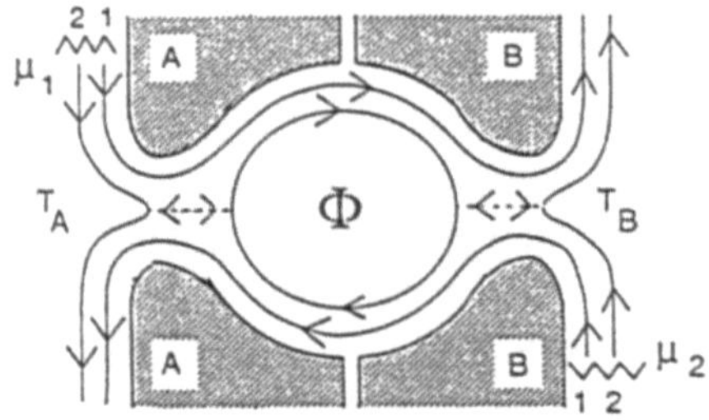

Figure 4. Schematic layout of the quantum dot with diameter of 1.5µm and two 300nm wide quantum point contacts. The electron flow in edge channels is shown when a high magnetic field B is applied. A 1D loop is formed which encloses a flux ϕ, when an edge channel is only partially transmitted by both quantum point contacts.

case, the conductance G_D of the quantum dot is given by: $G_D=(e^2/h)(M+T_2)$, in which M denotes the number of fully transmitted edge channels and T_2, given by eq.1b, is the transmission through the partially transmitted edge channel (see fig.4). The conductance is quantized at plateau values $M\cdot e^2/h$ whenever all edge channels are either completely transmitted or reflected ($T_2=0$). From eq.1b it is expected that G_D oscillates in the transition region between two quantized plateaus (where $T_2\neq 0$), when B is varied or the dot area is changed by means of the gate voltage V_g, both resulting in a change of the phase θ.

Fig.5a shows G_D versus B at constant V_g=-0.35V at 10mK in the transition from the second to the third plateau, which corresponds to the complete transmission of two (spin resolved) edge channels and the partial transmission of the third channel. Large oscillations are seen between the two plateaus with an amplitude modulation up to 40% of e^2/h. Fig.5b displays the oscillations on an expanded scale showing their regularity. The period B_o of the oscillations smoothly varies from B_o=2.5mT at B=2.5T to B_o=2.8mT at B=2.7T. This period cannot be deduced directly from the enclosed area A, because the location of the edge channels at the boundary also depends on B [4]. In fig.5c the region of low transmission or weak coupling is plotted. Here the conductance is $2e^2/h$ (the third edge channel is completely reflected here), except when the Fermi energy coincides with the energy of a 0D-state. The discrete conductance peaks demonstrate the resonant transmission through 0D-states. It is shown elsewhere [4], that a calculation of G_D from eq.1 with the measured conductances of the individual QPCs substituted for the transmissions T_A and T_B, gives good agreement with the measured G_D shown in fig.5.

In fig.6 it is demonstrated that the phase θ can also be varied by changing the area of the dot. Here G_D is measured as a function of gate voltage for a fixed B=2.5T. In fig.6a the voltage is varied on both gates (which affects the total area of the dot) and the oscillation period is 1mV. In fig.6b the voltage on one gate pair is kept fixed and varied on the other (which effects half of the dot area) and, as expected, the oscillation period is now 2mV.

The above measurements demonstrate that the conductance of a quantum dot in a high magnetic field can be strongly modulated by changing the magnetic flux enclosed by the current carrying electrons. If the Fermi energy coincides with the

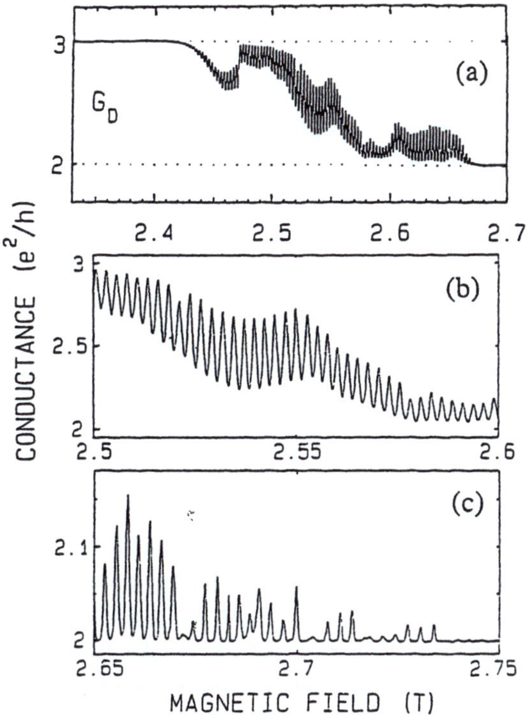

Figure 5. (a) Transition from the second to the third plateau of the dot conductance G_D, showing large oscillations with amplitudes up to 40% of e^2/h. (b) Enlarged oscillations from (a) showing their regularity (period B_0=2.5mT). (c) Region of low transmissions of the third edge channel (T_A, T_B≈0). The discrete conductance peaks demonstrate resonant transmission through 0D-states.

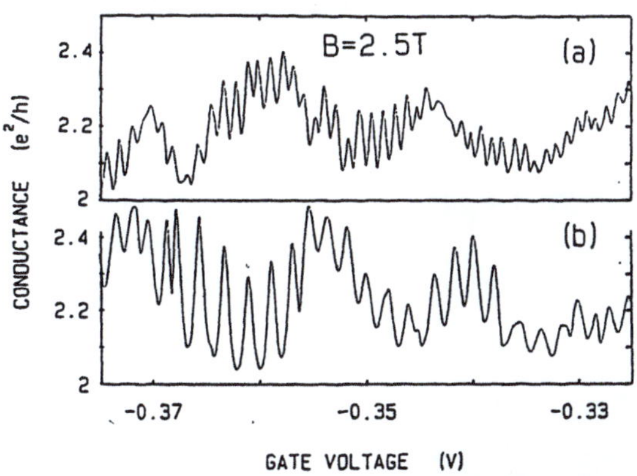

Figure 6. Conductance oscillations as a function of gate voltage for a fixed magnetic field B=2.5T. In (a) the voltage on both gate pairs is varied (period=1mV) and in (b) the gate voltage on one pair is kept fixed at -0.35 V and varied on the other gate pair (period= 2mV).

energy of a 0D-state, the transmission through the dot is at resonance. The rounded oscillations for large transmissions T_A and T_B and the sharp peaks for low T_A and T_B illustrate the effect of coupling of the discrete 0D-states with the continuous states in the 1D edge channel in the wide 2DEG regions outside the dot, which is in accordance with the results in fig.3 (there shown for N=16).

4. Transport through a finite 1D crystal of coupled quantum dots.

In this section, the principles of magnetically induced 0D-states in a single quantum dot will be extended to a series of 15 coupled quantum dots. The formation of a mini-band structure due to the periodic potential yields this device to be an artificial finite 1D crystal. The SEM photograph in fig.7 and the inset of fig.9 show the layout of a periodically corrugated gate geometry, which defines 15 quantum dots in series in the 2DEG. The ungated 2DEG of this sample has an electron density of $2.7 \cdot 10^{15} m^{-2}$ and a transport mean free path of 10μm. The voltage V_{g1} on the first gate defines the depletion region around the 16 "fingers" at a period of 200nm. The effect of making the voltage V_{g2} more negative is to increase the depletion region around the second gate. This reduces the area of each dot and lowers the Fermi energy in the conducting regions. The potential landscape in the 2DEG presumably resembles a periodic electrostatic confinement, with the asymmetric saddle-shaped maxima defined by the fingers.

The conductance measurements in this section are performed at 10mK and at a fixed magnetic field of B=2T. Fig.8 shows the first quantum Hall plateau for several fixed values V_{g1} and changing the voltage V_{g2} on the second gate. Large oscillations are seen below the first plateau (i.e. $G<e^2/h$), and in the plateau region deep downward peaks enclose smaller oscillations. Some of the deeper peaks are marked, to indicate their shift when the voltage V_{g1} is varied. The plateau region of the curve V_{g1}=-0.45V is measured and shown enlarged in fig.9. As can be seen, the two deeper peaks enclose 15 smaller oscillations, which corresponds exactly to the number of quantum dots.

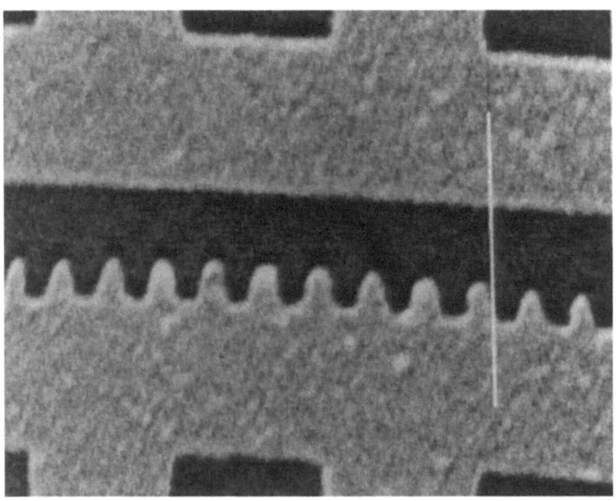

Figure 7. Scanning Electron Microscope photograph of the 1D crystal device. The corrugated gate contains 16 fingers in total. The white marker indicates 1μm.

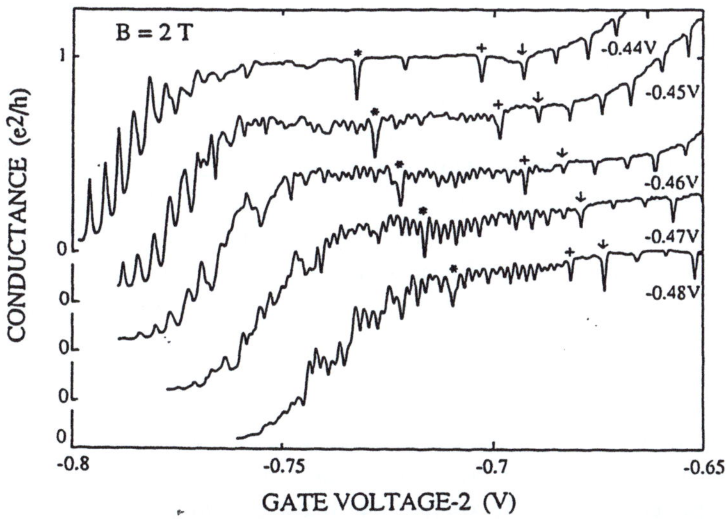

Figure 8. Conductance as a function of gate voltage-2 V_{g2} on the second gate at a fixed magnetic field of 2 T and for different values of gate voltage-1 V_{g1} on the first gate. Corresponding peaks are marked. The curves have been offset for clarity.

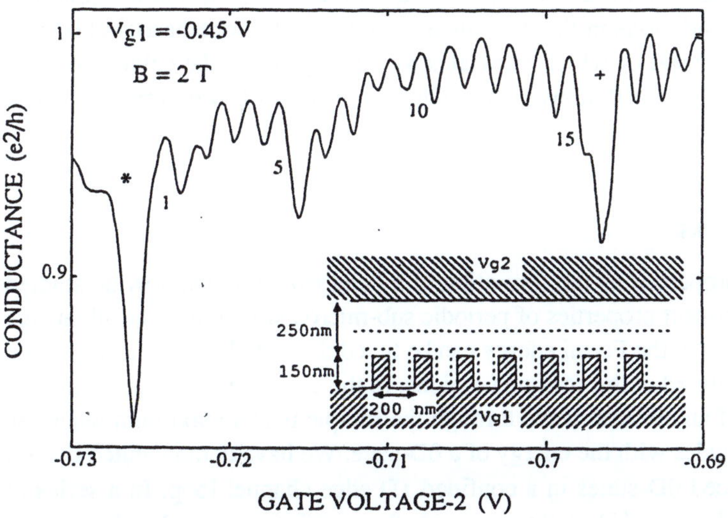

Figure 9. Conductance as a function of gate voltage-2 V_{g2} on the second gate at 2 T and V_{g1} = -0.45V on the first gate. The inset schematically shows the gate geometry; the dashed lines indicate the depletion regions in the 2DEG.

This simple counting comparison between the number of oscillations and the number of quantum dots, shows that the observed oscillations can be associated with the formation of a mini-band structure in the periodic 1D crystal device. The deeper peaks correspond to the energy gaps and the smaller oscillations with the discrete states, which form the mini-bands. This interpretation is supported by the calculations shown in fig.2 and 3. From going vertically through fig.3 one can also see that when the transmissions T are varied relatively fast compared to the phase θ, the Fermi energy moves through the mini-band structure. This situation corresponds to the large oscillations below the first plateau, where T changes from 0 (at pinch-off) to 1.

The effect of irregularities on the mini-band structure can also be seen in fig.8. The trace of V_{g1}=-0.46V shows 16 smaller oscillations between the marks "*" and "+", and in the curve of V_{g1}=-0.47V the gap marked by "+" is not seen. Note also the smaller oscillations on the left of the mark "*" indicating another mini-band, although clearly less regular. The 16 oscillations between two gaps cannot be explained by the number of quantum dots, not even when they are considered to be unequal. An additional scatterer somewhere in the device could be the cause of this extra oscillation. The disappearance of a gap is presumably due to irregularities, which introduces mixing between mini-bands. This also follows from further calculations [7] of the transmission T_N from eq.2 for N=16, in which irregularities are introduced by variations in the individual t's and θ's. With one barrier chosen in the middle of the crystal of which t deviates by a few percent, the calculations [5] show that the smaller oscillations group together in pairs of two, very similar to various parts of fig.8.

5. Concluding remarks.

Basic solid state properties as the formation of a band structure can now be studied from the linear transport properties of periodic sub-micron structures. The advantage of gated samples is that the Fermi energy can be tuned through the discrete electronic states, which are formed in the confined regions. In the case of a single quantum dot, the transmission of an electron wave through the sample is at a maximum when the Fermi energy coincides with the energy of a 0D-state. We have demonstrated this for magnetically induced 0D-states in a confined 1D edge channel loop. In a series of coupled quantum dots, the 0D-states group together and form a mini-band structure. The agreement between the observed number of oscillations in a band and the number of quantum dots, shows that the electron transport is phase coherent over the entire device. The fact that the mini-band structure originates from a modulation of the lowest 1D magnetic subband makes this device a true 1D crystal. The mini-band structure is reflected in the conductance, in a way that the 1D crystal is electrically insulating or conducting whether the Fermi energy is in a gap or a mini-band respectively. A simple model can account for the observed features in the conductance and illustrates the formation of a band structure in the solid state in terms of resonant transmission.

We thank C.W.J. Beenakker, J.E. Mooij, T.P. Orlando, G. Schön, and J.G. Williamson for valuable discussions, M. van Eijck for performing the calculations, J.R. Kraayeveld for assistance with the experiment, and A. van der Enden for his contributions to the fabrication of the devices, and the Stichting F.O.M. for financial support.

References

1. L. Esaki and R. Tsu, IBM J. Res. Dev. 14, 61 (1970).
2. L. Esaki and L.L. Chang, Phys. Rev. Lett. 33, 495 (1974).
3. see for a review on grown superlattices in a magnetic field: J.C. Maan, Festkörperprobleme 27, 137 (1987).
4. see also: B.J. van Wees et al., Phys. Rev. Lett. 62, 2523 (1989); L.P. Kouwenhoven et al., Proc. of the workshop on Science and Engineering of 1- and 0- dimensional Semiconductors, edited by C.M. Sotomayor-Torres and S.P. Beaumont (Plenum Press, 1990); L.P. Kouwenhoven et al., Proc. of the EP2DS8, to be published in Surface Science.
5. see also: L.P. Kouwenhoven, F.W.J. Hekking, B.J. van Wees, C.J.P.M. Harmans, C.E. Timmering, and C.T. Foxon, (to be published).
6. see for earlier theoretical work, but with different methods: R. Tsu and L. Esaki, Appl. Phys. Lett. 22, 562 (1973); D.J. Vezzetti and M.M. Cahay, J. Phys. D. 19, L53 (1986); H.W. Lee, A. Zysnarski, and P. Kerr, Am. J. Phys. 57, 729 (1989); S.E. Ulloa, E. Castano and G. Kirczenow (to be published).
7. F.W.J. Hekking, M. van Eijck, L.P. Kouwenhoven, and G. Schön, (in preparation).
8. C.G. Smith et al., J. Phys. Condens. Matter 1, 9035 (1989).
9. Y. Hirayama and T. Saku, Solid State Comm. 73, 113 (1990).
10. B.I. Halperin, Phys. Rev. B 25, 2185 (1982); P. Streda, J. Kucera, and A.H. MacDonald, Phys. Rev. Lett. 59, 1973 (1987); J.K. Jain and S.A. Kivelson, Phys. Rev. Lett. 60, 1542 (1988); M. Büttiker, Phys. Rev. B 38, 9375 (1988).
11. B.J. van Wees et al., Phys. Rev. Lett. 62, 1181 (1989).
12. A theoretical study of a 2D dot to which narrow leads are attached is given by: U. Sivan, Y. Imry, and C. Hartzstein, Phys. Rev. B 39, 1242 (1989); U. Sivan, and Y. Imry, Phys. Rev. Lett. 61,1001 (1988). Transmission through single quantum dots in the quantum Hall regime is treated also in a recent review by: M. Büttiker, volume of Semiconductors and Semimetals, entitled, "Nanostructured Systems" edited by M.A. Reed (Academic Press, Orlando, Florida, 1990).
13. M. Büttiker, Y. Imry, and R. Landauer, Physics Letters 96A, 365 (1983).
14. Conductance quantization at zero magnetic field has been observed in quantum point contacts by; B.J. van Wees et al., Phys. Rev. Lett. 60, 848 (1988); D.A. Wharam et al., J. Phys. C 21, L209 (1988).
15. L.P. Kouwenhoven et al., Phys. Rev. B 40, 8083 (1989); A similar structure has been studied experimentally by: P.C. Main, et al. Phys. Rev. B 40, 10033 (1989), and theoretically by: L.J. Glazman and M. Johnson, J. Phys. Condens. Matter 1, 5547 (1989).

Controllable Scattering, Fabry-Pérot States and Quantum Coupling in Ballistic Devices

C.G. Smith, R.J. Brown, M. Pepper, H. Ahmed, J.E.F. Frost, D.G. Hasko, R. Newbury, D.C. Peacock, D.A. Ritchie, and G.A.C. Jones*

Cavendish Laboratory, University of Cambridge, Cambridge CB3 0HE, UK
*Also at the GEC Hirst Centre, Wembley, Middlesex HA9 7PP, UK

1. Introduction

With the continued reduction in device dimensions it is now possible to explore the conduction processes of devices small in comparison to an electron mean free path. In this regime the notion of classical Boltzmann conductivity breaks down and we have to consider the ballistic nature of the electrons. If in addition the samples are small in comparison to λ_F, the Fermi wavelength of the electrons, and the temperatures are low enough, so that scattering is reduced and thermal smearing is minimised, then large quantum effects can be observed.

To maximise the chance of seeing quantum ballistic transport a combination of small sample size, large Fermi wavelengths, low temperatures and long electron mean free paths are required. For these reasons all the devices mentioned in this paper were fabricated using high resolution electron beam lithography to pattern sub-micron Schottky gates on modulation-doped GaAs-AlGaAs heterostructures. These substrates contain a high mobility two-dimensional electron gas (2DEG) less than 1μm below the surface which, because the donors are separated from the 2DEG by a spacer layer, can have an electron mean free-path greater, than 5μm at 4.2K. The application of a negative gate voltage depletes the carriers under them leaving a sub-micron conducting path defined in the 2DEG that can be varied in size, because the depletion of the 2DEG away from the gates is proportional to the gate voltage. Such a sample design was used by Thornton et al. [1] to look at the diffusive conduction properties of a 1-D path defined in a 2DEG between two 15μm long gates separated by 0.5μm. This principle was later used by van Wees et al. [2] and Wharam et al. [3] to look at ballistic transport in 1-D by shortening the length of the constriction to 0.3μm. In this paper we will be discussing three devices based on a ballistic 1-D channel.

2. Sample Fabrication and Measurement Techniques

The heterostructures used in all the devices mentioned in this paper have the same layer structure, and were grown by molecular beam epitaxy (MBE) on an undoped GaAs wafer. A superlattice buffer of 25 AlAs (2.4nm) GaAs (2.5nm) double layers was grown first followed by a 1μm undoped GaAs layer. On top of this an AlGaAs layer was grown as a spacer separating the 2DEG from the next layer of Si-doped AlGaAs, which is 40nm thick. Finally a 10nm capping layer is deposited on top. The resulting 2DEGs used with all three samples described subsequently had a carrier concentration of between $3\times10^{15}/m^3$ and $2.7\times10^{15}/m^3$ and a mobility of between $98m^2/Vs$ and $110m^2/Vs$.

The next stage in the processing was to pattern the device into a Hall bar with 8 Ohmic contacts and four Au gate pads, which overlap the mesa by 5μm between two sets of voltage probes, using optical lithography. In the last stage the fine metal gate pattern is produced on the mesa by electron beam lithography and an appropriate metallisation procedure connecting with the large metal bond pads. (See figure 1.)

All the measurements were made in a dilution refrigerator with a base temperature of 50mK. The conductance and resistance measurements were made using a low current, low frequency (70Hz) AC phase sensitive detector technique.

Springer Series in Solid-State Sciences, Vol. 97
Localization and Confinement of Electrons in Semiconductors
Editors: F. Kuchar · H. Heinrich · G. Bauer

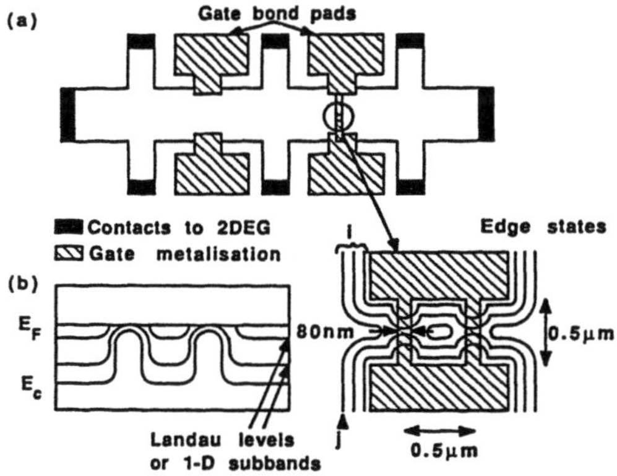

Figure 1. (a) A schematic diagram of the optically patterned heterostructure mesa showing the large gate bond pads and the Ohmic contacts. The circled region has an arrow pointing to an enlarged diagram of the detail of the quantum box. Three incident edge states, i, are marked as well as one reflected edge state j. (b). An energy diagram of the conduction band on going through the quantum box. The effect of the barriers on 1-D subbands or Landau levels is marked.

3. The Quantum Box

By defining a metal gate consisting of a 0.5μm wide gap between two 1μm long metal pads with two 80nm wide fine lines running between them, which have a centre to centre separation of 0.5μm (see figure 1 insert), a lateral quantum box can be defined. Because the narrow 80nm wide lines are 70nm above the 2DEG, capacitative edge effects result in the potential at the 2DEG being less under the fine lines than under the wider portions of the gate. This results in there being a range of gate voltages over which the 1-D channel is defined, but the barriers caused by the fine lines are lower than the Fermi energy.

3.1 Scattering and Tunneling Through Zero-Dimensional States

From the experiments of Wharam et al. [3] and van Wees et al. [2] in ballistic 1-D channels we know that the resistance at low temperature is inversely proportional to the number of 1-D subbands, i, within the device so that

$$R = h/2Ne^2T_N. \qquad (1)$$

The fact that the transmission coefficient T_N for the Nth subband remains unity is a consequence of the smooth potential at the entrance to and exit from the 1-D channel [4]. By putting barriers in the channel the transmission probabilities T_N no longer remain unity and we have introduced a controllable amount of scattering into the system. Changing the gate voltage has two effects, firstly the 1-D subbands with large transverse momentum are reflected from the barriers, resulting in the normal step-like variation of resistance with gate voltage, but in addition reflections between the barriers lead to the build up of quasi bound states above them every time an integer number of half Fermi wavelengths fit between the two. A small change in gate voltage moves these quasi bound states through the 1-D subband energies in the rest of the channel leading to large changes in the conductance of up to 60%, as the transmission probability is modulated [5,7], thus altering the conductance not by directly changing the carrier concentration, but by varying the transmission probability using quantum interference. In figure 2a the variation of conductance with gate voltage for such a devices is plotted showing three poorly quantized plateaux with well-defined large oscillations on them, which result from reflections off the quasi 1-D bound states. Such a potential in a 1-D channel has been modelled by Martin-Moreno and Smith [6], who calculated that the structure seen in the conductance versus gate voltage could be closely reproduced if the potential barriers are

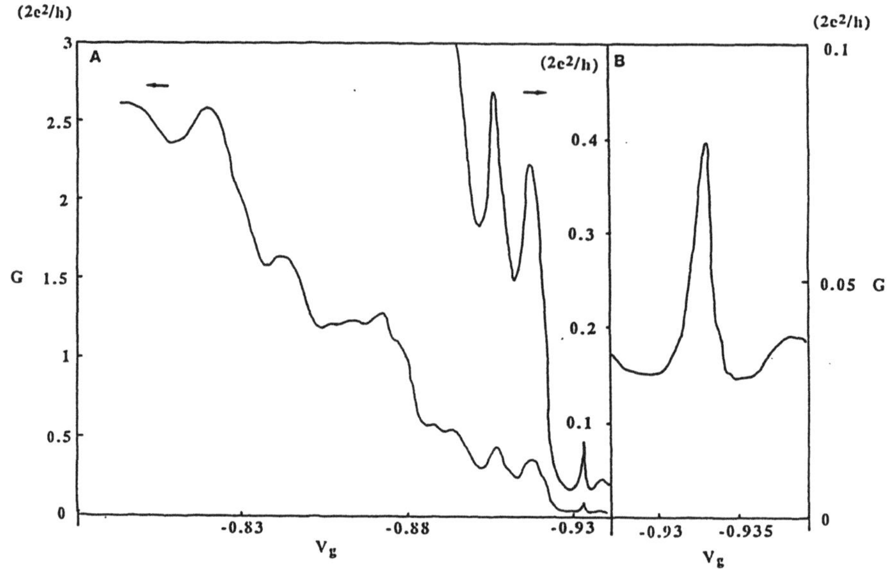

Figure 2. (A) A plot of the conductance versus gate voltage at 80mK with B=0 for the quantum box shown in figure 1. Three poorly quantized plateaux with oscillations can be seen. The vertical scale for the trace on the left is marked on the left hand side, while the scale for the enlargement of that trace is shown on the right hand side of figure 2A. Figure 2B shows an enlargement of the conduction peak associated with resonant tunnelling through a zero-dimensional state. The vertical scale for this is shown on the right of figure 2B.

assumed to have a rounding which extends over a distance of half the separation of the 2DEG from the gates. They also showed that the structure is greatly reduced if there are any differences in the heights of the two barriers.

At more negative gate voltages when the last 1-D subband has been reflected and the barriers are just higher than the Fermi energy the conductance drops sharply below $h/2e^2$. As the barriers are raised further a sudden sharp increase in the conductance of over 80% is observed as the zero-dimensional bound states within the device are pushed up past the Fermi energy (see figure 2b). By applying a DC bias voltage to the device when it is on resonance one can calculate the barrier heights to be 180μV and their width to be 80nm. The magnitude of the transmission probability suggests that the top of the potential is rounded [8].

3.2 Magnetic Edge States in a Quantum Box

When a magnetic field is applied to a 2DEG the energy levels are quantized into Landau levels that are forced to higher energies by the potential at the edge of the device, which means that even when the Fermi energy is between Landau levels in the middle of the device there are still states for conduction at the edge where the Landau levels are forced up through the Fermi energy. When the edge states carry all the current the electrons with forward momentum are confined to one edge of the device and electrons with reverse momentum are confined to the other side. As the edge states extend into the bulk of the sample by approximately one cyclotron orbit diameter L_c, where $L_c = (hk_F/2\pi eB\sqrt{2})$, then as long as the two edges of the device are further than L_c away from each other there will be no scattering of electrons with momentum in opposing directions. Where the edge states move through the potential caused by the gates in the quantum box sample they are reflected by the barriers in the same way as the 1-D subbands were. In this regime if the magnetic field is held constant and the gate voltage is swept to negative values the two-terminal resistance shows quantized plateaux at values of R given by

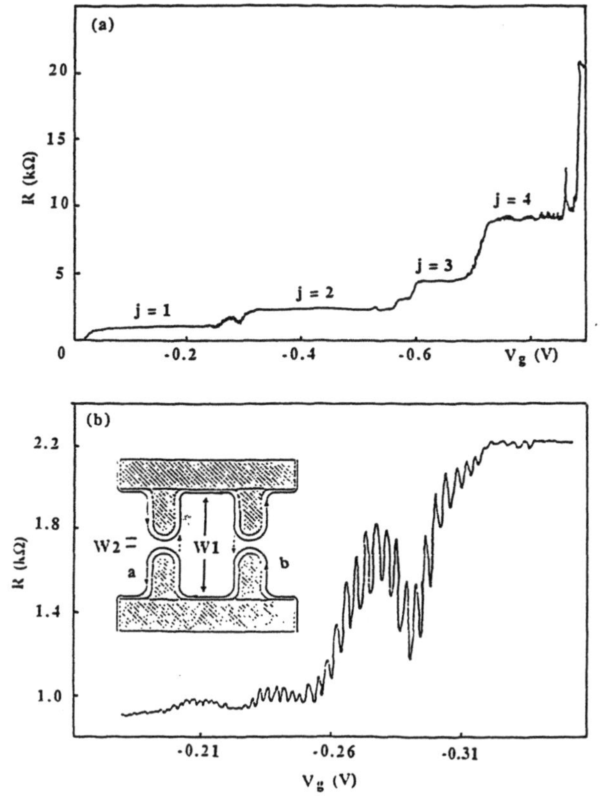

Figure 3. (a). The magneto-resistance of the quantum box at 2 Tesla and 100mK, showing the reflected Landau levels j = 1,2,3 and 4. (b) Detail of the magneto-resistance between -0.16 V and -0.36V showing the Aharonov Bohm oscillations, (small period) and the $\sin^2$ variation in amplitude (large period envelope). Insert. The full curve shows the relevant edge state, with the dashed curve indicating the tunneling paths. Taken from [10].

$$R = (h/2e2)[1/(i-j) - 1/i], \qquad (2)$$

where i is the number of incident Landau levels and j the number of reflected levels (see figure 3a). With a different arrangement of current and voltage probes the resistance is given by a different expression [8]. This was first observed for a single fine line gate over a 2DEG by Haug et al. and Washburn et al. [8,9]. In the quantum box the addition of the second barrier allows interference of the two opposite travelling edge states at two closely spaced points, because they are forced closer than L_c as they surmount the two barriers. This leads to a proportion of the current circulating round the closed edge states defined between the barriers, leading to the observation of the Aharonov Bohm effect [10]. These are shown clearly as the closely spaced oscillations in figure 3b. Both figure 3a and 3b were measured with a magnetic field of 2 Tesla, where changing the gate voltage changes the depletion width from the gate and therefore the width W1 and therefore the area A. As the area is changing in a constant magnetic field the flux contained within the central dot changes, causing an interference in the circulating edge state whenever

$$h/eB = dA. \qquad (3)$$

The insert in figure 3b shows how tunneling from one edge state to another at the barriers leads to a closed current-carrying loop. Such an effect was originally theoretically predicted by Sivan and Imry [11]. In addition to the Aharonov Bohm effect there is an additional envelope modulation of the amplitude which results from the variation of the overlap of the two oppositely travelling edge states as they go over the barrier. This modulation has been theoretically investigated by Yosefin and Kaveh [12] and is predicted by them to be proportional to $\sin^2(mW1/W2)$. Although both W1 and W2 vary with gate voltage they have a different constant of proportionality, resulting in a $\sin^2$ variation in the amplitude of the Aharonov Bohm effect.

4. The Fabry-Pérot Interferometer for Electron Waves

Extending the idea of a quantum box, a device was made using electron beam lithography and a metallisation lift-off technique consisting of a 1-D channel with two impenetrable barriers placed at the entrance to and exit from the channel with a separation of 1.02µm. A diagram of the central part of the metallisation is shown in figure 4. With a negative gate voltage on all four gates four voltage probes are defined, in the underlying 2DEG, inside a device with a variable width and variable length. Even when the reflecting gates B1 and B2 are grounded the Schottky potential creates enough band bending under the gates to lead to backscattering of the electrons leaving the 1-D channel, which destroys the well-defined plateaux in resistance by reducing the transmission probability for the electrons in those subbands [13].

From the variation of conductance with gate voltage for each pair of gates A1,A2 and B1,B2, the variation of the depletion width between each gate with gate voltage can be estimated. If Y is the width of the conducting region between gates A1 and A2, and U is the separation of the depletion boundaries around B1 and B2 then

$$Y = [3.27 + V_g(A1,A2)] \times 0.12\mu m \quad (4)$$

and

$$U = [5.92 + V_g(B1,B2)] \times 0.19\mu m. \quad (5)$$

These are calculated using the fact that the variation of conductance with gate voltage for each pair is linear and that when the gates just deplete the carriers under them, the channel width is that defined lithographically minus the distance from the 2DEG to the gates [14]. At the other extreme when the channels pinch off they have the width of one half Fermi wavelength.

The first observations showed that when the conduction width is defined and the sample length is altered by varying the voltage on the reflector gates B1 and B2, oscillations in the resistance are measured on a nearly constant background. Minima in resistance should be

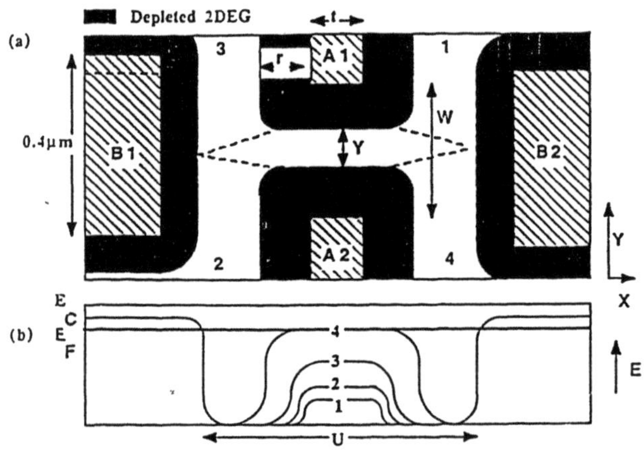

Figure 4. (a). A schematic diagram of the central gated region of the Fabry-Pérot interferometer. The black regions show the extent of the depletion of the 2DEG from the gates (diagonally shaded). The numbers 1,2,3 and 4 are the 2DEG regions that connect to the Ohmic contacts. (b) An energy diagram showing the 1-D subbands and the conduction band bending.

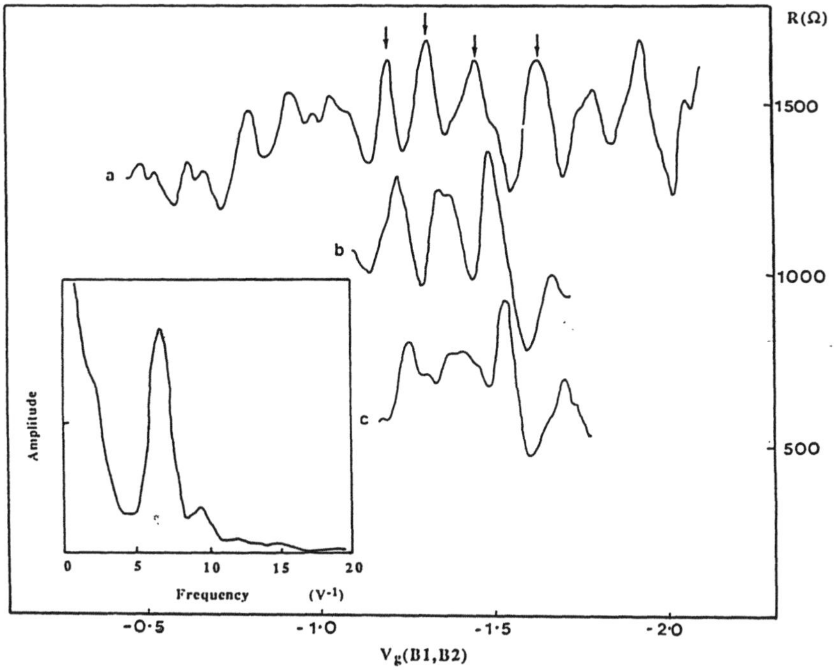

Figure 5. The resistance measured with current in contacts 1 and 3 and voltage measured off 2 and 4, while V_g(B1,B2) is varied. For curve a, Vg(A1,A2) = -2.225V; curve b, Vg(A1,A2) = -2.202V, while for curve c, Vg(A1,A2) = -2.157V. For clarity curve b has been shifted down 200Ω, and curve c has been shifted down a further 200Ω. The insert is a Fourier transform power spectrum of curve a. The amplitude has an arbitrary scale.

measured whenever there is an allowed state defined between the reflectors, which will occur when the following condition is satisfied:

$$K_x U + \text{Ø}_n = m\pi, \qquad (6)$$

where m takes integer values 0,1,2,3.. and Ø_n is the difference in phase felt by an electron passing through the nth 1-D subband defined between gates A1 and A2. When the channel between A1 and A2 is wide there are many 1-D subbands and the resulting structure in R versus Vg(B1,B2) is complicated. On the other hand, if the constriction is narrowed a simple geometric argument shows that electrons will only be reflected back through the constriction when

$$K_y < K_x \, Y/(U\text{-}L). \qquad (7)$$

The lowest 1-D subbands have the smallest transverse momentum, K_y, and are therefore most likely to resonate between the reflectors. Thus to see periodic oscillations the 1-D channel must not be so large that there are too many phase terms in equation (6) but must not be so narrow that equation (7) is not satisfied. When the device is set up so that there are three 1-D subbands in the channel periodic oscillations are observed as the length is varied using Vg(B1,B2), see figure 5. The Fourier transform power spectrum of the main figure is shown in the insert with a peak at a frequency of $7V^{-1}$, which from equation (5) corresponds to a change in sample length of 27nm. Shubnikov-de Haas measurements of the sample at these gate voltages indicate that the Fermi energy is reduced by the closely spaced negatively charged gates giving a value for the Fermi wavelength of 55nm. Thus the periodic oscillations in resistances of 30% occur every time the device length changes by one half Fermi wavelength, or in other words the device is behaving as a Fabry-Pérot interferometer for electron waves. The device is similar that suggested by Sols et al. [14] for a T-shaped 'stub' geometry.

5. Two 1-D Ballistic Resistors in Parallel

It has been shown by Wharam et al. [15] that two ballistic 1-D channels in series have a combined resistance equal to that of the narrowest 1-D channel. This result can be understood in terms of conservation of quantum number on passing from a 1-D subband in one channel into a subband in the other. To investigate how ballistic resistors add in parallel when the separation is less than the elastic and inelastic scattering length, the following device was devised. By using electron beam lithography a 150nm thick layer of PMMA (poly-methyl methacrylate) was exposed in the pattern shown in figure 6. The diagonal shaded area shows the region of PMMA which was exposed and developed away revealing the underlying GaAs heterostructure. Next three 100nm thick layers of Au are evaporated at 45° to the surface from three different directions to ensure that the Au conformally coats the device. This technique, which was first developed by Ford et al. [16], relies on that the fact that the surface states screen the 2DEG from the potential on the gate metal above the resist, however, where it touches the GaAs a Schottky gate is formed. Using this method small isolated gates of less than $0.2\mu m^2$ can be connected to the bond pads at the edge of the mesa. In this case a negative gate voltage depletes the 2DEG to define two 1-D ballistic channels separated by $0.5\mu m$.

The first measurements, which were made at 80mK, are shown in figure 7. The variation in resistance with gate voltage shows plateaux in resistance with values given by

$$R = h/4e^2N, \qquad (8)$$

where N is the number of 1-D subbands in each ballistic 1-D channel. This result implies that both 1-D channels have the same width and the same variation of width with gate voltage. The only plateau not well defined is that for N=1 which shows a double structure separated by 30mV in gate voltage. Using the same technique as that described in section 4 we can calculate the variation of depletion width r from the edge of the gates to get

$$r = 0.152(Vg+0.5V)\mu m. \qquad (9)$$

The double structure for N=1 implies a difference in width of 10nm, which corresponds to the resolution limit of PMMA.

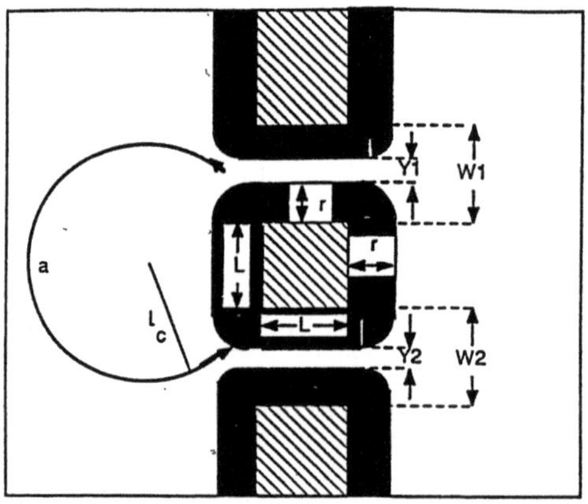

Figure 6. The gate metallisation is marked by the diagonal shading where it touches the GaAs substrate. The black shading shows the regions of the 2DEG when partially depleted. W1=W2 = 0.3µm while L = 0.2µm. The lengths Y1 and Y2 are the conducting widths of the two channels while Lc is the cyclotron orbit radius at 0.22 Tesla.

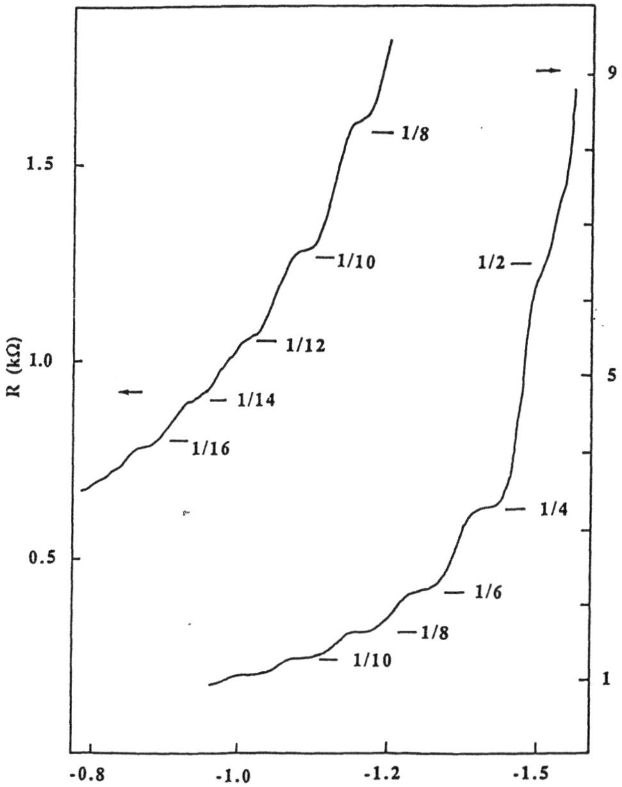

Figure 7. The resistance as a function of gate voltage for the two 1-D constrictions in parallel. The resistance scale for the curve on the left is marked on the left while that for the curve on the right is marked on the right. Taken from [17].

5.1 The Aharonov Bohm Effect Around the Central Dot

By applying a magnetic field perpendicular to the 2DEG we can again force the carriers to travel in edge states around the gated structure. In this case a circular path is defined around the central dot which shows Aharonov Bohm oscillations whenever a flux of h/e is threaded through the central area of $(L+2r)^2$. As each edge state has a width of a cyclotron radius L_c the accuracy of the Aharonov Bohm effect will be given by $2L_c(L+L+2r)/(L+2r)^2$, where this is the ratio of the conducting edge state area to the enclosed area. By keeping the magnetic field constant and changing the gate voltage, oscillations in the resistance are seen as the area of the central dot changes, in a similar manner as that seen in the quantum box. Using this result at different magnetic fields and for different ranges of gate voltage the accuracy of equation (9) can be tested, because the variation dA/dV_g can be measured from the Aharonov Bohm effect at a constant magnetic field. Figure 8 is a plot of the variation of gate voltage required to change the area to include one more flux quanta h/e. The resulting slope gives a value for dA/dVg of $2.3\times10^{-13}m^2V^{-1}$ at around Vg = -2.15V. This is close to that predicted from equation (9) which gives a value of $2.43\times10^{-13}m^2V^{-1}$.

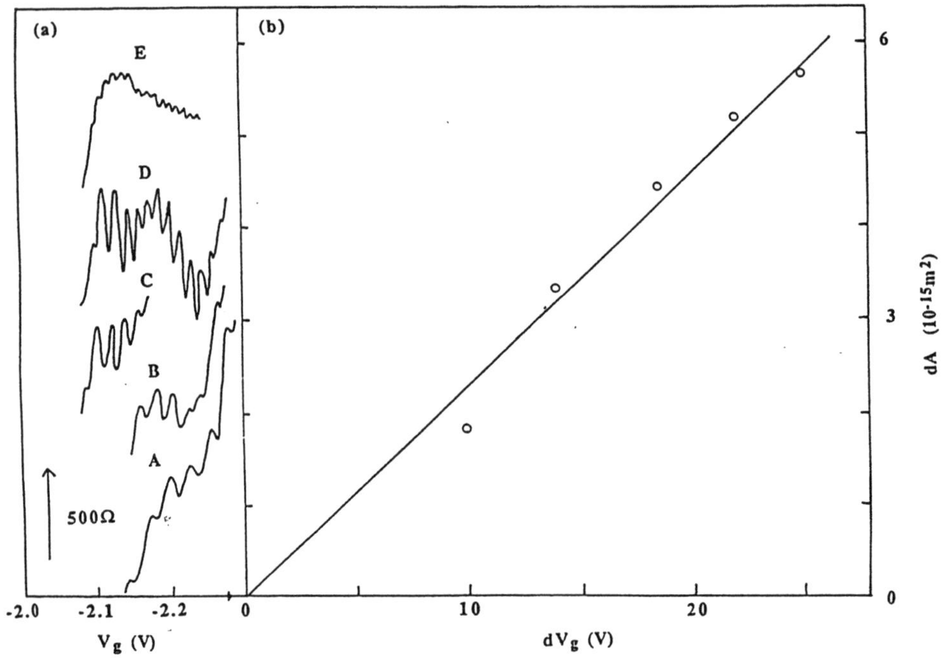

Figure 8. (a). The oscillations in resistance seen when the gate voltage is swept at T=80mK, for five different values of magnetic field. Each trace is displaced by 500Ω from the one above. Curve A is for B = 0.715 Tesla; curve B, B = 0.8 Tesla; curve C, B = 0.93 Tesla; curve D, B = 1.24 Tesla; curve E, B = 2.21 Tesla. (b) A plot of the change in area dA, at field B, required to allow one extra flux quantum (h/e) in the central dot, plotted against the period in gate voltage dV_g of the resulting oscillations. Taken from [17].

5.2 Locking of 1-D Subbands in the Two Channels

Although this device behaved in a repeatable manner when it was at helium temperatures, room temperature cycling, or sweeping the gate voltage past that required to deplete the channels, resulted in one 1-D channel pinching off at a different gate voltage to the other. This could be explained if there was a defect under the gate near to one 1-D channel which charged up differently on different occasions and therefore altered the width of one channel or the Fermi energy in that channel. For ease of representation various curves of conductance against gate voltage are represented in figure 9, where each curve has been displaced by 200mV from its neighbour. The curves were measured on different occasions, with increasing time as one goes from curve A to D. In curve A all the steps are quantized in units of $4e^2/h$ while curve B has a step of $2e^2/h$ followed by steps of $4e^2/h$. This can be explained if one 1-D channel is pinched off while there is still one subband in the other, and over the rest of the gate voltage range there are N subbands in one channel and N+1 subbands in the other. Curve C is similar, only now there are N subbands in one channel and N+2 in the other. For curve D the difference has increased further with N subbands in one corresponding to N+3 in the other. The fact that the slope of G versus V_g doubles as the steps go from $2e^2/h$ to $4e^2/h$ tells us that when the second 1-D channel opens the subbands in each channel depopulate at the same gate voltages. The probability of the subbands lining up over all these measurements is less than 10% [17], implying that there is a 90% chance that there is some locking mechanism between the 1-D subbands in the two channels which ensures that they depopulate at the same gate voltage.

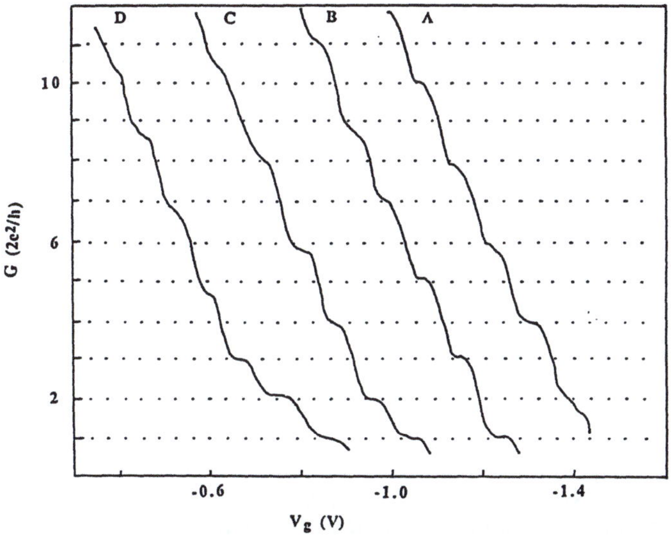

Figure 9. A plot of conductance versus gate voltage, taken on four different occasions, with increasing time going from curve A to D. Curves B,C and D are displaced 0.2V, 0.4V and 0.6V, respectively, to the left of curve A for clarity. The estimated differences in channel widths are 5.5nm for curve A, 11nm for curve B, 32nm for curve C and 45nm for curve D.

5.3 Possible Explanations for Locking of 1-D Subbands

From the low field Aharonov Bohm and mobility measurements we know that both the inelastic and elastic scattering lengths are longer than the circumference of the device, so each electron must pass through both 1-D channels simultaneously. The one-dimensional density of states in each subband means that the number of electrons in it is proportional to $(E_F-E_N)^{-1/2}$, where E_N is the energy of that Nth subband. Thus as a subband in one channel comes close to depopulation the number of electrons contained within that channel is divergent. Any electron-electron effects will then cause the energy of that subband to rise in comparison to the subbands in the neighbouring channels, resulting in an apparent locking mechanism. Alternatively, the close proximity of the two channels may result in space charge effects at the entrance of one constriction altering those at the entrance of the other, thus changing the current carried by the two.

6. Conclusions

By using electron beam lithography to pattern sub-micron metal gates on a high mobility GaAs GaAlAs heterostructure, we have made three different devices based on ballistic 1-D channels. The first, the quantum box, can be used to show scattering and resonant conduction through zero-dimensional states. With a magnetic field Aharonov Bohm oscillation in the circulating edge states of the quantum box can be observed, with a $\sin^2$ modulation of the amplitude. The second device has four closely spaced gates that define a device of variable width and length. This can be used as a Fabry-Pérot interferometer for electron waves showing 30% oscillations in resistance as the sample length is changed by one half Fermi wavelength. Finally the third sample, which defines two ballistic 1-D channels in parallel, shows steps in conductance of $4e^2/h$ as both channels change width. There is also an apparent locking mechanism which causes the 1-D subbands in the different channels to depopulate at the same gate voltage.

7. Acknowledgements

The authors would like to thank the S.E.R.C., the ESPRIT Basic Research Action, 3043 and the European Office of the U. S. Army for their financial support.

References

[1]. T. J.Thornton, M.Pepper, H.Ahmed, D.Andrews and G. J. Davies , Phys. Rev. Lett. **56**, 1198 (1986)
[2]. B. J.van Wees, H.van Houten, C. W. J. Beenakker, J. G.Williamson, L. P. Kowenhoven, D. van der Marel and C. T.Foxon, Phys. Rev. Lett. **60**, 848 (1988)
[3]. D. A. Wharam , T. J. Thornton , R.Newbury, M. Pepper, H.Ahmed, J. E. F.Frost, D. G.Hasko, D. C.Peacock, D. A.Richie and G. A. C. Jones, J Phys. C: Solid State Physics, **21**, L209 (1988)
[4]. L.I.Glazman, G.B.Lesovick, D.EKmelnitskii, R.I.Shekhter, Pys'ma v ZhETF, **48**, No 4, (1988)
[5]. C. G.Smith, M.Peppper, R.Newbury, H.Ahmed, D.G.Hasko, D. C.Peacock, J. E. F.Frost, D. A.Ritchie, G.A.C.Jones, J. Phys.: Condens. Matter **1**, 6763 (1989)
[6]. L. Martin-Moreno and C. G. Smith., J. Phys.: Condens. Matter, **1**, 5421 (1989)
[7]. C. G. Smith, M. Pepper, H. Ahmed, J.E.F. Frost, D.G. Hasko, R. Newbury, D. C. Peacock, D. A. Ritchie and G.A.C. Jones, Surface Science, **228**, 387 (1990)
[8]. R.J.Haug, A.H.MacDonald, P.Streda and K.von Klitzing, Phys. Rev. Lett. **61**, 2797 (1988)
[9]. S.Washburn, A.B.Fowler, H.Schmid and D.Kern, Phys. Rev. Lett. **51** 2801 (1988)
[10]. R. J. Brown, C.G. Smith, M. Pepper, M.J. Kelly, D.G. Hasko, H. Ahmed, D.C. Peackock, J.E.F. Frost, D. A. Ritchie, G.A.C. Jones, J. Phys.: Condens. Matter **1**, 6291 (1989)
[11]. U.Sivan and Y.Imry, Phys. Rev. Lett. **61**, 1001 (1989)
[12]. M.Yosefin and M.Kaveh, J. Phys.: Condens. Matter **1**, 10207 (1989)
[13]. C. G. Smith, M. Pepper, H. Ahmed, J.E.F. Frost, D.G. Hasko, R. Newbury, D. C. Peacock, D. A. Ritchie and G.A.C. Jones, J. Phys. Condens. Matter **1**, 9035 (1990)
[14]. F. Sols, M. Macucci, U. Ravaioli , and K. Hess, Appl. Phys. Lett. **54**, 350 (1989)
[15]. D. A.Wharam, M.Pepper, H.Ahmed, J. E. F. Frost, D. G.Hasko, D. C.Peacock, D. A.Ritchie , and G. A. C.Jones, J. Phys. C **21** L887 (1988)
[16]. C. J. B.Ford, T. J.Thornton, N.Newbury, M.Pepper, H.Ahmed, D. C. Peacock, D. A.Ritchi, J. E. F. Frost, and G. A. C.Jones, Appl. Phys. Lett. **54**, 21 (1989)
[17]. C. G. Smith, M. Pepper, R. Newbury, H. Ahmed, D. G. Hasko, D. C. Peacock, J. E. F. Frost, D. A. Ritchie, G. A. C. Jones and G. Hill. J. Phys.: Condens. Matter. **1**, 6763 (1989)

Part II

Localization and the Metal–Insulator Transition

Localization and Transport of Disordered 2D Electrons in Strong Magnetic Fields

B. Huckestein, W. Apel, and B. Kramer

Physikalisch-Technische Bundesanstalt, Bundesallee 100,
D-3300 Braunschweig, Fed. Rep. of Germany

Abstract. The frequency dependent conductivity is considered for disordered two-dimensional systems in a strong magnetic field at zero temperature. In the tails of the lowest Landau band the instanton technique can be applied. Rigorous results for the low frequency behaviour are derived in various asymptotic regimes and a vanishing dc conductivity is found.

Near the centre of the lowest Landau band the scaling properties of the exponential decay length of the modulus of the Green's function on a cylindrical surface of finite circumference and infinite length are investigated. A new version of the numerical finite-size scaling method is used which includes a quantitative statistical procedure of data evaluation. In contrast to other suggestions we find universal one-parameter scaling behaviour. The localization length diverges at the centre of the band with a critical exponent $\nu = 2.34 \pm 0.04$.

The results are compared with recent experimental findings concerning the temperature and frequency dependence of the Quantum Hall Effect. The temperature exponent of the phase coherence length is determined to be $p/2 = 1.0 \pm 0.1$.

1 Introduction

The quantized Hall effect (QHE) is one of the most remarkable quantum transport phenomena discovered during the past decade [1]. In spite of many experimental and theoretical efforts [2] a complete understanding of the physical mechanisms underlying the quantization of the dc Hall conductivity in integer multiples of e^2/h and the simultaneous vanishing of the dc magneto-conductivity as a function of the charge density is not yet achieved. One of the two widely accepted models is based on Anderson localization [3]. Here the phenomenon is explained in terms of localized, and extended states which pin the Fermi level, and carry the current, respectively. Up to now it was impossible to create a "predictive" theory by using this model. There is no controllable approximation which allows for calculation of the magneto-conductivity and the Hall conductivity as a function of the various parameters like temperature, magnetic field, impurity concentration etc., and to compare quantitatively with experimental data. Even a sound theoretical foundation of the localization model has not yet been developed.

Since localization is closely related to quantum mechanical transport at zero temperature, it is natural to start from quantum mechanical linear response theory in order to develop a localization model for the QHE. Although it is not possible to evaluate the expressions for the conductivity components explicitly, except in certain asymptotic limits, discussed in section 2, they can be used to derive a localization criterion which is, on the one hand, simple enough to be treated numerically, and which, on

Springer Series in Solid-State Sciences, Vol. 97
Localization and Confinement of Electrons in Semiconductors
Editors: F. Kuchar · H. Heinrich · G. Bauer

the other hand, still provides some information about the behavior of the conductivity throughout the whole band.

Having obtained quantitative information about the localization properties which consists essentially in certain scaling laws (section 3) one can even go beyond the scope of the zero temperature approach by adopting one additional assumption. This connects the zero-temperature localization properties to the finite temperature transport properties. It can also be extended to include finite frequency effects. The main idea is that instead of calculating the finite temperature and finite frequency conductivity of an infinite system, one can equally well consider the disorder averaged zero temperature, and zero frequency conductivity of a finite system with a size which is given by a phase coherence length L_Φ.

As L_Φ originates in phase randomizing inelastic processes as electron-phonon and electron-electron collisions it depends on temperature, frequency, and disorder simultaneously. Approaches to calculate L_Φ are therefore restricted to perturbation theory and phenomenological arguments. Nevertheless, as will be shown in section 4, the concept can be used to explain certain experimentally observed scaling features of the transport in the QH regime. In addition, the width of the Hall plateaus and the width of the maxima in the magneto-conductivity as a function of temperature and frequency can be quantitatively determined.

2 Frequency-dependent conductivity

The Kubo formula for the conductivity tensor at finite frequency ω and zero wavevector of the external electric field

$$\sigma_{ij}(\varepsilon_F,\omega) = e^2 \int dz_1 dz_2 \, \frac{i}{\hbar\omega + i0 + z_1 - z_2} \, \frac{f(z_1) - f(z_2)}{z_2 - z_1} \, K_{ij}(z_1, z_2), \tag{1}$$

$$K_{ij}(z_1, z_2) = \frac{1}{2} \overline{<\mathbf{r}|\dot{r}_i \delta(z_2 - H)\dot{r}_j \delta(z_1 - H) + \delta(z_2 - H)\dot{r}_j \delta(z_1 - H)\dot{r}_i|\mathbf{r}>}. \tag{2}$$

Here, H is the Hamiltonian of electrons moving in two dimensions in a perpendicular magnetic field and an impurity potential $V(\mathbf{r})$, $f(z)$ denotes the Fermi function, and the bar the average over the distribution of the impurity potential. The diagonal part of σ is related to the dielectric function (i.e., density-density response). Replacing the current- by the density operator via the equation of charge conservation, one gets for the real part

$$\operatorname{Re} \sigma_{xx}(\varepsilon_F, \omega) = \frac{e^2}{h} \int d\varepsilon \, \frac{f(\varepsilon - \hbar\omega/2) - f(\varepsilon + \hbar\omega/2)}{\hbar\omega} \, K(\varepsilon, \omega), \tag{3}$$

$$K(\varepsilon, \omega) = -\frac{\pi^2}{2}(\hbar\omega)^2 \int d^2r \, r^2 \, \overline{<\mathbf{r}|\delta(z_- - H)|0><0|\delta(z_+ - H)|\mathbf{r}>}, \tag{4}$$

where $z_\pm = \varepsilon \pm \hbar\omega/2$. We study this expression in two different regimes of the (Fermi) energy ε_F, using analytical and numerical methods, respectively. First, we sketch the analytical evaluation of the average conductivity for ε deep in the band tails [4]. Second, we explain the application of a numerical technique which works in the middle of the band.

Before describing the analytical approach we determine the parameters entering the conductivity. σ depends on the following four energies: the cyclotron energy $\hbar\omega_c$, the Landau level bandwidth in self-consistent Born approximation (SCBA) Γ, resulting from the impurity scattering, the external frequency ω, the energy ε_F, which is given by the Fermi energy as ω goes to zero. Thus, σ is e^2/h times a dimensionless function of three dimensionless ratios of these energies. Here, we deal with the strong field limit and keep only states of the lowest Landau level. That limit is reached by letting ω_c tend to infinity and keeping the three other energies finite. Since Γ is given by $\sqrt{\omega_c/\tau}$, the case considered here corresponds to scaling the elastic scattering time, τ, (which is defined independently from a magnetic field) proportional to the magnetic field. Measuring all the energies in units of the bandwidth Γ, the conductivity then depends on the two dimensionless parameters $E = |\varepsilon_F - \hbar\omega_c/2|/\Gamma$ and $\Omega = \hbar\omega/\Gamma$.

The evaluation of the impurity average K in Eq. (4) with the instanton method starts from the following observation. Only those potentials contribute to the average (i.e., the function-space integral), which yield z_+ and z_- as eigenvalues of H. Now one first finds the specific potential V^I, which gives the maximum weight in the average under these additional eigenvalue constraints. This emerges as a spatially localized solution of non-linear variational equations and is called intanton potential. Second, the contribution of the quadratic fluctuations around V^I is calculated. If these fluctuations are small, the probability distribution of the impurity potential is strongly peaked around the instanton potential V^I, which then mainly determines the average K. For the present model of an electron in the lowest Landau level moving in a Gaussian (white noise) impurity potential, both steps of the calculation can be rigorously done (see [4]). As it turns out, it is the parameter E which governs the size of the fluctuations. If $E \gg 1$, the fluctuations are of order E^{-2}, and, in the limit $E \to \infty$, the instanton method becomes asymptotically exact. This limit can be reached in two ways. First, consider the limit of weak impurity scattering, $\Gamma \to 0$. Then, $\Omega \gg 1$ and Ω/E stays finite. The result for the conductivity is given by

$$K(\varepsilon,\omega) \propto (\Omega E)^2 D(\frac{\Omega}{2E}) e^{-2[(E-\Omega/2)^2+\Omega^2]} \tag{5}$$

where D is a positive, slowly varying function (cf. [4]). Second, in the hydrodynamic limit, $|\varepsilon - \hbar\omega_c/2| \to \infty$, and hence $\Omega/E \to 0$, while Ω stays finite. Here, one finds a cross over between two different regimes.

$$K(\varepsilon,\omega) \propto 2\Omega E^3 \ln\frac{1}{\Omega^2} e^{-2[E-\Omega/2]^2}, \tag{6}$$

for $1 \gg \Omega^2 \gg E^{-2}$, and

$$K(\varepsilon,\omega) \propto 8\Omega^2 E^4 \ln\frac{1}{\Omega^2} e^{-2E^2}, \tag{7}$$

for $1 \gg E^{-2} \gg \Omega^2$.

In all cases, the conductivity is exponentially small. In the limit $\omega \to 0$ one obtains from Eq. (7)

$$\sigma_{xx}(\varepsilon_F,\omega) = \frac{e^2}{\hbar} 4\pi^2 z(\varepsilon_F)^2 \omega^2 \ln\frac{1}{\Omega^2}, \tag{8}$$

where $z(\varepsilon)$ is the density of states. Up to a numerical prefactor, and apart from the ω-scale in the argument of the logarithm, this agrees with the result of a phenomenological argument given earlier by Ono [5].

Near the centres of the Landau bands the instanton approximation becomes invalid. At present no method is available for evaluation of the Kubo formula which is able to produce quantitative results to be compared with experimental data, and which, on the other hand, is based on controllable approximations only. To get some insight into the transport properties, one can resort to numerical calculations. A numerical calculation of the conductivity via the Kubo formula is difficult due to problems related to the finite size of the systems and the ensemble averaging [6,7].

3 Scaling behaviour in the centre of the lowest Landau level

A quantity that is accessible to a thorough numerical investigation is the localization length $\xi(E)$ that can be defined by using the one-particle Greens function $G(E;\mathbf{r},\mathbf{r}')$,

$$\xi^{-1}(E) = -\lim_{|\mathbf{r}-\mathbf{r}'|\to\infty} \frac{1}{2|\mathbf{r}-\mathbf{r}'|} \ln |G(E;\mathbf{r},\mathbf{r}')|. \tag{9}$$

We note that as long as $\xi(E)$ is finite, the Greens function falls off exponentially and the dc conductivity vanishes as can be seen from Eqs. (3) and Eq. (4).

Numerical calculations can only be performed for finite systems. To obtain $\xi(E)$ from these calculations it is necessary to perform an extrapolation from finite to infinite system size. This extrapolation is not trivial and is done in two steps. The systems we used were quasi-onedimensional, i.e. their width M is much smaller than their length L. This geometry allows for a recursive calculation of the exponential decay length of the Greens function on the strip [8,9,10]

$$\lambda_M^{-1}(E) = -\lim_{k\to\infty} \frac{1}{2kl_c^2} \ln |G(E;0,k)|. \tag{10}$$

Due to the one-dimensional character of the system, $\lambda_M(E)$ is finite for every energy, even if $\xi(E)$ would diverge in the infinite two-dimensional system. The desired $\xi(E)$ can be obtained from $\lambda_M(E)$ by taking the limit $M\to\infty$.

The two limits involved in the calculation of $\xi(E)$ have to be treated numerically quite differently. The first limit $k\to\infty$ is straightforward, since $\lambda_M(E)$ is self-averaging and converges as a function of k to its limiting value. From $\ln|G(E;0,k)|$ for a large but finite value of k we get an estimate for the limit $k\to\infty$. Repeating the calculation for different realizations of the disorder potential produces a distribution of values for $\lambda_M(E)$ from which the uncertainty of the estimate of the limiting value can be estimated. With the present method, and the computer capacity available to us, it is possible to obtain uncertainties of less than 1 %.

The limit $M\to\infty$ is more difficult to perform. This arises from the fact that $\xi(E)$ may be divergent, as is the case in the present situation. For localized states λ_M converges to $\xi(E)$ for $M\to\infty$. However, the convergence will occur only for $M \gg \xi(E)$, so that very large systems may be needed [11]. If this condition is not fulfilled one has to extract for every energy E the asymptotic behaviour of the unknown functions $\lambda_M(E)$ from a finite range of M-values. This can be done, e.g. by Padé approximation, but the results are not very accurate [12] due to the large number of parameters involved.

This problem can be avoided, if the dependencies of $\lambda_M(E)$ on M and E are known. Although they are in general unknown, under certain assumptions the functional form of the asymptotic behaviour can be evaluated. These assumptions have to be justified by checking their consistency with the numerical data. Let us look at the asymptotic behaviour of λ_M as a function of M. For extended states, i.e. for those energies, where $\xi(E)$ diverges, $\lambda_M(E)$ is at least proportional to M in the limit $M \to \infty$. The renormalized exponential decay length $\Lambda(E,M) \equiv \lambda_M(E)/M$ takes at least a non-zero value in this limit, while for localized states $\Lambda(E,M) \to \xi(E)/M$. The central assumption of the data evaluation is that near the critical energy E_c, where $\xi(E_c)$ diverges and $\Lambda(E_c,\infty)$ is finite, changes in the length scale can be related to changes in the energy scale by a single parameter $\xi(E)$. Then, $\Lambda(E,M)$ has to be asymptotically a function of the dimensionless variable $M/\xi(E)$,

$$\Lambda(E,M) = \Lambda\left(\frac{M}{\xi(E)}\right). \tag{11}$$

Due to homogeneity requirements under combined transformations of the length and energy scales, the scaling parameter $\xi(E)$ must diverge at the critical point

$$\xi(E) = \xi_0|E - E_c|^{-\nu}, \ \xi_0 = \text{const}. \tag{12}$$

Since for localized states, $\Lambda(E,M) \to \xi(E)/M$ for $M \to \infty$, $\xi(E)$ can be identified with the localization length.

The scaling behaviour of Eq. (11) is only asymptotically correct for large M, there will be deviations of the data from it for small system sizes, say $M < M_{sc}$. Furthermore, it will only be correct in a critical region near the critical energy E_c, so there will also be deviations for large energies, $E > E_{sc}$.

Fig. 1(a) shows the renormalized exponential decay length $\Lambda(E,M)$ for various energies and system widths. Only the data for $E = 0.01$ show no M-dependence for large M so that E_c can be estimated to be $\lesssim 0.01$ and is taken as zero. For the data evaluation we took data for $M \geq M_{sc}$ and $E \leq E_{sc}$ and tried to fit them to the scaling law Eq. (11) and the power law Eq. (12). At this stage it is indispensable to check whether the numerical data are really compatible with the assumed scaling behaviour [10]. This can be done by comparing the estimated uncertainties of the numerical data with their deviation from the assumed scaling behaviour. A quantitative statistical procedure that follows this approach shows that for $M_{sc} = 16$ (measured in units of $\sqrt{2\pi}l_c$, $l_c^2 = \hbar/eB$) and $E_{sc} = 0.5$ the data are compatible with the assumption of a one-parameter scaling of the renormalized exponential decay length Λ. The statistical test is a very sharp instrument to analyze numerical data. For instance, it proves $E_{sc} = 1$ to be about *ten* orders of magnitude less probable than $E_{sc} = 0.5$ [10].

The resulting scaling curve and the power law dependence of the localization length are shown in Fig. 1(b). The critical exponent is obtained to be $\nu = 2.34 \pm 0.04$. This result is in remarkable agreement with $\nu = 7/3$ obtained for a semi-classical approximation [13]. In that calculation the correlation length σ of the random potential was taken to be large compared to the magnetic length l_c, while in our case the potential was δ-correlated. Furthermore, the data for the network model [14], where again $\sigma \gg l_c$, fit to the same scaling curve as our data. This suggests, that the scaling behaviour of the exponential decay length Λ is independent of the microscopic details of the random potential.

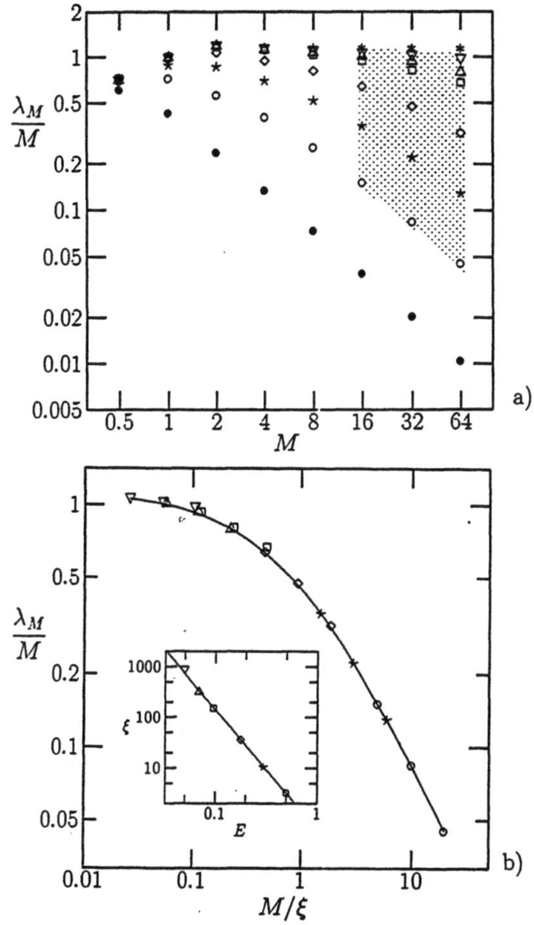

Figure 1: (a) The renormalized exponential decay length λ_M/M as a function of system width M for energies $E = 0.01$ ($*$), 0.05 ($\triangledown$), 0.07 ($\triangle$), 0.1 ($\square$), 0.18 ($\diamond$), 0.3 ($\star$), 0.5 ($\circ$) and 1 ($\bullet$). Data in the shaded area show scaling behaviour (Fig. 1(b)).
(b) The renormalized exponential decay length λ_M/M as a function of the scaling variable $M/\xi(E)$. The data plotted are taken from the shaded area of Fig. 1(a). The solid curve approximates the scaling function Eq. (11). The inset shows the localization length $\xi(E)$. (All lengths in Figs. (a) and (b) in units of $\sqrt{2\pi}l_c$).

4 Comparison with Experimental Data

In this chapter we apply the above described results of the scaling study to understand the experimentally observed temperature behaviour of the Quantum Hall effect [15,16], and its frequency induced breakdown [17,18,19].

4.1 Temperature dependence in the band centres

In order to obtain information about the temperature dependence of the magneto-transport we shall assume that temperature dependent phase breaking processes introduce an additional length $L_\Phi(T) \propto T^{-p/2}$, the phase breaking length, which plays the role of the size of phase coherent regions [20,21,22]. Since in our model the system width M limits the size of the phase coherent regions, we identify M with $L_\Phi(T)$. In this way, the scaling quantity $\Lambda(M/\xi(E))$ becomes a temperature dependent function. The variable $L_\Phi(T)/\xi(E)$ is a function of $ET^{-\kappa}$ with $\kappa = p/2\nu$. The n-th derivative of Λ with respect to E at $E = 0$ is proportional to $T^{-n\kappa}$, if it is non-zero. Since in the centre of a Landau band, E is proportional to the magnetic field B, the derivatives with respect to B show the same temperature dependence. This is exactly the behaviour observed experimentally by Wei et al. in measurements of the first three derivatives of the magneto-resistivity and the Hall resistivity in the Quantum Hall regime [15]. From the experimentally observed value $\kappa = 0.42 \pm 0.04$, we obtain $p/2 = 1.0 \pm 0.1$ for the exponent of the phase coherence length $L_\Phi(T)$. In the diffusive regime $L_\Phi(T)$ will be given by $L_\Phi(T) = \sqrt{D\tau_i}$. τ_i is the temperature dependent inelastic scattering time, and D is the diffusion constant. If D is independent of temperature, $\tau_i \propto T^{-p}$, i.e. $\tau_i \propto T^{-2}$.

4.2 Temperature Dependence of the Plateau Width

For a discussion of the temperature dependent plateau width, we have to take into account the density of states. We assume that it scales according to

$$\rho(B, E) = \frac{1}{2\pi l_c^2 \Gamma} \varrho(E). \tag{13}$$

Near the centre of the Landau bands $\varrho(E)$ may be expanded, $\varrho(E) = \varrho_0 - \varrho_1 x^2 + \ldots$. The constants may be determined either by comparing with theoretically determined densities of states [23,24,25,26] or used as fitting parameters for experimental data. The scaling assumptions may be justified when the disorder broadened Landau bands are well separated, i.e. $\hbar\omega_c \gg \Gamma$ (see section 1). This assumption is fulfilled in the experimental situation described below.

The width of the Hall plateaus as well as the width of the maxima in the magneto-conductivity may be estimated by using the following argument. In the energy regions where the localization length is small compared with the phase breaking length the states will not contribute to transport since the electrons are localized in a finite region within the coherence time. Here the longitudinal (magneto-)conductivity is vanishingly small. The Hall conductivity adopts its plateau value. On the other hand, if the localization length is large compared with the phase breaking length, the electrons are frequently scattered inelastically between the states such that they do not feel localization and transport becomes possible. The magneto-conductivity is finite, and the Hall conductivity deviates from the ideal plateau value. The condition

$$L_\Phi(T) = \xi(\pm E_c(T)) \tag{14}$$

defines quasi-mobility edges. They separate regions of quasi-extended states (due to inelastic scattering) from energy regions of localized states. If the Fermi energy lies between the mobility edges corresponding to a given Landau band $\sigma_{xx} \neq 0$, $\sigma_{xy} \neq ie^2/h$ (i integer). If the Fermi energy lies between the upper and lower mobility edges of two successive Landau bands, E_c^N and $E_{c'}^{N+1}$, respectively, $\sigma_{xx} = 0$, and $\sigma_{xy} = (N+1)e^2/h$ (Fig. 2)(a). The width of the plateaus is given by the distance between the quasi-mobility edges via the condition

$$\int_{-\infty}^{E_{c'}^{N+2}} \rho(B_2, E)dE = \int_{-\infty}^{E_c^{N+1}} \rho(B_1, E)dE = n, \tag{15}$$

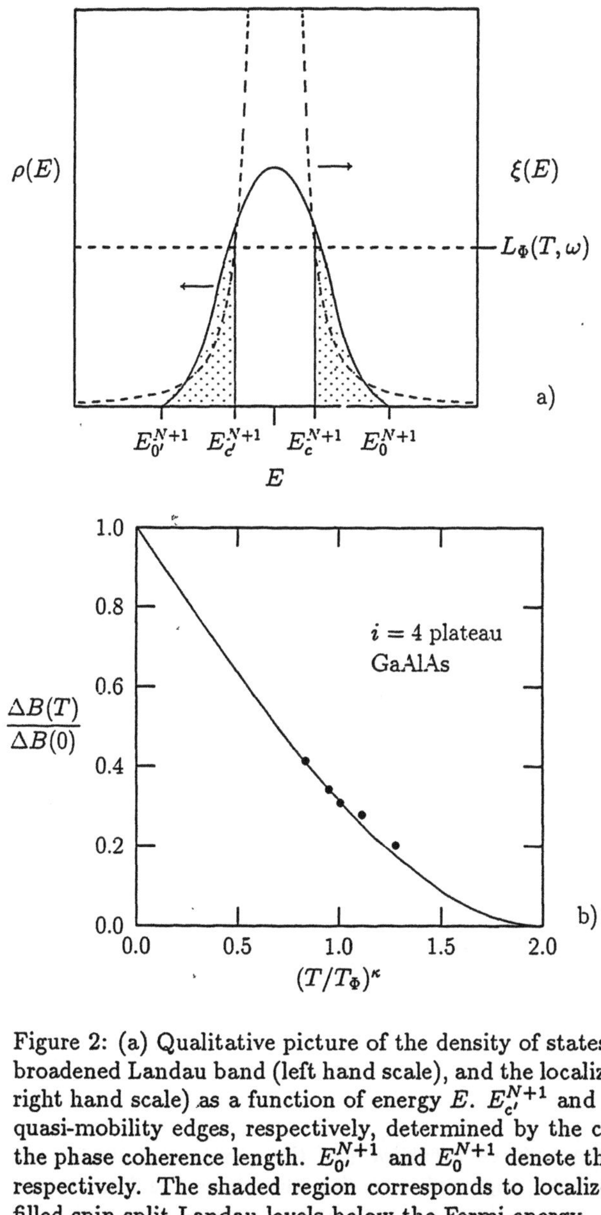

Figure 2: (a) Qualitative picture of the density of states $\rho(E)$ (solid line) of a disorder broadened Landau band (left hand scale), and the localization length $\xi(E)$ (dashed line, right hand scale) as a function of energy E. $E_{c'}^{N+1}$ and E_c^{N+1} are the lower and upper quasi-mobility edges, respectively, determined by the condition $\xi(E_c^{N+1}) = L_\Phi(T,\omega)$, the phase coherence length. $E_{0'}^{N+1}$ and E_0^{N+1} denote the lower and upper band edges, respectively. The shaded region corresponds to localized states. N is the number of filled spin split Landau levels below the Fermi energy.
(b) Width $\Delta B(T)/\Delta B(0)$ as a function of $(T/T_\Phi)^\kappa$. Dots correspond to experimentally obtained data for the fourth Hall plateau of a GaAlAs sample characterized by an electron density $n = 5.2 \cdot 10^{15}\,\mathrm{m}^{-2}$ and a mobility $\mu = 300\,000\,\mathrm{m}^2\mathrm{s}^{-1}$. $T_\Phi = 2.3\,\mathrm{K}$ and $\kappa = p/2\nu = 0.42$. As the magnetic field B is about $5\,\mathrm{T}$, $\hbar\omega_c \approx 7\,\mathrm{meV}$. From the mobility and electron density $\Gamma \approx 1\,\mathrm{meV}$ is obtained.

n is the electron density. Using the above discussed assumptions for the localization length and the density of states one obtains

$$\Delta B(T) \approx \frac{nh}{e}\left(\frac{1}{N+1/2+F(T)} - \frac{1}{N+3/2-F(T)}\right), \tag{16}$$

with $F(T) = 0.37((T/T_\Phi)^\kappa - 0.08(T/T_\Phi)^{3\kappa})$. Here, we have inserted $E_c(T) = (T/T_\Phi)^\kappa$ as obtained from the condition Eq. (14) into the integrated density of states. T_Φ is an adjustable constant containing D and ξ_0, $\kappa = p/2\nu$, N is the number of completely filled spin split Landau bands. Fig. 2(b) shows a fit to experimental data obtained for the fourth Hall plateau of a GaAlAs heterostructure [16]. The agreement between theory and experiment is rather good given the crude approximation of the density of states and in view of the fact that any dependence of T_Φ on the band index is neglected. More experiments, especially at lower temperatures and for different samples, are necessary to investigate the dependence on the various parameters systematically.

With a similar argument one can discuss also the width ΔB_{xx} of the peaks in σ_{xx} as function of the temperature. One obtains straightforwardly that $\Delta B_{xx}^{N+1}(T) \propto (N+1)^{-1}T^{-\kappa}$ in agreement with the results of Wei et al. discussed above [15]. For the ratio between the width of the two spin split components of the second Landau band the theoretical argument gives $\Delta B_{1\downarrow}/\Delta B_{1\uparrow} = 5/7$ ($N = 3$, 2, respectively) in agreement with experiment (Fig. 2 of [15]).

4.3 Frequency Dependence of the Plateau Width

With the same argument the width of the Hall plateaus as a function of the frequency can be obtained. A finite frequency will introduce an additional length

$$L_\Phi(\omega) = \sqrt{D/\omega}. \tag{17}$$

The phase breaking length $L_\Phi(T,\omega)$ is now given by

$$L_\Phi(T,\omega)^{-2} = L_\Phi(T)^{-2} + L_\Phi(\omega)^{-2}. \tag{18}$$

As before, with Eq. (14) and (16) $\Delta B(T,\omega)$ is obtained. Unfortunately, only a few experimental data are presently available, such that a fit to the data would not be very significant.

4.4 The Frequency Induced Breakdown of the Quantized Hall Effect

In order to discuss the frequency induced breakdown of the quantum Hall plateaus [17,18] observed recently, the above argument involving the phase breaking length cannot be used since one would have to take into account strong changes in the diffusion constant with energy. It is not clear yet, in how far this would still be consistent with the model assumption of the phase breaking length at all.

However, the physical mechanism that leads to the vanishing of the Hall plateaus with increasing frequency can still be discussed within the localization picture. The experimental situation is qualitatively as depicted in the Figure 2(a). The Landau bands are well separated. The density of states in the middle of the quasi-gap is Gaussian small as compared to the density of states in the centres of the bands. In this situation the centre of the Hall plateau corresponds to the energy region between two successive Landau bands, $E_0^N < E < E_{0'}^{N+1}$ (cf. Fig. 2(a)). The width of the plateau as a function of the magnetic field is almost completely due to the localized states in the upper ($E_c^N < E < E_0^N$) and lower ($E_{0'}^{N+1} < E < E_{c'}^{N+1}$) parts of the Nth and the $(N+1)$th bands, respectively. Deep in the localized tails the conductivity may be described by the asymptotic instanton method discussed in section 2. The magnetoconductivity σ_{xx} is here vanishingly small due to the rather small density of states as long as the frequency is small (cf. Eq. (8)).

Let us now assume, for instance, that the Fermi energy is at the lower edge of the upper band, near $E_{0'}^{N+1}$. This corresponds roughly to the centre of the plateau. If now the frequency of the electric field is such that a transition can be made from the localized states near $E_{0'}^{N+1}$ to the quasi extended state region near the band centre, the Hall conductivity will suddenly increase considerably. As similar argument can be made for the upper edge of the lower band, i.e. the lower half of the plateau. The critical frequency for the vanishing of the plateau width is then given by the condition

$$h\nu_0 \approx \Gamma. \tag{19}$$

For the sample mentioned above $\Gamma \approx 1\,\mathrm{meV}$ which corresponds roughly to $\nu_0 \approx 240\,\mathrm{GHz}$. This values is in respectable agreement (in view of the rather crude estimate) with the experimental result $\nu_0 \approx 160\,\mathrm{GHz}$ [19]. The experiments of the russian group [17] cannot be discussed with this simple argument since the mobility of the samples used were rather low. The widths of the Landau bands are of the order of $\hbar\omega_c$ such that the independent band approximation cannot be applied.

Acknowledgements

It is a pleasure to acknowledge enlightening discussions with Friedl Kuchar and Klaus von Klitzing.

References

[1] K. v. Klitzing, G. Dorda, and M. Pepper, Phys. Rev. Lett. **45**, 494 (1980).

[2] *The Quantum Hall Effect*, edited by R. E. Prange and S. M. Girvin, Graduate Texts in Contemporary Physics, Springer-Verlag, Berlin, 1987.

[3] H. Aoki, Rep. Prog. Phys. **50**, 655 (1987).

[4] W. Apel, J. Phys.: Condens. Matter **1**, 9387 (1989).

[5] Y. Ono, J. Phys. Soc. Jpn. **51**, 237 (1982).

[6] T. Ando, J. Phys. Soc. Jpn. **53**, 3101 (1984).

[7] T. Ohtsuki and Y. Ono, J. Phys. Soc. Jpn. **58**, 956 (1989).

[8] A. MacKinnon and B. Kramer, Phys. Rev. Lett. **47**, 1546 (1981). Z. Phys. B **53**, 1 (1983).

[9] B. Huckestein and B. Kramer, Solid State Comm. **71**, 445 (1989). Submitted to Phys. Rev. Lett..

[10] B. Huckestein, in *Proceedings of the International workshop on Anderson transition and mesoscopic fluctuations, PTB Braunschweig, Germany, 9–12 January, 1990, Physica A (to be published).*

[11] B. Kramer, A. MacKinnon, and D. L. Weaire, Phys. Rev. B **23**, 6357 (1981).

[12] H. Aoki and T. Ando, Phys. Rev. Lett. **54**, 831 (1985). T. Ando and H. Aoki, J. Phys. Soc. Jpn. **54**, 2238 (1985).

[13] G. V. Mil'nikov and I. M. Sokolov, JETP Lett. **48**, 536 (1988).

[14] J. T. Chalker and P. D. Coddington, J. Phys. C **21**, 2665 (1988).

[15] H. P. Wei, D. C. Tsui, M. Paalanen, and A. M. M. Pruisken, Phys. Rev. Lett. **61**, 1294 (1988). H. P. Wei, S. W. Hwang, D. C. Tsui, and A. M. M. Pruisken, in *Proceedings of the Eighth International Conference on Electronic Properties of Two-Dimensional Systems, Grenoble, France, 4–8 Sept, 1989, Surf. Sci. (to be published).*

[16] F. Kuchar, private communication.

[17] L. A. Galchenko, I. M. Grodnenskiĭ, M. V. Kostovetskiĭ, and O. R. Matov, JETP Lett. **46**, 542 (1987).

[18] K. Y. Lim, I. Auer, F. Kuchar, G. Weimann, W. Schlapp, and A. Forchel, in *Proceedings of the Eighth International Conference on Electronic Properties of Two-Dimensional Systems, Grenoble, France, 4–8 Sept, 1989, Surf. Sci. (to be published).*

[19] F. Kuchar, private communication.

[20] D. J. Thouless, Phys. Rev. Lett. **39**, 1167 (1977).

[21] P. W. Anderson, E. Abrahams, and T. V. Ramakrishnan, Phys. Rev. Lett. **43**, 718 (1979).

[22] B. L. Altshuler and A. G. Aronov, in *Electron-Electron Interactions in Disordered Solids*, edited by A. L. Efros and M. Pollak (North Holland, Amsterdam, 1985), p. 1.

[23] T. Ando, in *Anderson Localization*, edited by Y. Nagaoka and H. Fukuyama (Springer-Verlag, Berlin, 1982), p. 176.

[24] F. Wegner, Z. Phys. B—Condensed Matter **51**, 279 (1983).

[25] L. Schweitzer, B. Kramer, and A. MacKinnon, J. Phys. C **17**, 4111 (1984).

[26] E. Brezin, D. J. Gross, and C. Itzykson, Nucl. Phys. B **235[FS11]**, 24 (1984).

Critical Behaviour of the Metal–Insulator Transition

A. MacKinnon

Blackett Laboratory, Imperial College of Science, Technology and Medicine, Prince Consort Rd., London SW7 2BZ, UK

Abstract The critical behaviour at the disorder induced metal–insulator transition (MIT) has been the subject of considerable work over the years. Analytical Theory has produced critical exponents ranging from $\frac{1}{2}$ to about 2 with a concentration around 1 and $\frac{1}{2}$ and experiment has tended to give 1 except for uncompensated semiconductors which have $\frac{1}{2}$. Numerical results have also yielded indices ranging from about $\frac{1}{2}$ to about 2.

New calculations will be presented based on the finite size scaling method, but carried out with much higher accuracy than hitherto. Results will be presented for the the case of normal scattering as well as magnetic and spin–orbit scattering in 2D and 3D.

1. Introduction

In spite of the large volume of work published over the years the values of the critical exponents for the disorder induced Metal–Insulator transition remain controversial. Since our ability to predict these values is a measure of our understanding of the underlying physics, it cannot yet be claimed that the problem has been solved.

One important advance was the prediction of the existence of 3 Universality classes [1], which I shall term *Normal*, *Magnetic* and *Spin–Orbit*. These classes are more usually called *Orthogonal*, *Unitary* and *Symplectic*, especially by field theorists, but these names are not very informative to non–specialists. Within each class any model should yield the same critical behaviour. This assumption is actually crucial to any attempt to compare theory and experiment. In practice no theory is yet capable of working with realistic Hamiltonians, so the idea of universality provides the justification for comparing the results of very simplified model calculations with experiments.

The *Normal* Class contains all Hamiltonians with time–reversal symmetry and without spin effects. Thus all real Hamiltonians without spin, including the Anderson Hamiltonian and simple cases of off–diagonal disorder are in this class. Hence this class represents the normal case of scattering from non–magnetic impurities etc.

Springer Series in Solid-State Sciences, Vol. 97
Localization and Confinement of Electrons in Semiconductors
Editors: F. Kuchar · H. Heinrich · G. Bauer

The *Magnetic* Class contains all Hamiltonians without time–reversal symmetry, including spinless models with random phases in off–diagonal elements and models with spin but with real matrix elements. Physically this case might be a system with random magnetic fields due to magnetic impurities. It is not clear whether systems with constant magnetic fields should be included here (e.g. Quantum Hall Effect).

The *Spin–Orbit* Class contains Hamiltonians with random spin–orbit coupling *and* time–reversal symmetry. For this model there is believed to be a metal–insulator transition in both 2D and 3D, but it has proved difficult to make any reliable estimates of the critical exponents.

It should be noted that these classes are idealisations. Real systems probably do not fall squarely into one particular class but may change behaviour as a function of certain parameters.

2. Tight–Binding Models

We shall consider 6 different models based on the general tight–binding Hamiltonian

$$\mathcal{H} = \sum_i |i\rangle \varepsilon_i \langle i| + \sum_{i \neq j} |i\rangle V_{ij} \langle j| \tag{1}$$

where

$$V_{ij} \begin{cases} \neq 0 & \text{if } i \text{ and } j \text{ are nearest neighbours} \\ = 0 & \text{otherwise} \end{cases}$$

2.1 Normal Models

We have considered 3 distinct models which should all fall into the normal class [2], namely

Model	Hamiltonian		Accuracy
Anderson–1	$-\frac{1}{2}W < \varepsilon < \frac{1}{2}W$	$V = 1$	0.2%
Anderson–2	as above[3]		1%
Off–Diagonal–1	$\varepsilon = 0$	$V = \pm 1$	0.1%
Off–Diagonal–2	$\varepsilon = 0$	$-\frac{1}{2} < V < \frac{1}{2}$	0.2%

where, in each case the 3rd column gives the statistical accuracy with which the calculations were carried out.

It should be noted that the Anderson Model has 2 independent parameters, energy E, and disorder W, whereas the off-diagonal models only have the energy. Here data will be presented in which the disorder is varied for a fixed energy. Data for varying energy will be published later.

2.2 Other Universality Classes

For the *Magnetic* and *Spin–Orbit* classes models are chosen which have *Local Gauge Invariance*[1]: they have purely off–diagonal disorder, where the mean value of the matrix elements, $< V >$ is zero. In this case the ensemble–averaged 1–particle Green's function is diagonal and the phase coherence length is zero. It is expected that such models will show scaling behaviour for smaller system sizes while accurately reflecting the Universality class to which they belong.

3 Models were considered

Model	Hamiltonian	Accuracy
Magnetic 3D	$V_{ij} = \exp(i\theta_{ij})$	0.2%
Spin–orbit 3D	$V_{ij}^{\sigma\sigma'} = \exp(i\theta_{ij}^{\sigma\sigma'})$	1%
	$V_{ij}^{-\sigma,-\sigma'} = \sigma\sigma' \left\{ V_{ij}^{\sigma\sigma'} \right\}^*$	
Spin–Orbit 2D	as above	0.2%

where in each case $\sigma = \pm 1$ and θ is completely random ($0 \le \theta < 2\pi$) subject only to the constraint of hermiticity. In all 3 cases energy is the only parameter in the model.

3 Calculations

3.1 Transfer Matrices

Calculations were carried out by the now well established *Transfer Matrix* method[4]. Essentially the method calculates the probability of transmission of an electron through a long narrow system of length L and cross–section M^{d-1} where d is the dimensionality of the system. For long lengths L this will be an exponential function of L of the form

$$T(L, M) \approx T_0 \exp(-2L/\lambda_M) \tag{2}$$

where λ_M is the localisation length in the quasi–1D system.

It should be noted that many of the results presented here have an accuracy of 0.1% or 0.2% rather than the traditional 1%. This was done to check whether contradictions in previous results were due to systematic propagation of errors.

3.2 Scaling

In practice[4] it is found that the *renormalised localisation length* $\Lambda = \lambda_M/M$ obeys a scaling relationship of the form

$$\frac{\mathrm{d}\ln\Lambda}{\mathrm{d}\ln M} = \chi(\ln\Lambda) \tag{3a}$$

$$\Lambda = f(M/\xi) \tag{3b}$$

where the length scale ξ contains all information about energy, disorder, etc. This relationship is usually termed *one–parameter scaling*.

Consideration of the limiting cases allows us to identify the physical meaning of the parameter ξ. For strongly localised states we expect

$$\Lambda = \xi/M \qquad \chi = -1 \tag{4}$$

so that ξ is clearly the localisation length.

On the other hand, in the case of weak scattering, perturbation theory[4,5] gives a result like weak scattering in 1D with the disorder, W, averaged over the cross section of the strip or bar namely

$$W'^2 = W^2/M^{d-1} \tag{5a}$$

$$\lambda_M \sim W'^{-2} = W^{-2}M^{d-1}. \tag{5b}$$

When combined with (3b) this gives

$$\Lambda \sim (M/\xi)^{d-2} \qquad \chi = d-2. \tag{6}$$

Close to the Metal–Insulator transition $\chi(\Lambda) \approx 0$ this can be approximated by

$$\ln\Lambda = \ln\Lambda_c + A(\tau-\tau_c)M^{\alpha} \tag{7}$$

where $\alpha = \mathrm{d}\chi/\mathrm{d}\Lambda|_{\chi=0}$ is the derivative of χ at $\chi = 0$, and τ is some parameter (e.g. disorder, W, or energy E) which can be varied in order to move through the transition. In more conventional phase transitions τ would usually represent the temperature.

By comparing (3b) and (7) we see that

$$\xi \sim |\tau-\tau_c|^{-1/\alpha} \tag{8}$$

so that we can identify α with the reciprocal of the critical exponent, $s = \nu = 1/\alpha$ for the conductivity and localisation length respectively.

3.3 Analysis

In order to derive a value for the critical exponent it is necessary to analyse the data for λ_M as a function of M and the parameter τ on the basis of (7). Here τ represents either the disorder W or the energy E This has been done using the following procedure:

1) Calculate the gradient g of $\ln\Lambda$as a function of τ for fixed system size M. The gradient is taken at the point where the curves for different M cross

one another. From (7) these gradients are expected to obey

$$g = AM^{\alpha} + \text{corr.} \tag{9}$$

where "corr." refers to corrections for small M, i.e. deviations from one–parameter scaling.

2) By rearranging (9) we obtain the result

$$\frac{\ln g}{\ln M} = \alpha + \frac{A}{\ln M} + \text{corr.} \tag{10}$$

By plotting $\ln g/\ln M$ vs. $1/\ln M$ we should be able to extrapolate linearly to $M = \infty$ and obtain a value for α. It is also possible to form an opinion about the reliability of this extrapolation.

In practice it is found to be better to plot

$$\frac{\ln\left[(g/g_{\text{min}})M_{\text{min}}^{\alpha^*}\right]}{\ln M} \quad \text{vs.} \quad \frac{1}{\ln M} \tag{11}$$

where we have chosen $M_{\text{min}} = 4$ and $\alpha^* = 1$.

The calculation of the gradients g is the crucial step in this procedure. In the models with higher accuracy (0.1% or 0.2%) it was found that very few points were required to obtain an accurate estimate, whereas in those with lower accuracy (1%) the data was so noisy that a much larger number of points was required. This fact partly offsets the additional expense of the higher accuracy calculations.

4. Results and Discussion

Figure 1 summarises the results. The most striking aspect is the difference between the calculated exponents for the Anderson and Off–Diagonal models. This is all the more striking since the data otherwise fall on good straight lines; a fact which tends to confirm the validity of the one–parameter scaling approach.

The Anderson Model with Rectangular Disorder *Anderson 1 & 2* tends to have $\alpha \approx 0.6 \Rightarrow \nu \approx 1.5$ whereas the Off–Diagonal Models *Off–Diag 1 & 2* have $\alpha \approx 1$. It is difficult to believe that the curves for the Anderson Model are going to turn round and come back to $\alpha = 1$. In fact closer inspection of the high accuracy Anderson Model data *Anderson 2* suggests a slight curvature towards even smaller values of α.

Since it appears that the Anderson and Off–Diagonal Models have different critical exponents, we conclude that there are different physical processes responsible for the transition in the 2 cases. There is some evidence [6] that the Anderson Model with Gaussian disorder has an exponent of about $\nu \approx 0.9 \pm 0.3$. It would therefore be rash at this stage to make any comparison with experiment.

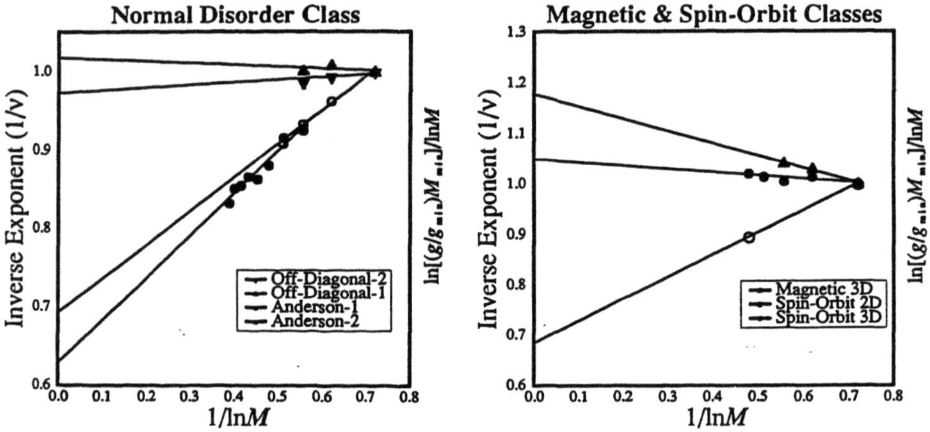

Figure 1 Extrapolations for the critical exponents versus $1/\ln M$ as discussed in the text, for the Normal (orthogonal) models (*left hand figure*) and for the magnetic and Spin–Orbit (unitary and symplectic) models (*right hand figure*). The meaning of the various symbols is given in the legend in the figures themselves. The lines are linear extrapolations and the arrows point to the extrapolated values for the inverse critical exponents.

Only a single model has been presented for the other universality classes. In some cases the results tabulated below represent the first attempt to calculate these exponents. However, these results too should be treated with caution in the light of the above discussion.

Model	Exponent
Anderson	1.5
Off–Diagonal	1.0
Magnetic 3D	0.84
Spin–Orbit 3D	0.97
Spin–Orbit 2D	1.45

References

1. F. Wegner: Phys. Rev. **B35**, 783 (1979)
2. P. W. Anderson: Phys. Rev. **109**, 1492 (1958)
3. Data kindly provided by M.Schreiber.
4. A. MacKinnon and B. Kramer: Zeitschrift für Physik **B53**, 1 (1983)
5. A. A. Abrikosov and I. A. Ryzhkin: Adv. Phys. **27**, 148 (1978)
6. B. Kramer, private communication

Photo-induced Metal–Insulator Transition in a Semiconductor

S. Katsumoto

Department of Physics, Faculty of Science, University of Tokyo,
7-3-1 Hongo, Bunkyo-ku, Tokyo 113, Japan

Abstract. Experimental results of the metal-insulator transition in $Al_{0.3}Ga_{0.7}As$: Si are presented. In the experiments, the effective electron concentration was finely controlled by using persistent photoconductivity. The electrical conductivity, and the tunneling density of states in the metallic phase, and the dielectric constant and the temperature dependence of the variable range hopping conductivity in the insulating phase were measured. The results were compared with the scaling theory and the theory by Efros and Shklovskii. Most of the results are in fairly good agreement though there remain several unexplained points.

1.Introduction

1-1 Metal-Insulator Transition

We present and discuss here the experimental results obtained in the vicinity of the metal-insulator transition (MIT) in $Al_{0.3}Ga_{0.7}As$: Si. The MIT is essentially caused by disorder in the lattice potential. The current physical picture of the MIT has been established through theoretical works by Mott, Anderson and others [1]. In this picture, the electron interference effect (localization effect) and the Coulomb interaction effect play important roles. In this section, we briefly review the present understanding of the phenomena around the MIT. In the latter part of this section we explain why we chose this material for the experiment.

In the metallic phase, the effect of disorder can be treated perturbatively from the metallic limit. Various phenomena caused by disorder (weak localization effects) in dirty metals have been quantitatively explained by the perturbation theory [2].

In the vicinity of the transition, the only theory which gives quantitative predictions on the critical behavior is the scaling theory which was introduced by Abrahams et al.[3] and was extended by Finkel'stein [4] and Castellani et al.[5] to take account of the interaction effect, though its mathematical bases have not been thoroughly established yet [6].

In the insulating phase, the theory of disordered insulators introduced by Austin and Mott [7], and extended by Efros and Shklovskii [8], can be applied to many experimental results, though at present it is not as powerful as the perturbation theory in the metallic phase. For the temperature dependence of the conductivity, Mott's variable range hopping conduction (VRHC) is observed in a wide range of materials.

Since the appearance of the scaling theory, the experiments need to be more sophisticated than before. In order to obtain precise values of the critical exponents, one should approach very close to the critical point. The theory also predicts some relations between the metallic phase and the insulating phase. Measurements in the insulating phase are as important as those in the metallic phase.

However, it is not very easy to approach the critical point. The electron concentration n is most frequently adopted as the relevant parameter for the MIT. When a sample is specified, n cannot be varied in usual three dimensional systems, thus we must prepare many samples. Various "tuning" techniques of the MIT have been developed in order to avoid this difficulty [9]. For example, the MIT in Si:P system has been extensively studied by the stress tuning method. Unfortunately that system is considered to belong to a universality class for which the present theory cannot give unambiguous predictions.

1-2 Persistent Photoconductivity

In $Al_xGa_{1-x}As:Si$ $(0.2 < x < 0.45$, Si can be replaced with other donors), photoconductivity has unique characteristics [10]. Because all donors in these alloys have deep ground states, all the electrons are trapped in donors at low temperatures. Once these electrons are excited to the conduction band or shallow states by light illumination, they seldom fall into the deep states. Therefore the photoconductivity continues after the light is turned off. This phenomenon is called persistent photoconductivity (PPC).

When the relaxation time of the PPC is long enough compared with the time for measurements, we can treat the system as in equilibrium. The concentration of electrons free from the deep states is a smooth function of the total light exposure time. Thus the concentration can be finely controlled over a wide range by increasing the total exposure time. In order to satisfy the requirement for the experiments mentioned in 1-1, we used PPC to adjust the electron concentration.

2.Experiment

2-1 Samples

All samples used here were $Al_{0.3}Ga_{0.7}As$ single crystal films doped with Si grown by molecular beam epitaxy on GaAs substrates. We prepared a simple uniform film for the conductivity measurement, and two special structures which are shown in Fig.1(a),(b).

The structure (a) is for the dielectric constant measurement in which an $Al_{0.3}Ga_{0.7}As$ layer was inserted between two highly doped GaAs layers. These GaAs layers are metallic and were used as "metal electrodes". We treated the whole structure as a capacitor and measured the capacitance to obtain the dielectric constant.

The structure (b) is for the tunneling density of states measurement. We adopted here a heterostructure single barrier to avoid band bending effects which are inevitable in usual Schottky type tunneling probes.

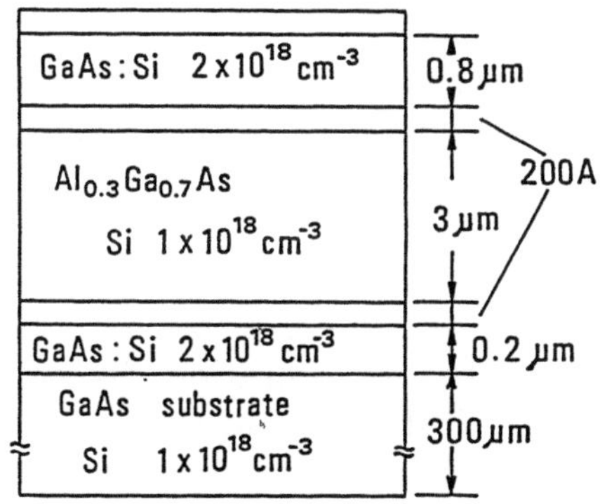

(a) Sample for the dielectric constant measurement.

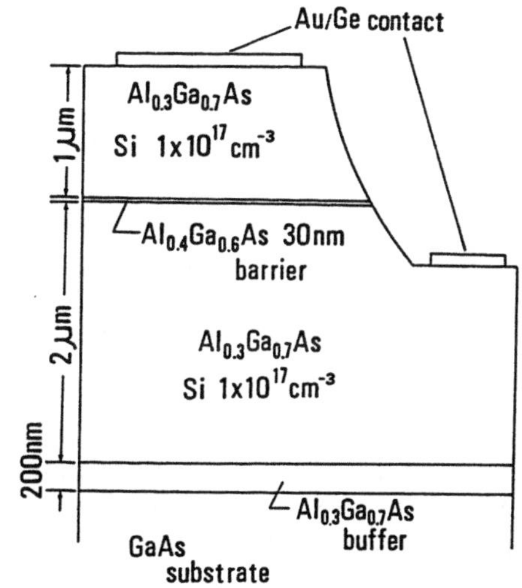

(b) Sample for the tunneling conductance measurement.

Fig.1 Cross sectional views of the samples used in the experiment.

2-2 Measurement

The samples were placed in the mixing chamber of a dilution refrigerator. They were illuminated by light emitting diodes (LEDs) which have the intensity peak at 9200A in wavelength.

The sample conductance was measured by DC and AC 0.2Hz$\sim$60Hz in the temperature range from 30mK to 800mK. The capacitance was measured by a three-probe capacitance bridge at a frequency of 103kHz. At the lowest temperatures the Joule power loss in the samples was kept below 10^{-12}W.

3.Results and Discussion

3-1 Universality Class of the System

To compare the results with the scaling theory for interacting electrons (STIE), the universality class of the system i.e. the symmetry of the system Hamiltonian, is essential. Factors which determine the universality class are external magnetic field, spin-orbit scattering (SOS) and spin flip scattering (SFS) by magnetic centers.

We classify the symmetry of the system to one of four limited cases;

1) general case in which the time reversal symmetry (TRS) is conserved;

2) spin-orbit case in which the spin rotation symmetry is modified by strong SOS;

3) magnetic impurity case in which the TRS is broken by strong SFS;

4) magnetic field case in which the TRS is broken by the external magnetic field.

To estimate the SOS and the SFS rates, the effective concentration was increased so as to bring the sample deep in metallic region in which the perturbation theory is reliable. We applied the theory to the data of magnetoresistance and obtained the SOS time as 5×10^{-10}s and the SFS time as 5×10^{-9}s.

In the following, we refer to these estimates. However we have to keep in mind that these rates may vary with the concentration. Recent static magnetic susceptibility measurements [11] showed that the ground state of the deep donor in $Al_{0.3}Ga_{0.7}As$ has a free spin. This means that the concentration of paramagnetic centers is represented by (donor doping concentration)−(effective electron concentration). In this case the SFS rate may decrease with an increase in the electron concentration. Recently we also measured the static susceptibility and obtained a controversial result. The problem has not been settled yet.

3-2 Metallic Phase

We performed a Hall measurement at 77K and confirmed that the effective electron concentration is proportional to the total exposure time when the concentration is relatively low ($< 2 \times 10^{17}cm^{-3}$). The electron mobility varies little in this region i.e. the degree of disorder is nearly constant.

Cooled in the dark, the sample resistance was very high ($\rho > 10^{10}\Omega$cm). Below 1K the relaxation time of the PPC was immeasurably long.

To determine the critical exponents the parameter dependence at $T = 0$ is needed, therefore we must use a reasonable temperature dependence for the conductivity and extrapolate it to $T = 0$. The STIE gives the temperature dependence just on the transition as

$$\sigma(T) = \left(\frac{12z_0}{\pi}\right)^{1/2} \frac{e^2}{\hbar} \left(\frac{k_B T m^*}{2\pi^2\hbar^2}\right)^{1/2} \tag{1}$$

for the spin-orbit case and as

$$\sigma(T) = \frac{(8\sqrt{6}\pi^2 z_0)^{2/5}}{2\pi^2} \frac{e^2}{\hbar} \left(\frac{m^* k_F^{1/2}}{2\pi^2\hbar^2} k_B T\right)^{2/5} \tag{2}$$

for the magnetic impurity case where m is the effective mass and z_0 is a parameter of the order of 1 [12].

We first adopt the spin-orbit case and check the consistency. In Fig.2, the conductivity is plotted against square root of T. The temperature dependence is well represented by

$$\sigma(n, T) = \sigma(n, 0) + aT^{1/2}, \tag{3}$$

where a and $\sigma(n, 0)$ are fitting parameters. We can fit formula (1) to the data with $z_0 = 6$. The agreement is satisfactory.

In the theory, the behavior of the conductivity near the MIT is described by

$$\sigma(n, 0) = \sigma_0(\frac{n}{n_c} - 1)^\nu \tag{4}$$

where ν is the critical exponent of the transition.

Figure 3 shows $\sigma(n, 0)$ as a function of the total exposure time (i.e. the effective concentration). Negative $\sigma(n, 0)$ are replaced with 0. With increasing n, the conductivity increases linearly, hence $\nu = 1$ is obtained. This is again in agreement with the STIE.

We should check the consistency between the results and the assumption of the spin-orbit case. The question is whether the SOS affects the critical behavior exhibited in the experiment. This can be judged from whether the spin-orbit scattering length $\sqrt{D\tau_{so}}$ (D : diffusion constant, τ_{so} : SOS time) cuts off the characteristic length or not. In the metallic phase, the characteristic length is the coherence length which is estimated to be $e^2/\hbar\sigma$. Substituting the parameters, we obtain the condition for

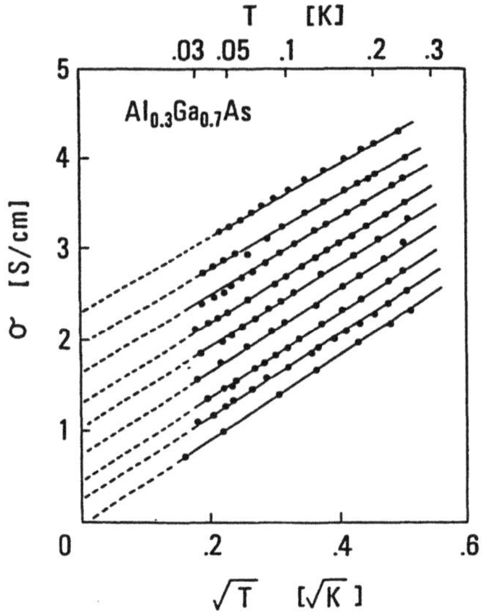

Fig.2 $\sqrt{T}$ dependence of the conductivity below 300mK in the vicinity of the MIT.

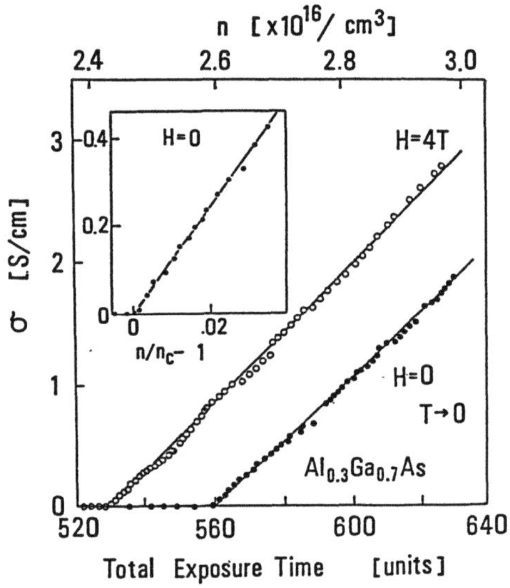

Fig.3 Zero-temperature conductivity near the MIT in magnetic fields of 0T and 4T. The behavior in the vicinity of the MIT is shown in the inset.

which SOS is effective, as $\sigma < 10(\mathrm{S/cm})$. Clearly our experiment satisfies this condition. Through similar analysis on the SFS, it turns out that SFS is ineffective in the experiment.

The analysis mentioned above is based on the parameters obtained in 3-1. We notice in 3-1, that SFS might be more frequent near the critical point. The difference in the exponent of the temperature between eq.(1) and eq.(2) is so small that the experiment cannot distinguish them. Using the exponent 2/5, we again obtain the critical exponent $\nu = 1$ and this also agrees with the theory.

Figure 3 also shows the response to the external magnetic field of 4T. The magnetoresistance is negative and n_c shifts to a slightly lower concentration. According to the theory, n_c should shift to a much higher concentration in the spin-orbit case, or should not shift in the magnetic impurity case. To explain the negative magnetoresistance, we should consider some other factors such as the reduction of the spin-dependent scattering by the magnetic field which results in decrease of the degree of disorder.

The most characteristic feature in the interacting electron picture is the anomaly in the density of quasi-particle states at the Fermi level

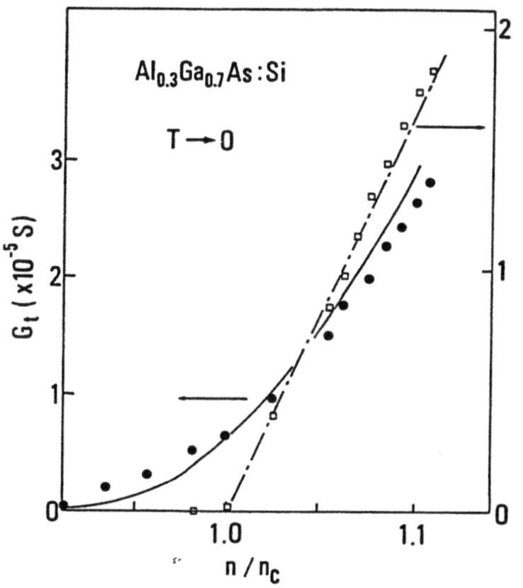

Fig.4 Zero-temperature tunneling conductance in the vicinity of the MIT.

[13]. That is the density of quasi-particle states at E_F vanishes at the critical point together with the conductivity. We can examine this by the tunneling measurement.

We used T-linear dependence to extrapolate the vertical (tunneling) conductance in the structure (c) to $T = 0$. The result is shown in Fig.4. The vanishing point is $0.91n_c$ and qualitative agreement is obtained. The reason for the appearance of the density of states at a slightly lower concentration is not clear now.

3-3 Insulating Phase

On the insulating side of the MIT, the dielectric constant corresponds to the conductivity in the metallic phase. The critical behavior of the dielectric constant near the MIT is described by

$$\epsilon(n,0) = \chi_0(1 - \frac{n}{n_c})^{-\nu'} \tag{5}$$

where ν' is the critical exponent.

The dielectric constant has strong temperature dependence below 100mK. When the effective concentration is low, the temperature dependence is T^2. The coefficient of T^2 becomes larger with the increasing concentration. Further increase of the concentration changes the dependence from T^2 to T^4. Thus the temperature dependence is fitted by the formula

$$\epsilon(n,T) = \epsilon(n,0)(1 + a(n)T^2 + b(n)T^4) \tag{6}$$

in these regions. Although the temperature dependence of the dielectric

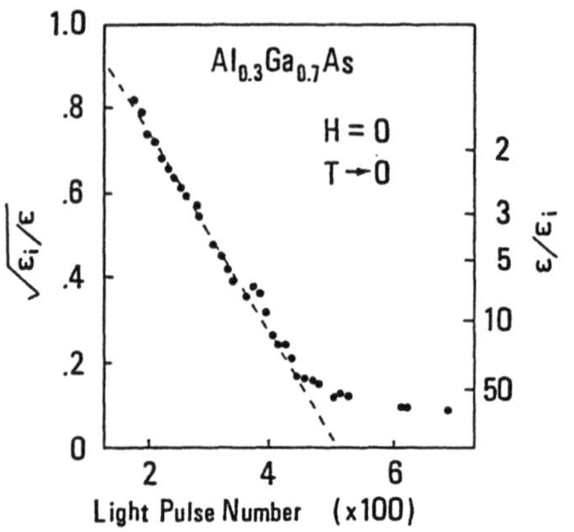

Fig.5 Square root inverse of the zero-temperature dielectric constant as a function of the total exposure time. $\epsilon_i = 13\epsilon_0$.

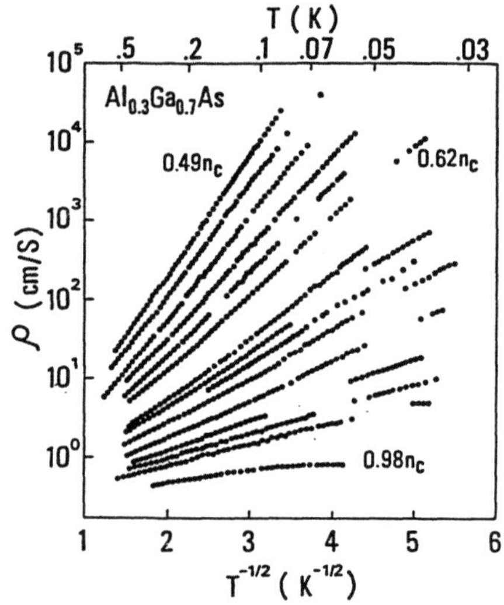

Fig.6 Resistivity plotted as a function of $T^{-1/2}$ for various effective electron concentrations.

constant has not been given by the scaling theory, this is in qualitative agreement with the theoretical expression based on the Efros and Shklovskii (ES) model of disordered insulators [14]. We adopted formula (6) for the extrapolation of the dielectric constant to $T = 0$.

In Fig.5, $(\epsilon_i/\epsilon(n,0))^{1/2}$ is plotted against the total exposure time. As shown by the broken line, the data can be fitted by relation (5) with $\nu' = 2$ over a wide range. However the analysis of the universality class similar to that in the metallic phase, shows that neither the SOS nor the SFS are effective in the experiment. In other words we should approach closer to the critical point in order to treat the system as the spin-orbit case or as the magnetic impurity case. Indeed the scaling theory gives $\nu' = 1.5$ and 1 for the spin-orbit and the magnetic impurity cases respectively, and neither agrees with the result [12].

In the insulating phase, the conductivity at finite temperature is characterized by the VRHC. In this regime, the temperature dependence of the conductivity is represented as

$$\sigma(T) = B\exp(-\frac{T_1}{T})^{\gamma}, \tag{7}$$

where T_1 is a parameter and $\gamma = 0.25, 0.5$ for non-interacting and interacting electrons respectively [8]. As shown in Fig.6, $\gamma = 0.5$ fits better

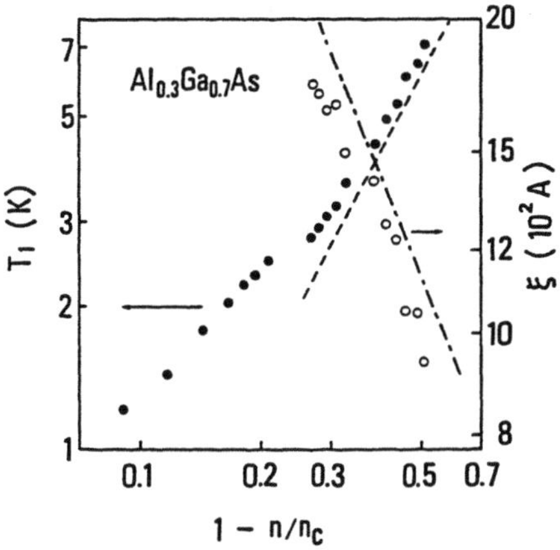

Fig.7 Effective concentration dependence of T_1 (solid circles) and of the localization length ξ (open circles).

to our data, this means that the Coulomb repulsion is important in the conduction.

Combining the data in Fig.5 and Fig.6, and applying the ES theory, we can calculate the localization length. The result is shown in Fig.7. The broken line and the dotted one indicate the predicted concentration dependence of T_1 and the localization length respectively. The ES theory works well deep in the insulating phase.

4.Conclusion

We have described the experiment on the MIT in $Al_{0.3}Ga_{0.7}As$: Si. Many of the results are explained by the scaling theory. Some of them remain as open questions. Theoretical work to understand the experiment is needed and more sophisticated experiments, especially in the insulating phase, are also required.

Acknowledgment

This work was done in collaboration with Prof. S. Kobayashi, Dr. F. Komori and Prof. N. Sano. I would like to thank Prof. A. Kawabata, Prof. C. Di Castro, and Prof. A. G. Aronov for theoretical advice.

References

[1] N. F. Mott, *Conduction in Non-Crystalline Materials,* (Clarendon, Oxford, 1987).

[2] P. A. Lee and T. V. Ramakrishnan, Rev. Mod. Phys. **57**, 2887 (1985).

[3] E. Abrahams, P. W. Anderson, D. C. Licciardello and T. V. Ramakrishnan, Phys. Rev. Lett. **42**, 673 (1979).

[4] A. M. Finkel'stein, Zh. Eksp. Teor. Fiz. **84**, 168 (1983) [Sov. Phys. JETP **57**, 97 (1983)].

[5] C. Castellani, C. Di Castro, P. A. Lee, M. Ma, S. Sorella and E. Tabet, Phys. Rev. **B30**, 1596 (1984).

[6] F. Wegner, Nucl. Phys. **B316**, 663 (1989); B. Kramer, in *Anderson Localization*, ed. T. Ando and H. Fukuyama (Springer, 1988) p.84.

[7] I. G. Austin and N. F. Mott, Adv. Phys. **18**, 41 (1969).

[8] A. L. Efros and B. I. Shklovskii, J. Phys. **C8**, L49 (1975).

[9] M. A. Paalanen, T. F. Rosenbaum, G. A. Thomas and R. N. Bhatt, Phys. Rev. Lett. **48**, 1284 (1982).

[10] R. J. Nelson, Appl. Phys. Lett. **31**, 351 (1979).

[11] K. A. Khachaturyan, D. D. Awschalom, J. R. Rozen and E. R. Weber, Phys. Rev. Lett. **63**, 1311 (1989).

[12] A. Kawabata in *Anderson Localization*, ed. T. Ando and H. Fukuyama (Springer, 1988) p.119.

[13] B. L. Altshuler and A. G. Aronov, Zh. Eksp. Teor. Fiz. **77**, 2028 (1979) [Sov. Phys. JETP **50**, 968 (1979)].

[14] R. N. Bhatt and T. V. Ramakrishnan, J. Phys. **C17**, L639 (1984).

Localization in Diluted Magnetic Semiconductors

T. Dietl, M. Sawicki, J. Jaroszyński, J. Wróbel, T. Wojtowicz, and A. Lenard
Institute of Physics, Polish Academy of Sciences,
Al. Lotników 32/46, PL-02688 Warszawa, Poland

Various models of the metal-insulator transition [MIT] in doped semiconductors are surveyed vis-a-vis our millikelvin experimental results for diluted magnetic semiconductors. Particular emphasis is put on: (i) magnetic field and temperature dependencies of conductivity in metallic n-CdSe and n-$Cd_{0.95}Mn_{0.05}Se$; (ii) critical behavior of conductivity tensor components and dielectric susceptibility at the field-induced MIT in p-type and n-type $Hg_{1-x}Mn_xTe$. It is shown that most of the experimental results are compatible with the model that the MIT is a result of quantum localization driven by disorder-modified electron-electron interactions. At the same time some results seem to suggest that, in addition to electrons forming the Fermi liquid and undergoing localization at the MIT, there is a certain concentration of local electron moments, even on the metallic side of the MIT. The formation of these moments can presumably be described in terms of a disordered Hubbard–Mott model.

1 Introduction

It has been known for a long time that when the concentration of dopants in a semiconductor decreases, a metal-to-insulator transition takes place [1]. In other words, there is a critical impurity concentration n_c, below which the zero-temperature conductivity vanishes and the material becomes an insulator. It is an empirical rule that for hydrogenic-like impurities n_c depends primarily on one material parameter, i.e., the effective Bohr radius a_B: $n_c^{1/3}a_B \approx 0.26$ [1]. Hence, n_c is as low as $\sim 10^{14}\ \mathrm{cm}^{-3}$ for n-InSb, but as high as $\sim 3 \cdot 10^{18}\ \mathrm{cm}^{-3}$ for n-Si. In terms of the wave function of electrons at the Fermi level, the MIT marks a transformation from extended states into localized states. Thus, as the kinetic energy involves space derivatives of the wave function, the MIT is obviously associated with an increase of the kinetic energy of electrons. Therefore, quite generally, the MIT will occur provided that the localization results in such a drop of the potential energy which overcompensates the above mentioned increase in the kinetic energy.

A considerable experimental and theoretical effort has been devoted over the recent years to establish the dominant driving mechanisms of the MIT in bulk doped semiconductors [1,2]. In this paper some of the proposed models are briefly described, and their predictions are compared with selected experimental results for diluted magnetic semiconductors. A novel aspect of these

Springer Series in Solid-State Sciences, Vol. 97
Localization and Confinement of Electrons in Semiconductors
Editors: F. Kuchar · H. Heinrich · G. Bauer

materials is the exchange interaction between effective-mass electrons and localized Mn d-spins. This interaction gives rise to giant exchange spin-splitting of electronic states, spin-disorder scattering, and formation of bound magnetic polarons [3].

2 Models of the MIT in Doped Semiconductors

2.1 Mott–Hubbard transition

Numerical studies of electronic states in a crystal built up of hydrogen atoms suggest that the MIT occurs for $n_c^{1/3} a_B = 0.22$ [4]. The driving force of this transition is the energy of the Coulomb interaction between electrons on the same site, U. For an appropriately large lattice constant the increase of the kinetic energy due to localization becomes smaller than the decrease of the potential energy related to the fact that electrons can no longer appear on the same site. In the insulating phase there is one electron localized around each atom, and thus the model explains how local magnetic moments can be formed. The MIT is, presumably, discontinuous in the sense that at n_c extended states undergo transformation to localized states with the localization radius $\sim a_B$. In contrast, in doped semiconductors, because of a random distribution of impurities and the presence of minority impurity charges, the localization could occur gradually. Within this scenario, the MIT marks the end-point of the transformation of the extended states with energies $\epsilon \leq \epsilon_F$ into the singly occupied local states characterized by the localization radius $\sim a_B$.

2.2 Wigner crystallization

In this model the MIT is also driven by the Coulomb interactions among electrons, but the impurities are assumed to act only as a uniformly charged neutralizing background. To a good accuracy this situation is accomplished in the recently designed modulation doped heterostructures [5]. It has been also argued that such a model should apply if the conducting electrons originate from resonant donors with bound states above the Fermi energy ϵ_F [6]. Standard arguments demonstrate that the critical concentration of the Wigner condensation should scale with a_B^{-3} [6,7]. The proportionality factor is anticipated to be, however, at least one order of magnitude smaller than that observed in doped semiconductors.

2.3 Percolation transition

Imagine a degenerate liquid of electrons to be placed in a randomly fluctuating potential produced by impurities. If the rms amplitude of the potential fluctuations increases, the volume occupied by electrons shrinks. Once the electron droplets turn to occupy less than $\sim 16\%$ of the sample volume, they cease to form a percolation cluster, and the materials turns into an insulator [8]. Such an MIT is continuous. Experimental studies [9] of a mixture of conducting and nonconducting materials imply that the conductivity vanishes with the critical exponent $t = 1.8 \pm 0.2$. When approaching the MIT from the insulating phase, the dielectric susceptibility diverges with the critical exponent $s = 0.7 \pm 0.1$. Obviously, the percolation model disregards effects of electrón-electron (e-e) interactions as well as quantum phenomena, such as tunneling and interference, which tend to delocalize and localize electrons, respectively.

2.4 Anderson localization

Anderson's MIT is the quantum localization of the Fermi liquid induced by scattering. It occurs due to a collective action of many scattering centers and would take place even if the potential of a single impurity, or defect, had no bound states. In macroscopically homogeneous systems, the Anderson localization is thought to set-in before the percolation transition. An interesting exception is probably the localization in 2d systems under the quantum Hall effect conditions [10].

It appears, by now, that there are two main mechanisms by which scattering may give rise to the localization [2,11,12]. One of them is the quantum interference of scattered waves. It is known that, when the positions of scattering centers are spatially correlated (as, e.g. , atoms in a crystal) the interference enhances or diminishes the transition probability between two points in space depending on the electron wavelength. If, on the other hand, the scattering centers are distributed randomly, the interference terms will vanish after averaging over all possible configurations of scattering centers. In other words, the mean transition probability between two points is well approximated by the sum of probabilities of all possible electron paths, in agreement with the classical Boltzmann description of diffusion. This is no longer true, however, if we consider the probability of returning to the starting point. This is because the "optical paths", and hence the phases of the transition probability amplitudes, are identical for the clockwise and counterclockwise trajectories of an electron wave along the same sequence of scattering centers. Since such interference is constructive, it increases the probability of return and thus, diminishes the conductivity.

Another consequence of scattering by impurities is a substantial modification of the interactions between electrons. A simplified picture behind it is that the diffusion motion allows two quasiparticles to meet and interact several times. If the two quasiparticles have similar energies they will acquire a small phase difference between the successive meetings. In such a case the successive interaction events cannot be regarded as independent. This invalidates several conclusions of the standard Fermi liquid theory and, in particular, leads to corrections to the Boltzmann conductivity.

The role played by the above quantum effects grows with the increase of $\hbar/\epsilon_F\tau$, where τ is the momentum relaxation time. Since, according to the Brooks-Herring formula (and its modification by Moore [13]) $\epsilon_F\tau$ for ionized impurity scattering depends only on $n^{1/3}a_B$, the model in question is compatible with the proportionality of n_c to a_B^{-3}. An interesting aspect of these quantum corrections is their unusual sensitivity to the temperature and to the so-called symmetry lowering perturbations. The corrections depend on the inelastic scattering processes (appearing at $T > 0$, $\tau_{in} \sim T^{-p}$, where $p \geq 1$) as well as on the magnetic field, spin-orbit- and spin-disorder scattering, because these change the phase of the wave functions. The contribution from the electron-electron interactions is affected additionally by thermal broadening of the Fermi–Dirac distribution function, and by the spin-splitting, as the latter influences the time-evolution of the phase difference if the two quasiparticles come from the different spin subbands. Depending on the symmetry lowering perturbations present (magnetic field, spin-splitting, spin-orbit- or spin-disorder scattering) various universality classes have been identified [12], each characterized by a different set of critical exponents controlling the behavior of relevant variables in the immediate vicinity of the MIT.

The models of the MIT listeded above by no means exhaust all the possibilities. In particular, an interesting question arises whether it is justified to treat, say, phonons or a subsystem of interacting localized spins only as a source of scattering. While for the inelastic processes the corresponding scattering rates vanishes at $T \rightarrow 0$, it seems plausible that in this limit the dynamic aspects of these couplings could show up, affecting the localization.

3 Discussion of Experimental Results

3.1 Weakly Localized Regime

As it was mentioned above, the quantum correction to σ resulting from the interference is expected to depend on the magnetic field (vector potential), while that due to *e-e* interactions should be additionally affected by the spin-splitting of electronic levels. Therefore, diluted magnetic semiconductors, such as $Cd_{1-x}Mn_xSe$, offer a novel opportunity to separate these two contributions. This is because the *s-d* interaction gives rise to an additional spin-splitting $\hbar\omega_S$ of electronic states. This spin-splitting is proportional to the field-induced magnetization of Mn spins, $M(T,H)$, available from independent measurements. In n-$Cd_{0.95}Mn_{0.05}Se$ $\hbar\omega_S(H)$ corresponds at $\sim$ 1.5 K to an effective Landé factor as large as 200, and therefore $\hbar\omega_S$ is more than ten times greater than the cyclotron energy $\hbar\omega_C$. In contrast, in non-magnetic CdSe, the orbital effects prevail, as $\hbar\omega_S$ is about thirty time smaller than $\hbar\omega_C$. Magnetoresistance studies across the MIT in n-CdSe and n-$Cd_{0.95}Mn_{0.05}Se$ have indeed shown a very different behavior of these two materials [14,15]. A quantitative description of the experimental results is possible in the metallic phase when $\epsilon_F\tau/\hbar > 1$. It is the so-called weakly localized regime, where the existing theory [11], developed to lowest order in $(\epsilon_F\tau)^{-1}$, should be valid.

A comparison between the measured and calculated dependencies of σ on the magnetic field is shown in Fig. 1 for metallic CdSe:In with $n \approx 3n_c$ [14]. The magnetoresistance (MR) is seen to be negative, reflecting the destructive effect of the magnetic field on the interference of scattered waves. Below 1 K,

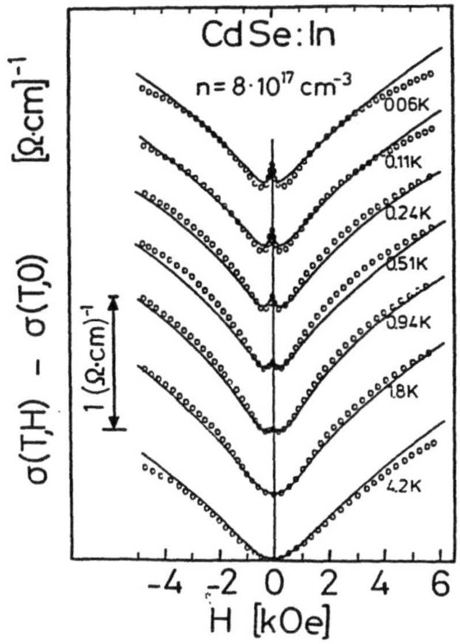

Fig. 1. Conductivity changes vs. magnetic field for a metallic sample of n–CdSe. Solid lines are calculated taking quantum corrections [11] into account.

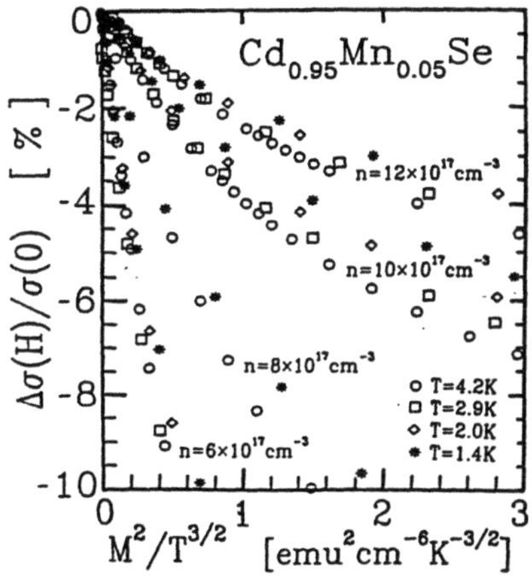

Fig. 2. Field–induced relative changes of conductivity for metallic samples of n–$Cd_{0.95}Mn_{0.05}Se$ plotted as a function of $M^2/T^{3/2}$, where $M(T,H)$ is the magnetization of d–spins and T is the temperature.

a small positive MR becomes visible in this material. This MR turns out to reveal the influence of spin-flip scattering upon the interference. In the case of wurzite-structure CdSe, efficient spin relaxation is brought about by the randomly oscillating magnetic field that is experienced by electrons diffusing in an acentric crystal. This effect appears when the temperature is low enough to make the inelastic scattering time longer than the characteristic time of this spin-orbit interaction.

In the case of n-$Cd_{0.95}Mn_{0.05}Se$:In, the MR stemming from the interference is found to be totally masked by the positive MR resulting from the influence of the giant spin-splitting upon the e-e interactions [14–16] . In weak magnetic fields, $\hbar\omega_S < 2\pi k_B T$, this MR is predicted to be proportional to $\omega_S^2/T^{3/2}$, that is in our case to $M^2(T,H)/T^{3/2}$. As it is shown in Fig. 2, the experimental results [16] fulfil this expectation. Furthermore, the data point out that the relative change of σ induced by ω_S increases with decreasing electron concentration. This convincingly demonstrates that the contribution of the quantum corrections to the total conductivity grows when approaching the MIT.

3.2 Metal-insulator transition

Especially accurate information on the critical region can be obtained if the MIT occurs under the influence of an external perturbation. Figure 3 presents

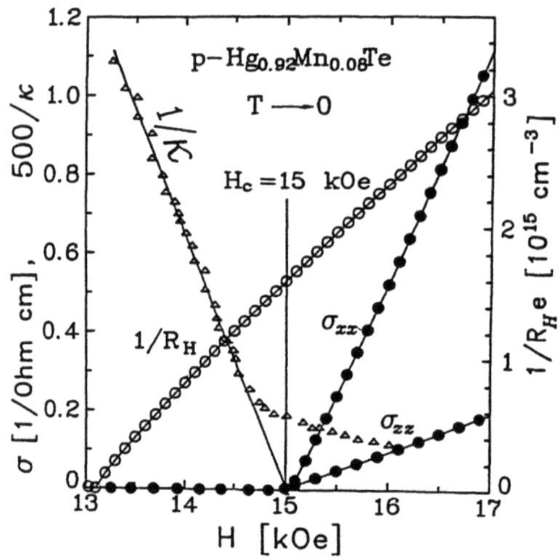

Fig. 3. Conductivity tensor components σ_{xx} and σ_{zz}, the inverse values of the Hall coefficient R_H and of the dielectric constant κ_{xx} extrapolated to zero temperature for p-$Hg_{0.92}Mn_{0.08}Te$. The solid lines indicate critical behavior. The finite values κ_{xx} in the metallic phase result from the Schottky barrier capacitance.

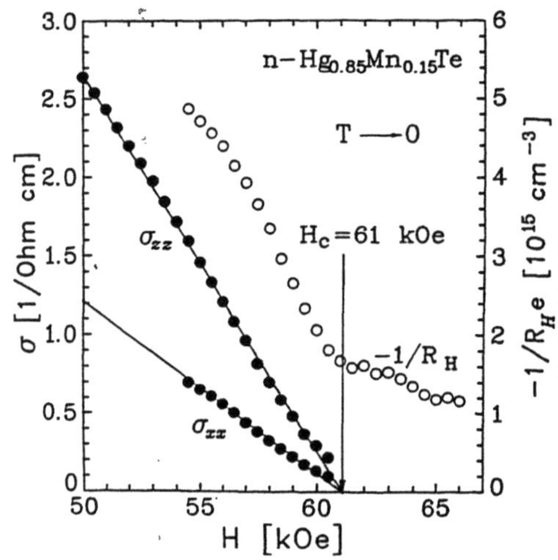

Fig. 4. Conductivity tensor components and the inverse Hall coefficient for n-$Hg_{0.85}Mn_{0.15}Te$ near the field-induced localization. Solid lines indicate critical behavior. Above H_c the conductivity seems to be perturbed by accumulation layers.

zero-temperature transverse and longitudinal conductivity, as well as the inverse Hall coefficient and dielectric constant, as functions of the magnetic field for p-$Hg_{0.92}Mn_{0.08}Te$ [17]. Similar data for n-$Hg_{0.85}Mn_{0.15}Te$ [18] are shown in Fig. 4. The data points were obtained from resistivity and capacitance measurements in the temperature range 30 mK $\leq T \leq$ 0.5 K. It is seen that p-HgMnTe undergoes an insulator-to-metal transition when the magnetic field gets stronger. This transition is primarily caused by the large exchange spin-splitting. The spin-splitting, if comparable to ϵ_F, leads to a redistribution of carriers between spin-subbands. The redistribution increases the kinetic energy of majority-spin carriers at ϵ_F, and hence diminishes their tendency towards localization. The effect is particularly strong in p-type semimagnetics, as the redistribution of holes between Γ_8 subbands is connected with a change of the transverse hole mass from a heavy mass to a light one [3]. Accordingly, σ_{xx} of p-HgMnTe is greater than σ_{zz}. The narrow-gap n-HgMnTe undergoes, in turn, a traditional field-induced metal-to-insulator transition. Such a transition takes place because of the diamagnetic effects, which dominate when the electrons are confined at the lowest Landau level, causing the kinetic energy to decrease with increasing magnetic field.

Several important conclusions can be drawn from the data displayed in Fig. 3 and 4. Firstly, in both cases the transition is continuous. This underlines the leading role of randomness, as MITs occurring in ordered systems (Mott–Hubbard transition, Wigner crystallization) would be, presumably, discontinuous in the 3d case. Secondly, the Hall coefficient R_H is seen to remain finite in the magnetic fields at which the conductivity vanishes. This contradicts the simple freeze-out picture in the framework of which the carrier concentration should vanish at the same field as the conductivity. The third conclusion comes from the fact that the diagonal components of the conductivity tensor tend to exhibit critical behavior of the form $\sigma_{ii} \sim |H - H_c|^t$, where $t = 1.0 \pm 0.2$. Such a value of t, observed commonly in doped semiconductors [2], is in apparent disagreement with $t = 1.8 \pm 0.2$ expected from the percolation theory. This indicates that the transition is caused neither by long-range fluctuations of the impurity potentials nor by macroscopic inhomogeneities in the impurity distribution. At the same time $t = 1$ is in accord with theoretical predictions for Anderson's MIT, provided that effects of scattering-modified electron-electron interactions are taken into account [12]. This conclusion is supported by the temperature dependence of the conductivity near the MIT [19], as well as by the character of the divergence of the localization length ξ at H_c, evaluated from the hopping conductivity [20] and dielectric constant [17] measurements on the insulating side of the MIT. The large values of ξ indicate also that the localization of electrons at the Fermi energy is caused by many scattering centers and must *not* be regarded as a freeze-out of carriers on individual impurities. Finally, in spite of the large field-induced anisotropy, visible in Figs. 3 and 4, the same values of H_c and t are observed for both σ_{xx} and σ_{zz}. This finding confirms the theoretical expectations concerning the quantum localization in anisotropic systems [21].

3.3 Formation of Local Electron Moments

The experimental results discussed above substantiate the picture of the MIT as the quantum localization of the Fermi liquid. According to this model, a Fermi-liquid-type description of electronic states remains valid as long as the size of the system is smaller than the localization length ξ. On the other hand, it is well known that deep in the insulating phase the electrons are bound to individual impurities forming, owing to the Hubbard energy U, localized magnetic moments. Obviously, in the latter case the electronic states cannot be described within the Fermi-liquid formalism at any length scale. A question then arises when does a transformation of the Fermi liquid electrons into local moments occur. There is a growing number of evidences that this transforma-

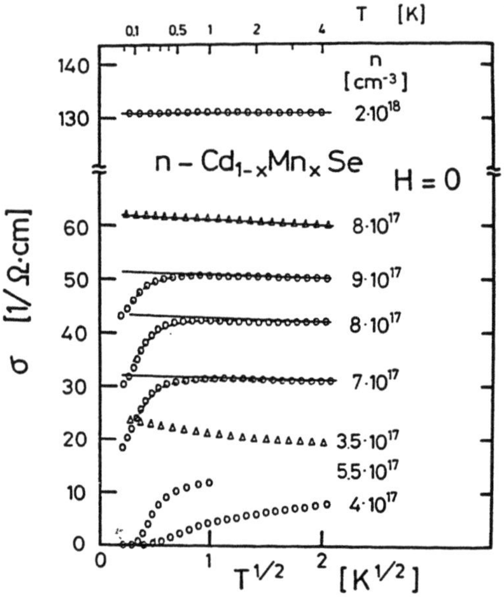

Fig. 5. Zero–field conductivity of n–CdSe (△) and n–$Cd_{0.95}Mn_{0.05}Se$ (○). Solid lines are calculated taking quantum corrections [11] into account but neglecting the presence of bound magnetic polarons.

tion begins already on the metallic side of the MIT, leading to the existence of local moments in metallic samples [22]. At the same time, it has been speculated [18] that the divergence of the Hall coefficient, visible in Figs. 3 and 4, marks the point where the Fermi-liquid electrons disappear totally.

The presence of local moments in metallic samples has been inferred, in particular, from the unusual temperature dependence of the conductivity discovered in $Cd_{0.95}Mn_{0.05}Se$:In below 0.5 K [14], as shown in Fig. 5 [23]. In magnetic materials the local moments can polarize, via *s-d* interaction, the neighboring Mn spins (there are about 200 Mn ions within the Bohr orbit in $Cd_{0.95}Mn_{0.05}Se$). The ferromagnetic bubbles (bound magnetic polarons) formed in this way constitute centers of spin-dependent scattering for the itinerant electrons. The efficiency of this scattering increases rather steeply with decreasing temperature, as the degree of the bubble polarization is proportional to the magnetic susceptibility of Mn spins, $\chi(T)$. It has been demonstrated that the drop of conductivity below 0.5 K can be semiquantitatively explained by taking into account the influence of this scattering mode on the quantum corrections to the conductivity [23]. It is also probable that the drop of σ is partly caused by the temperature dependent transformation of the Fermi-liquid electrons into local moments, as the binding energy of bound magnetic polarons becomes greater when the temperature is lowered.

ACKNOWLEDGMENT: This work is supported in part by the Institute for Low Temperature and Structural Research, Polish Academy of Sciences.

References

[1] for review see, e.g., *The Metal Non–Metal Transition in Disordered Systems*, eds. L.R. Friedman and D.P. Tunstall (SUSSP Publ., Edinburgh 1978).

[2] see, e.g., *Anderson Localization*, eds. T. Ando and H. Fukuyama, (Springer, Berlin 1988).

[3] for general properties of Mn-based semimagnetics see, e.g., *Diluted Magnetic Semiconductors*, volume eds. J.K. Furdyna and J. Kossut, in: *Semiconductors and Semimetals*, vol. **25**, series eds. R.K. Willardson and A.C. Beer (Academic Press, Boston 1988); for review of transport properties see, J. Kossut, ibid. p. 183; N.B. Brandt and V.V. Moshchalkov, *Adv. Phys.* **33**, 193 (1984); Y. Shapira, in: *Diluted Magnetic (Semimagnetic) Semiconductors*, Symposium Proceedings of Material Research Society, vol. **97**, eds. S. von Molnar et al. (MRS, Boston 1987) p. 209; T. Dietl, M. Sawicki, T. Wojtowicz, J. Jaroszyński, W. Plesiewicz, L. Świerkowski and J. Kossut, in [2], p. 58; T. Dietl, *J. Crystal Growth*, in press.

[4] see, e.g., T.M. Rice, *Phil. Mag.* **B42**, 419 (1985).

[5] see, e.g., M. Shayegan, this volume.

[6] G. Nimitz, in: *High Magnetic Fields in Semiconductor Physics*, ed. G. Landwehr (Springer, Berlin 1987) p. 491, and references therein; R.R. Gerhardts, ibid., p. 482.

[7] see, e.g., C.M. Care and N.H. March, *Adv. Phys.* **24**, 101 (1975); for Wigner condensation within the system of resonant-donor charges, see W. Jantsch and Z. Wilamowski, this volume.

[8] B.I. Shklovskii and A.L. Efros, *Electronic Properties of Doped Semiconductors*, (Springer, Berlin 1984); R. Zallen, *The Physics of Amorphous Solids* (John Willey & Sons, 1983).

[9] see, e.g., Y. Song, T.W. Noh, S.–I. Lee and J.R. Gaines *Phys. Rev.* **B33**, 904 (1986).

[10] see, e.g., S. Luryi, in [6], p. 16.

[11] for review see, e.g., B.L. Al'tshuler, A.G. Aronov, D.E. Khmelnitskii and A.I. Larkin, in: *Quantum Theory of Solids*, ed. I.M. Lifshits (Mir Publishers, Moscow 1982) p. 130; P.A. Lee and T.V. Ramakrishnan, *Rev. Mod. Phys.* **57**, 287 (1985); B.L. Al'tshuler and A.G. Aronov, in: *Electron-Electron Interactions in Disordered Systems*, eds. A.L. Efros and M. Pollak, (North-Holland, Amsterdam 1985) p. 1; H. Fukuyama, ibid. p. 155.

[12] A.M. Finkel'stein, *Zh. Eksp. Teor. Fiz.* **86**, 367 (1984) [*Sov. Phys. JETP* **59**, 212 (1984)]; C. Castellani, C. Di Castro, P.A. Lee and M. Ma, *Phys. Rev.* **B30**, 527 (1984); C. Di Castro, in [2], p. 96; G. Kotliar, in [2], p. 107; T. Dietl, M. Sawicki, J. Jaroszyński, T. Wojtowicz, W. Plesiewicz and A. Lenard, in: *Proc. 19th Int. Conf. on Physics of Semiconductors, Warsaw 1988*, ed. W. Zawadzki (Institute of Physics, Warsaw 1988) p. 1189.

[13] E.J. Moore, *Phys. Rev.* **160**, 607 and 618 (1967).

[14] M. Sawicki, T. Dietl, J. Kossut, J. Igalson, T. Wojtowicz and W. Plesiewicz, *Phys. Rev. Lett.* **56**, 508 (1986).

[15] T. Dietl, L. Świerkowski, J. Jaroszyński, M. Sawicki and T. Wojtowicz, *Physica Scripta* **T14**, 29 (1986).

[16] M. Sawicki and T. Dietl, in: *Proc. 19th Int. Conf. on Physics of Semiconductors, Warsaw 1988*, ed. W. Zawadzki (Institute of Physics, Warsaw 1988) p. 1217.

[17] J. Jaroszyński, T. Dietl, M. Sawicki, T. Wojtowicz and W. Plesiewicz, in: *High Magnetic Fields in Semiconductor Physics II*, ed. G. Landwehr (Springer, Berlin 1989) p. 514.

[18] J. Wróbel, T. Dietl, G. Karczewski, J. Jaroszyński, W. Plesiewicz, A. Lenard, M. Dybiec and M. Sawicki, *Semicond. Sci. Technol.*, in press.

[19] T. Wojtowicz, T. Dietl, M. Sawicki, W. Plesiewicz and J. Jaroszyński, *Phys. Rev. Lett.* **56**, 2419 (1986).

[20] T. Wojtowicz, T. Dietl, M. Sawicki, J. Jaroszyński and W. Plesiewicz, *Physica* **B155**, 357 (1989).

[21] R.N. Bhatt, P. Wölfle and T.V. Ramakrishnan, *Phys. Rev.* **B32**, 569 (1985).

[22] R.N. Bhatt, M.A. Paalanen and S. Sachdev, *J. Physique* **49–C8**, 1179 (1988), and references therin; M. Milovanović, S. Sachdev and R.N. Bhatt, *Phys. Rev. Lett.* **63**, 82 (1989); M. Lakner and H.v. Lohneyson, *Phys. Rev. Lett.* **63**, 648 (1989); D. Romero, M.-W. Lee, H.D. Drew, M. Shayegan and B.S. Elman, in: [2], p. 53.

[23] T. Dietl, M. Sawicki, T. Wojtowicz, J. Jaroszyński, W. Plesiewicz, L. Świerkowski and J. Kossut, in [2], p. 58.

Spatially Correlated Occupancy of Resonant Donors

W. Jantsch[1] *and Z. Wilamowski*[2]

[1]Institut für Experimentalphysik, Johannes Kepler Universität, A-4045 Linz-Auhof, Austria
[2]Institute of Physics, Polish Academy of Sciences, PL-02668, Warsaw, Poland

Abstract. For high concentrations of resonant donors, the Fermi level becomes pinned and both ionized and neutral donors coexist. Coulomb interaction among the ionized donors causes their spatial ordering. Models and experimental evidence for HgSe:Fe and n-GaAs are reviewed.

1. Introduction.

Usually, the mobility of free carriers in a semiconductor [1] decreases for high doping concentration. As a consequence, both the conductivity and the dielectric relaxation rate are limited for a given semiconductor, which affects the achievable power of high frequency devices. For systems with dominating ionized impurity scattering the mobility can be improved substantially by modulation doping, i.e., the spatial separation of dopants and a two-dimensional (2D) current carrying channel [2]. In this paper, we discuss another effect, which shows up the possibility for a mobility enhancement in 3D- and 2D systems. This effect has been observed for the first time in bulk HgSe:Fe [3,4]. In contrast to Ga-doped samples the low temperature mobility of HgSe:Fe increases by a factor of 3 to 4 when the iron concentration N_{Fe} is increased from $5x10^{18}cm^{-3}$ to about $10^{19}cm^{-3}$ [3,4] (see Fig.1). There is no explanation of this effect within the usual picture of randomly distributed, "uncorrelated" ionized donors.

In the same Fe-concentration regime the concentration of free carriers n_{FC} saturates at a value of about $4.5x10^{18}cm^{-3}$ [3-6]. The latter observation is attributed to the resonant nature of the iron donor state. For iron concentrations less than $5x10^{18}cm^{-3}$, iron is self-ionizing: each substitutional iron atom supplies one electron to the conduction band. As a consequence, the Fermi level increases with increasing N_{Fe} and finally it reaches the $Fe^{2+/3+}$-donor level (ionic notation) where it becomes pinned. Any additional Fe-atoms are incorporated in their neutral state. In

Springer Series in Solid-State Sciences, Vol. 97
Localization and Confinement of Electrons in Semiconductors
Editors: F. Kuchar · H. Heinrich · G. Bauer

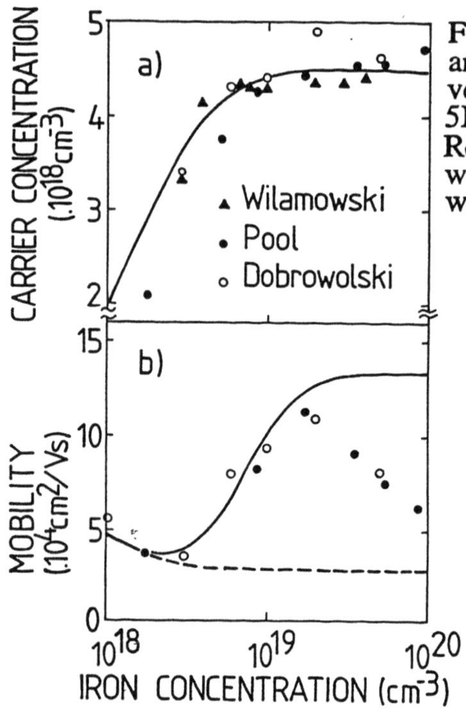

Fig.1: Carrier concentration (a) and mobility (b) of HgSe:Fe versus iron concentration at 5K. Experimental points from Refs.3-5, solid line: calculated with correlations, dashed line without (Ref. 13).

this concentration regime, ionized and neutral iron donors coexist, they are said to constitute a "mixed valence system".

In the mixed valence regime, there is a large number of different possible arrangements of the $N_{Fe}{}^{2+} = N_{Fe} - n_{FC}$ electrons among the N_{Fe} donors, which are thought to be randomly distributed within the cation sublattice, and we may ask for the ground state of the system. Coulomb interaction among the remaining $N_{Fe}{}^{3+} = n_{FC}$ charged donors is sizable: for an average distance of 60Å, the screened Coulomb energy per pair is about 10meV, much larger than the effective Rydberg of direct gap semiconductors or the thermal energy kT in the considered temperature range.

Many-body Coulomb-interaction among the ionized donors was considered first by J. Mycielski [7], who predicted the existence of a 3D charge superlattice consisting of Fe^{3+}-ions: the "Madelung energy" for the ionized donors becomes minimal if the $N_{Fe}{}^{2+} = N_{Fe} - n_{FC}$ electrons are distributed among the Fe donors in such a way that the remaining $N_{Fe}{}^{3+} = n_{FC}$ ionized donors form a superlattice. As a consequence, ionized impurity scattering is much less effective than in the disordered state [7]. Resonant scattering, which may be expected because of the pinning of the Fermi-level at the donor level is also quenched [8] because of the existence of a Coulomb gap [9]: a threshold energy is required to put an electron to a charge-superlattice site.

In this article, we review recent realistic models for the ground state of this system, in which short range correlation is explicitly taken into account (section 2). In section 3, we discuss investigations of dynamic aspects by means of electron-spin-resonance (EPR) of the Fe^{3+}-state. The observed small EPR line width demonstrates the extraordinary long life time, implying weak hybridization of the Fe^{3+} state with the conduction band, which is necessary for localizing the charge at the superlattice. In metals, hybridization of magnetic impurities leads to some homogeneous fractional charge distribution as it is evident from the disappearance of local magnetic moments [10]. The observed temperature dependence of the Fe^{3+}-EPR amplitude provides further support for the Coulomb-gap model. In section 4, we review recent experimental work both on n-type bulk and two dimensional δ-doped GaAs which also indicate correlation and Coulomb superlattice effects.

2. The Correlation Model.

The ground state of the mixed valence subsystem of ionized donors is determined by the total inter-ion Coulomb energy [11]:

$$E^{Coul.} = \frac{1}{2} \sum_{i}^{N_{Fe}} \eta_i eV_i \; ; \quad \text{with:} \quad eV_i = \sum_{ij} \frac{\eta_j e^2}{\varepsilon r_{ij}} \exp(-r_{ij}/\lambda) \; . \qquad (1)$$

Here r_{ij} is the distance between the impurities i and j, ε the dielectric constant and λ the screening radius. η_j is an occupation factor, equal to 1 for ionized donors and 0 for neutral donors. The treatment of η_j requires self-consistency- it depends on the relative position of the one-electron level with respect to the Fermi energy:

$$E_j = E_D + eV_j - eV_e \; ; \qquad \text{with:} \; V_e = 4\pi \, \lambda^2 n_{FC} e^2 / \, \varepsilon \; . \qquad (2)$$

(E_D: donor level, V_e: screened free carrier potential).

In the first treatment of this problem J.Mycielski [7] showed, that the total energy is lower by 10meV per Fe^{3+}-ion for a hypothetical regular superlattice of Fe^{3+}-ions as compared to a random arrangement. In real crystals, however, we cannot assume that the random matrix of substitutional donors really allows the formation of a perfect lattice. Instead, the charge superlattice will exhibit more or less disorder, which will be smaller for higher Fe-concentrations. More rigorous treatments of the disorder involve a pair correlation function g(r) which represents the probability of finding a second ionized donor within a distance r. For an analytical approach, a step-like correlation function is assumed [12]:

$$g(r) = \Theta(r-r_c) \qquad (1 \text{ for } r > r_c, \quad 0 \text{ for } r < r_c) \quad , \tag{3}$$

which means that no other ionized centers exist within the correlation radius r_c. For larger distances, $r > r_c$, the distribution is assumed to be random. Assuming a Poisson distribution for the Fe-donors, the correlation radius can be obtained from statistical arguments [11-13]. All quantities of interest can be calculated in terms of the correlation radius. The total energy of the system includes also electron-electron and electron-ion interactions, the kinetic energy of the electrons, and the neutral donor energy. The ground state of the system is obtained considering $dE_{tot}/dn_{FC}=0$, which provides a condition for the equilibrium carrier concentration $n_{FC}= N_{Fe^{3+}}$. The mobility can be calculated also from the correlation function making use of the structure factor:

$$S(q) = 1 + 4\pi\, N_{Fe^{3+}} \int [g(r) - 1] \,\frac{\sin\,(qr)}{q}\, r\, dr \quad , \tag{4}$$

which is required to calculate the momentum relaxation time for ionized impurity scattering.

This analytical model has been shown to explain the experimental data for the carrier concentration, their mobility [13] and the temperature dependence of the Fe^{3+}-EPR [11,12]. A more rigorous, but less transparent treatment of this problem employs a numerical simulation [11,13]. In this method, a random distribution of Fe-atoms is simulated within a finite volume and some fraction of them is assumed to be ionized. The minimum total energy now is obtained by probing charge transfer processes and allowing only for those which reduce the total energy. The one-electron density of impurity states is obtained by calculating the one-electron energies according to Eq.(2). Results for their distribution in energy are given in Fig.2. The results clearly show the existence of a Coulomb gap. The shaded regions indicate occupied states. The Fermi level is located right at the minimum of the defect density of states, as it is necessary for the quenching of the resonant scattering. The number of empty states (area of the unshaded peak-

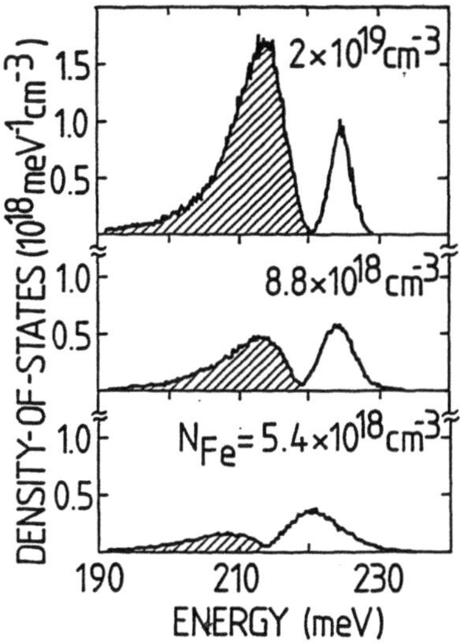

Fig.2: Results of the numerical simulation for the one-electron donor density of states for 3 different Fe concentrations (after Ref. 13).

like structures in Fig.2) is practically constant reflecting the experimental finding of a saturated carrier density.

Numerical results for the mobility are given in Fig.1 with and without correlation effects (solid and dashed line, respectively). For $N_{Fe}<2x10^{19}cm^{-3}$ the correlation model describes the experimental data very well. For higher Fe-concentrations, however, obviously additional effects become important. As a possible mechanism, the delocalization of the donor states can be considered as will be discussed in the following section in the context of EPR results.

3. Stability and Dynamics of the Charge Superlattice.

So far, we have considered only a static distribution of charges among impurity states assuming highly localized states. If, however, the hybridization of the resonant donor states with conduction band states (s-d) and/or with other impurities (d-d) becomes large, it will eventually lead to a more or less complete delocalization of charge, and ordering will disappear. In EPR, the Fe^{3+}ionized donor produces 5 lines due to the fine structure of the $3d^5$ state [6]. Surprisingly at first, the well resolved spectrum can be observed up to the highest concentrations of $N_{Fe}= 4x10^{19}cm^{-3}$ without appreciable broadening. This observation can be explained in terms of weak s-d hybridization which is necessary for the existence of a charge superlattice and hence, the Coulomb gap [12].

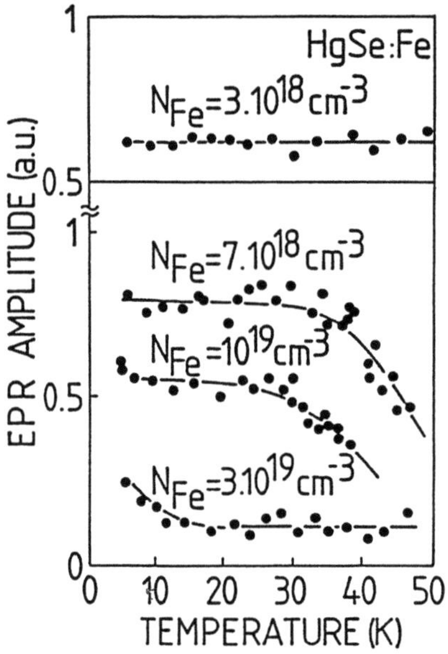

Fig.3: Fe^{3+}- EPR intensity as a function of temperature for 4 Fe concentrations. (After Ref. 5)

Typically, the observed EPR-line width is 30 G, which imposes a lower limit of 10^{-9}s for the life time of the Fe^{3+} state. With increasing iron concentration, the EPR line width remains practically constant but the integral intensity decreases, as can be seen in Fig.3. The EPR intensity exhibits also a rather peculiar temperature dependence: in the mixed valence regime, it is constant at low temperatures and starts to decrease only above 30-40K. Only for the highest concentration which is already in the region of decreasing mobility, the EPR amplitude decreases already at low temperatures.

In order to understand the anomalous temperature dependence of the EPR amplitude in the mixed valence regime, we consider inhomogeneous line broadening which results from the peculiar defect density of states in the presence of a Coulomb gap (Fig.2). In this situation, carrier capture by Fe^{3+}-ions is a thermally activated process: only Fe^{3+} states closer than a few kT to the Fermi level can capture an electron. Their life time will be much shorter than that of higher lying states. The resulting inhomogeneous EPR line broadening practically leaves the apparent line width unaffected, but the excessive broadening of lower states leads to a decrease in the total amplitude, their contribution to EPR is quenched. Qualitatively, the existence of the flat part in the temperature dependence of the EPR amplitude indicates low density of states close to the Fermi-energy, as it is expected for a Coulomb gap.

The quantitative treatment requires that the lifetime broadening of donor states at the Fermi-level is less than 0.01 meV. This value is smaller than the Coulomb gap by almost 3 orders of magnitude, which explains the stability of the charge superlattice with respect to s-d hybridization [5,11,12].

At present, there is no quantitative explanation for the drop in mobility for high iron concentrations and the strong decrease of the EPR amplitude with increasing temperature. Qualitatively, the decrease of the EPR amplitude has been attributed to donor-donor hybridization, which was shown to produce a line broadening if the distance of an Fe^{2+} donor to the next Fe^{3+} ion becomes less than 30Å [5,11]. It is still not clear, however, if this effect suffices to affect the correlation, and hence the mobility [5,11]. For the latter, additional scattering mechanisms have to be considered which have not been identified so far.

4. Correlation Effects due to DX-centers in GaAs.

Doping of $Al_xGa_{1-x}As$ or other III-V compounds with group IV or group VI donors inevitably leads to the appearance of a deep level (the "DX-state" [14,15]) in addition to the expected shallow effective mass states. These DX states are believed to originate from the same defect and there is evidence for large lattice relaxation which is held responsible for various persistent effects [15-17]. In $Al_xGa_{1-x}As$ the DX-level is located in the energy gap only for $x > 0.22$, suggesting that for lower x the state may be resonant. A resonant state has been observed in heavily Si-doped GaAs by capacitance transient spectroscopy at a level energy of 0.2-0.3 eV above the Γ conduction band edge, which corresponds to an extrapolated saturation of the carrier concentration at n_{FC} = $1.5\text{-}2 \times 10^{19} cm^{-3}$ with increasing donor concentration [18]. The same level apparently has been observed in high pressure experiments which show carrier freeze-out [19]. Extrapolation to atmospheric pressure leads to comparable level energies, although there is some unexplained variation of this energy with doping concentration [19]. Both groups attribute the resonant state to the DX state [18,19].

Now it is tempting to look for correlation effects in n-GaAs. In the high pressure experiment by Maude et al. [19], (recently refined by Suski et al. [20], see Fig.4), there is clear evidence for an increase in mobility under mixed valence conditions, which cannot be explained solely by the reduction of the density of ionized impurities due to freeze-out. Theoretical treatment along the

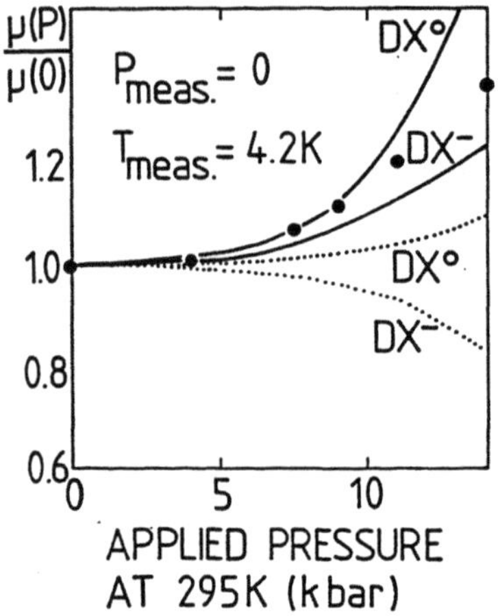

Fig.4: Mobility of n-GaAs:Si. In this experiment, DX-states are populated by cooling under high hydrostatic pressure (abscissa) which is released at low temperature. The calculations were performed both for a simple donor DX-state (solid curve marked DX^0) and for a hypothetical negative-U state ($2DX^0 \rightarrow DX^+ + DX^-$). (After Ref. 20)

analytical model described in section 2 yields at least good qualitative explanation (s. Fig.4), which indicates the existence of correlation effects also in GaAs [20,21].

There are clear indications also for the correlation induced mobility enhancement in 2D δ-doped GaAs [22,23] under pressure. In such layers, the two dimensional carrier density saturates as a function of the doping density which is attributed to the beginning population of the resonant DX-states. In order to test this hypothesis, Zrenner et al. performed Shubnikov-de Haas (SdH) experiments under hydrostatic pressure [22]. By using a Fourier transform, the contributions of the different electrical subbands could be clearly separated. At atmospheric pressure three subbands are occupied. The lowest (i=0) subband appears only as a wide, weak structure indicating lower mobility than in the higher electric subbands. Considering the wave function of this state this is not surprising: the i=0 wave function has a single maximum in the center of the channel, where the donor density is the highest. Therefore, ionized impurity scattering is expected to be most effective for this state. Under hydrostatic pressure the amplitude of the i=0 peak increases strongly, much stronger than the i=1 peak. Qualitatively, this effect again can be explained in terms of the spatial ordering of the ionized donors in the δ-plane: due to the hydrostatic pressure the DX-level moves down with respect to the electronic states, the Fermi-level is pinned at the DX-level and the mixed valence situation is attained. Spatial correlation affects particularly the i=0 mobility since these carriers suffered most from ionized impurity scattering [23]. It will be most interesting to discuss quantitative mobility data.

5. Conclusions

An enhancement of the electron mobility has been observed in such different systems as the zero gap semiconductor HgSe:Fe and n-GaAs under heavy doping conditions. In both cases, this effect can be explained quantitatively in terms of spatial correlation of well localized ionized donors in the mixed valence situation. This many-body effect produces a natural 3D (or 2D in the case of δ-doped layers) superlattice which is likely to stimulate a new area both in basic and applied research.

Acknowledgements: Herewith we express our gratitude to T. Dietl, G. Hendorfer, F. Koch, and J. Kossut for their valuable contributions. Work supported by the "Jubiläumsfonds der Österreichischen Nationalbank", Vienna, and by the project CPBP 01.04 in Poland.

References:

1. S. M Sze: *Physics of Semiconductor Devices,* 2nd Ed. (J. Wiley, New York, 1981) p 29
2. H. L. Störmer, Proc.15th ICPS, Kyoto 1980, J. Phys. Soc. Japan **49**, Supl. A, 1013 (1980)
3. W. Dobrowolski, K. Dybko, A. Mycielski, J. Wrobel, S. Piechota, M. Palczewska, H. Szymczak and Z. Wilamowski, Proc. 18th ICPS, O. Engstrom ed., (1987), p 1743.
4. F. Pool, J. Kossut, U. Debska, and R. Reifenberger, Phys. Rev. B**35**, 3900 (1987)
5. Z. Wilamowski, W. Jantsch and G. Hendorfer, Proc. Int. Conf. Phys. Narrow Gap Semic., Gaithersburgh 1989, in: Semicond. Sci. Technol., in print.
6. Z. Wilamowski, A. Mycielski, W. Jantsch and G. Hendorfer, Phys. Rev. B**38**, 3621 (1988)
7. J. Mycielski, Solid State Commun. **60**,165 (1986)
8. T. Dietl, Jap. J. Appl. Phys. **26** Suppl. 26-3, 1907 (1987)

9. B. I. Shklovski and A. L. Efros: *Electronic Properties of Doped Semiconductors,* (Springer, New York, 1984)
10. D. Riegel, L. Buermann, L.D. Gross, M. Luszik-Bhandra and S.N. Mishra, Phys. Rev. Lett. **68**, 316 (1989)
11. Z. Wilamowski, Acta Phys. Pol., in print.
12. Z. Wilamowski, A. Mycielski, W. Jantsch and G. Hendorfer, Proc. 19th ICPS, W. Zawadzki ed., (1988) p 1225, and: Superlattices and Microstructures 5, 157 (1989)
13. Z. Wilamowski, K. Swiatek, T. Dietl, and J. Kossut, Solid State Commun., in print, and: Ref.12, p 1571
14. D. V. Lang, in: *Deep Centers in Semiconductors,* S. Pantelides ed., (Gordon & Breach, New York 1986), p 489
15. P. M. Mooney, in: *Festkoerperprobleme* **29**, 215 (1989)
16. J. E. Dmochowski, J. M. Langer, J. Raczynska and W. Jantsch, Phys. Rev. B**38**, 3276 (1988)
17. G. Brunthaler, K. Ploog and W. Jantsch, Phys. Rev. Lett. **63**, 2276 (1989)
18. T. Theis, P. M. Mooney and S. L. Wright, Phys.Rev.Lett. **60**, 361 (1988)
19. D. K. Maude, J. C. Portal, L. Dmowski, T. Foster, L. Eaves, M. Nathan, M. Heiblum, J. J. Harris and R. B. Beall, Phys. Rev. Lett. **59**, 815 (1987)
20. T. Suski, P. Wisniewski, E. Litwin-Staszewska, J. Kossut, Z. Wilamowski, T. Dietl, K. Swiatek, K. Ploog and J. Knecht, Semicond. Science Techn., in print
21. T. Dietl, L. Dmowski, J. Kossut, E. Litwin-Staszewska, R. Piotrzkowski, T. Suski, K. Swiatek and Z. Wilamowski, Acta Phys. Pol., in print
22. A. Zrenner, F. Koch, R. L. Williams, R. A. Stradling, K. Ploog and G. Weimann, Semic. Sci. Technol. **3**,1203(1988)
23. F. Koch and J. Kossut, private communication

The High-Frequency Properties of Mesoscopic Junctions

V.I. Fal'ko and D.E. Khmel'nitskii

Institute of Solid State Physics, Chernogolovka,
Moscow District, 142432, USSR

The mesoscopic photovoltaic effect and generation of even and odd harmonics in microjunctions are predicted and evaluated. It is shown that frequency dependences of these effects and the mesoscopic part of an ac-conductance in the microwave regime can serve as a "frequency-fingerprint" of a sample. The microwave photoconductivity of mesoscopic junctions is also discussed.

We have examined linear and nonlinear high-frequency properties of small conductors (the so called mesoscopic conductors) of dimensions less than the inelastic length, the length of phase breaking and the temperature length $L_T=(hD/T)^{1/2}$. In such systems, due to interference effects, details of impurity distribution manifest themselves in kinetic coefficients of a specific sample. For example, the fine structure of magnetoresistance of a mesoscopic junction can serve as its "magneto-fingerprint" [1]. We argue that frequency-dependences of all ac-coefficients represent a set of "frequency-fingerprints" of a sample.

This suggestion is based on detailed calculations of correlation functions of mesoscopic fluctuations of linear and nonlinear terms in currents induced in microjunctions by microwave electromagnetic fields of different frequencies. All analysed correlation functions can be derived from expressions, corresponding to diagrams shown in Fig.1, by expanding in weak external ac-field **E** of frequency ω.

Our results about the mesoscopic part of the current in a microjunction can be symbolically represented in the form of a series

$$\delta I = e\sqrt{\frac{\omega}{\tau_f}}\ \{\ a_1(eLE/h\omega) + a_2(eLE/h\omega)^2 + a_3(eLE/h\omega)^3 + \ldots\} \qquad (1)$$

where $\tau_f=L^2/D$ is the time of diffusion through the junction. The first term of this series describes the mesoscopic fluctuation of the ac-conductance with the mean square value

$$<|\delta G_\omega|^2> \sim \left(\frac{e^2}{h}\right)^2 \frac{V_d}{L^d} f_d(\omega)/(\omega\tau_f)\ ; \qquad (2)$$
$$f_1=1;\ f_2=\ln(\omega\tau_f);\ f_3=\sqrt{\omega\tau_f}\ .$$

This value appears to be smaller than an amplitude of dc-conductance fluctuation by the factor $(\omega\tau_f)^{1/2}$.

Springer Series in Solid-State Sciences, Vol. 97
Localization and Confinement of Electrons in Semiconductors
Editors: F. Kuchar · H. Heinrich · G. Bauer

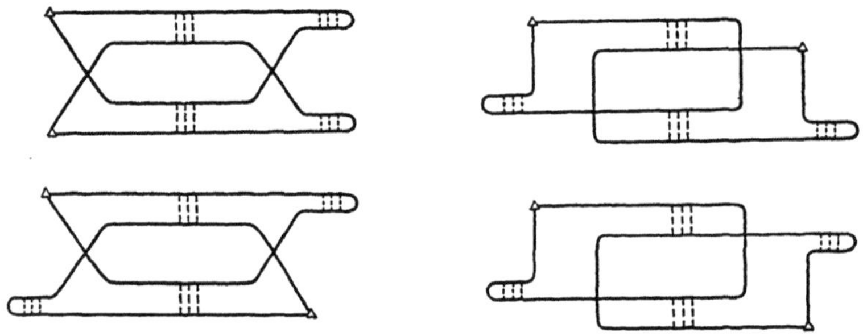

Fig.1. The diagrams corresponding to the correlation function of mesosopic fluctuations of current. Triangles correspond to current vertices, appendices with dashed lines correspond to Keldysh Green functions connected with the distribution function. Solid lines are connected with advanced or retarded functions and dashed lines mean an average over a random potential

The last expression allows to describe the frequency and field dependences only of the amplitude of these fluctuations. To imagine the behavior of the mesoscopic part of an ac-conductance in a specific sample we have analysed the correlation function

$$K_1(\omega,\Delta\omega) = \langle \delta G_{\omega+\Delta\omega/2} \delta G^*_{\omega-\Delta\omega/2} \rangle$$

of two mesoscopic conductances at different frequencies $\omega\pm\Delta\omega/2$. It appears to be

$$K_1(\omega,\Delta\omega) = \langle |\delta G_\omega|^2 \rangle / (\omega\tau_f)^{1/2} \qquad (3)$$

in one-dimensional conductor, $d=1$, and it depends on $\Delta\omega$ as

$$K_1(\omega,0) - K_1(\omega,\Delta\omega) \sim (\Delta\omega/\omega) \langle |\delta G_\omega|^2 \rangle \qquad (3')$$

for $d=2,3$. That shows the suppression of correlations of δG_ω upon increasing the deviation in frequency $\Delta\omega$ above the value of the inverse diffusion time π^2/τ_f. Therefore, δG_ω is a random function of frequency (with the characteristic scale $\Delta\omega\sim\pi^2/\tau_f$) and can serve as a "fingerprint" of a sample, as it was announced above. One can subtract this "frequency-fingerprint" from ac-magnetoresistance measurements. An additional possibility to amplify individual features of a sample in a related "frequency-fingerprint" consists in investigation of the derivative $dG/d\omega$. This value is practically determined by the mesoscopic part of the conductance

$$\left\langle \left|\frac{d\delta G_\omega}{d\omega}\right|^2 \right\rangle \sim (\omega\tau_f/8\pi)\,(V_d/L^d)\,\frac{e^2}{\hbar}\,(\tau_f/\pi^2)^2 \qquad (4)$$

because the frequency derivatives of the classical Drude conductance and weak localization corrections are comparatively small:

$$\frac{dG_{Drude}}{d\omega} \Big/ \frac{d\delta G_\omega}{d\omega} \sim \frac{\varepsilon_f \tau}{h} [1/L]^2 (\omega\tau_f)^{-1/2} (L^d/V_d)^{1/2}$$

$$\frac{dG_{w\,loc}}{d\omega} \Big/ \frac{d\delta G_\omega}{d\omega} \sim (\pi^2/\omega\tau_f)^{(5-d)/2} (8L^d/\pi V_d)^{-1/2} .$$

The second term in the symbolic series (1) describes the characteristic values of the photogalvanic effect [2] and the second harmonic generation [3] caused by breaking of inversion symmetry in the mesoscopic conductor due to a random distribution of scatterers. Similar to the coefficient a_1 in the first term, a_2 is a random function of frequency. The corresponding correlation function

$$K_{0,2}(\omega,\Delta\omega) = \langle \delta I_{0,2}(\omega+\Delta\omega/2)\ \delta I^*_{0,2}(\omega-\Delta\omega/2)\rangle$$

of two currents induced by ac-fields of different frequencies is also suppresed in the region $\Delta\omega > \pi^2/\tau_f$

$$K_{0,2}(\omega,\Delta\omega) \sim \langle |I_{0,2}|^2\rangle / (\Delta\omega\tau_f/\pi^2)^{2-d/2} \tag{5}$$

Both the latter property and the simple way to estimate the amplitude $\langle I_0^2\rangle^{1/2}$ of the photovoltaic effect can be clarified by the following reasoning. The ac-field effect on a junction brings about excitation of a number of nonequilibrium carriers, which amounts to the number $N \sim VE^2\sigma/h\omega$ of field quanta absorbed in the volume of the sample per unit time. After expiration of the diffusion time $\tau_f = L^2/D$ these excited carriers leave the junction and reach to the bulk electrodes, where they relax. Due to asymmetry of a junction, different number of carriers will diffuse to the left and right sides of the bridge. The degree of this asymmetry

$$\alpha \sim \frac{e^2/h}{\langle G\rangle}$$

can be estimated as a ratio of the value of the characteristic mesoscopic fluctuation of conductance to its average value. Because of special statistical properties of electron spectrum in a junction [4] this parameter randomly depends on electron energy varying on the scale $\Delta\varepsilon \sim h/\tau_f$. Therefore in the case of high frequencies $\omega > \pi^2/\tau_f$ a great number $n = \omega\tau_f/\pi^2$ of uncorrelated energy intervals take part in current formation. Thus a partial self-averaging leads to an additional small factor $n^{-1/2}$ in the amplitude of the total current and the latter takes the form

$$I_0 \sim eN\alpha/\sqrt{n} \sim e\sqrt{\omega/\tau_f}\ [eEL/h\omega]^2 \tag{6}$$

The random frequency dependences of the rectified current I_0 and the second harmonic I_2 in the current can be explained in the same way. Upon increasing the frequency above the value $\Delta\omega > \pi^2/\tau_f$ some new uncorrelated regions of the spectrum get involved into the play, so that the total photovoltaic effect can change its value unexpectedly: it can change both the magnitude and the sign (the second harmonic can change its initial phase).

Other higher terms in the series (1) describe the generation of higher even and odd harmonics in the current with correlation properties almost analogous to those discussed above:

$$K_N(\omega,\Delta\omega) \sim \langle|I_N(\omega)|^2\rangle/(\Delta\omega\tau_f/\pi^2)^{N-d/2},$$
$$\langle|I_N(\omega)|^2\rangle \sim 8\pi^2 \frac{\omega}{\tau_f} e^2 [2^{3/2} eEL/h\omega]^{2N} \tag{7}$$

The complete set of functions $a_i(\omega)$ contains full information about electron spectrum in a specific sample. It is interesting to note that the correlations between even and odd harmonics are sufficiently suppressed, as these two parts of the current originate from different features of electron spectrum. Even harmonics arise due to the inversion symmetry breaking by a random impurity distribution and the odd ones originate from an electron-hole asymmetry near the Fermi level. Moreover, the correlations between different even (or odd) current components, which are maximal at low frequencies, gradually vanish in the microwave region

$$\langle I_{2N}(\omega) I^*_{2N'}(\omega)\rangle \sim$$
$$\sim \langle|I_{2N}(\omega)|^2\rangle^{1/2}\langle|I_{2N'}(\omega)|^2\rangle^{1/2}(\pi^2/\omega\tau_f)^{N-N'}/(N-N')! \ . \tag{8}$$

The point is that different modes are connected with electron transitions between states within spectral intervals of different width $Nh\omega$ and $N'h\omega$, Therefore all functions $a_i(\omega)$ are practically independent.

All discussed results on the photovoltaic effect and higher even and odd harmonics generation are described by the formulae (1)-(8) in the limit of weak, $eEL/h\omega \ll 1$, high-frequency, $\omega\tau_f/\pi^2 \gg 1$, fields. An estimation of the low frequency limit of these effects in weak fields can be obtained by replacing ω by π^2/τ_f in these formulae (as can be obtained from the analysis of nonlinear mesoscopic conductance [5]).

Another effect we want to discuss is the microwave mesoscopic photoconductivity, i.e. the effect of microwave irradiation on the dc-conductance of the mesoscopic junction. It is known from the weak localization theory [6] that the high-frequency electromagnetic field E_ω stimulates the conductivity through the dynamical suppression of quantum corrections to conductivity. The magnitude of this effect induced by low intensity irradiation in the mesoscopic junction,

$$\langle\Delta G\rangle = \langle G(E_\omega)-G(0)\rangle =$$
$$= \frac{e^2}{h} 8\chi_d \zeta(5-d)\ (3-d)!!\ [eEL/h\omega]^2\ V_d/L^d,\ \chi_1=2/\pi,\ \chi_{2,3}=1, \tag{9}$$

is measured in fractions of e^2/h ($\zeta(x)$ denotes the Riemann ζ-function). Therefore the total value of the photoconductivity is determined by an interplay between the regular (always positive) part (9) and the variation of the mesoscopic part of the conductance. The latter arises from weak electron-hole asymmetry near the Fermi level, and the calculations [3] show that it has the same order of magnitude as (9), but a random frequency dependence. These two parts can be separated by high magnetic field $H>\Phi/S$ suppressing weak localization corrections in dirty conductors of area S. The mesoscopic part of photoconductivity depends randomly on H but does not change its amplitude. In the high magnetic field region it can be described by the correlation function of photoconductances induced by fields E_ω and $E_{\omega'}$ of different frequences ω and ω', $\Delta\omega=\omega-\omega'$. If $\Delta\omega\tau_f<<1$, it can serve to estimate the magnitude of effect,

$$K_G(\omega,0) = <[\Delta G(E_\omega)]^2> = (e^2/h)^2\{F_1+F_2\}\,\zeta_d V_d/L^d . \qquad (10)$$

If $\Delta\omega\tau_f/\pi^2>>1$, then

$$K_G(\omega,\Delta\omega) = (e^2/h)^2\{F_1+(\pi^2/\Delta\omega\tau_f)^2 F_2\}\,\zeta_d V_d/L^d , \qquad (11)$$

where

$$F_1 = (9/8)\,|eE_\omega L/\hbar\omega|^2\,|eE_{\omega'}L/\hbar\omega'|^2 ,$$
$$F_2 = |e^2L^2E_\omega E_{\omega'}/\hbar^2\omega\omega'|^2$$

and $\zeta_1=0.21$, $\zeta_2=0.1$, $\zeta_3=0.055$. As a result, the total photoconductivity induced in the mesoscopic junction by weak microwave field also can serve as a "frequency-fingerprint" of the sample.

In conclusion, we hope that further experimental investigation of microwave properties predicted above can give us new interesting information about the statistics of electron spectrum in disordered conductors.

References

[1] S.Washburn and R.A.Webb, Adv. Phys. 35 (1986) 375

[2] V.Fal'ko and D.E.Khmmel'nitskii, ZETF 95 (1989) 349

[3] V.Fal'ko, Europhys. Lett., 8 (1989) 785

[4] B.L.Al'tshuler and B.I.Shklovskii, Sov.Phys JETP 64 (1986) 127

[5] D.E.Khmel'nitskii and A.I.Larkin, Sov.Phys JETP 64 (1986) 1075

[6] B.L.Al'tshuler, A.G.Aronov and D.E.Khmel'nitskii, Solid State Commun. 39 (1981) 619

Part III

The Integer and Fractional Quantum Hall Effect and Electron–Electron Interactions

Metal–Insulator Transition in the Integer Quantum Hall Effect

H.P. Wei[1], *D.C. Tsui*[1], *and A.M.M. Pruisken*[2]

[1]Department of Electrical Engineering, Princeton University, Princeton, NJ 08544, USA
[2]Institute for Theoretical Physics, University of Amsterdam, The Netherlands

Abstract: We review a series of recent experiments, in $In_{0.53}Ga_{0.47}As/InP$ heterostructures, which unambiguously demonstrate that there is scaling in the integer quantum Hall effect.

1 Introduction

The temperature (T) dependent transport of electrons of a two dimensional electron gas (2 DEG), formed at a semiconductor interface, showed deviations at low T from that described within the framework of the Boltzman equation [1]. In particular, in a Si–MOSFET and in the absence of any magnetic field (B), Bishop, Tsui and Dynes [2] found that the conductivity σ has a T dependence of the form: $\sigma_0 + C\ \ln(T/T_0)$, where σ_0 is the Drude conductivity due to impurities, and C is a constant. Since the microscopic effective sample size is determined by a T dependent inelastic scattering length, $l_{in}(\sim T^{-p/2})$ [3], one can conceptually cast the T into a length scale. The above result can, then, be viewed as a demonstration of scaling of the electronic transport in a disordered system. Moreover, the scaling theory [4] predicts that all the electronic states are localized in the limit of very large sample size and there is no metal–insulator transition, for 2 DEG at $B = 0$.

The physics changes under high B, due to the existence of a series of impurity broadened, highly degenerate Landau levels, and due to the breaking of time reversal symmetry by B. More specifically, when T is much less than the energy separation between two neighboring Landau levels, the transport coefficients exhibit the integer quantum Hall effect (IQHE) [5,6,7]. For illustration, we plot ρ_{xx}, the longitudinal resistance per square, and the Hall resistance ρ_{xy} of an InGaAs/InP heterostructure as a function of B at $T = 53\,\mathrm{mK}$ in Fig.1. There exists a series of regions in B where ρ_{xy} is accurately quantized according to h/ie^2 (where i is an integer, h the Planck constant and e the electron charge). Concomitantly, ρ_{xx} is zero. Because the Landau level degeneracy increases with increasing B, the Fermi level E_F of the system is in effect sweeping through succesive Landau levels. When E_F is close to the center of a Landau level, where the density of states has a peak, ρ_{xx} has a maximum value; whereas when E_F is in the tails of a Landau level, ρ_{xy} and ρ_{xx} are in the plateau regions. The fact that the transport is dissipationless in the plateau regions indicates that the electronic states in the Landau level tails are localized. The appearance of finite ρ_{xx} and of a sharp transition in ρ_{xy} between two adjacent

Springer Series in Solid-State Sciences, Vol. 97
Localization and Confinement of Electrons in Semiconductors
Editors: F. Kuchar · H. Heinrich · G. Bauer

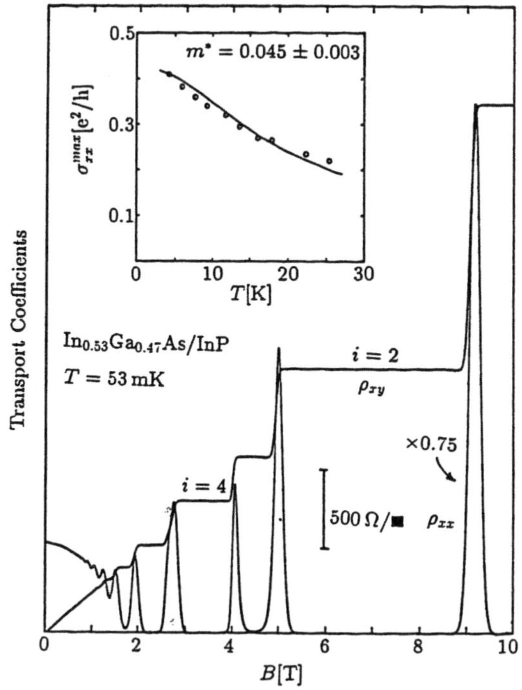

Fig.1 Transport coefficients ρ_{xx} and ρ_{xy} as a function of B at $T = 53\,\mathrm{mK}$. The inset is σ_{xx}^{max} vs. T at $B = 9.42\,\mathrm{T}$. The solid line is from a model calculation based on SCBA, with parameter $m^* = 0.045 \pm 0.003$.

quantum Hall plateaus further indicates that there is only a very narrow range in energy in each Landau level, where the electronic states are not localized [8]. Consequently, for 2 DEG under high B, not all states can be localized and the transport data in Fig.1 is a display of a series of metal–insulator transitions.

A theoretical investigation in Ref.[9] predicts that the electron transport in the IQHE regime shows scaling. Both the longitudinal conductance σ_{xx} and the Hall conductance σ_{xy} are length scale dependent, i.e. $\sigma_{\alpha\beta} = \sigma_{\alpha\beta}(L)$ and follow a specific renormalization group flow diagram on the σ_{xx} vs. σ_{xy} plane [9]. There are two different types of fixed points: the so called "quantum Hall fixed points" at $\sigma_{xx} = 0$ and $\sigma_{xy} = ie^2/h$, and the "delocalization fixed points" at $\sigma_{xy} = (i + \frac{1}{2})e^2/h$ and σ_{xx} finite. This latter describe a diverging electron localization length ξ according to:

$$\xi \sim |E - E_c|^{-\nu} \tag{1}$$

Here, E represents the Fermi energy of the 2 DEG and E_c corresponds to the singular values of E_F where the Hall resistance steps become infinitely sharp in the limit of large samples and low temperatures. Moreover, the renormalization theory predicts specific scaling relations for the transport coefficients ρ_{xx} and

ρ_{xy} near the delocalization critical points, which are amenable for experimental verification.

In this paper, we review a series of our recent experiments designed to investigate the nature of the transport phenomenon in the IQHE and to test out the predictions of scaling. The experiments focus on the singular structure of the transport coefficients and examine in depth how the Hall steps sharpen themselves and how the ρ_{xx} peaks, in between two adjacent Hall plateaus, narrow their widths as T decreases. We have found that there is a characteristic temperature, T_{sc}, below which the extrema of $d\rho_{xy}/dB$, $d^2\rho_{xx}/dB^2$, and $d^3\rho_{xy}/dB^3$ diverge like $T^{-\kappa}$, $T^{-2\kappa}$, and $T^{-3\kappa}$, respectively. The exponent $\kappa = 0.42 \pm 0.04$, independent of the Landau levels [10]. The result shows that for B sufficiently close to B^*, at which $E_F = E_c$, the transport coefficients are a differentiable functions of a single parameter, $T^{-\kappa}(B - B^*)$. That is, $\rho_{\alpha\beta}(T, B) = \rho_{\alpha\beta}(T^{-\kappa}(B - B^*))$ [11]. By writing the phase-breaking length as $l_{in} \sim T^{-p/2}$ and $\kappa = p/2\nu$, this single parameter is simply $(l_{in}/\xi)^{1/\nu}$. The experiments, thus, demonstrate explicitly and quantitatively that there is scaling in the IQHE.

2 Sample and Experimental Method

The sample used in this work is an $In_{0.53}Ga_{0.47}As/InP$ heterostructure as descirbed in Ref.[12]. The 2DEG is formed near the interface and in the $In_{0.53}Ga_{0.47}As$ layer, with an effective mass $m^* = 0.047$ (in units of free electron mass) [13]. Because the $In_{0.53}Ga_{0.47}As$ is an alloy, short range potential fluctuations is expected to be dominant [14]. Our sample has an electron density $n = 3.3 \times 10^{11}\,\mathrm{cm}^{-2}$ and a mobility $\mu = 35{,}000\,\mathrm{cm}^2/\mathrm{Vs}$ at 4.2K. The data presented in this paper are a collection of data taken in a variety of cryostats, including one standard variable–temperature dewar, two ^{3}He systems and two dilution refrigerators, for T ranging from 50 K down to 40 mK. The magneto-transport coefficients ρ_{xx} and ρ_{xy} were measured simultaneously using two lock–in amplifiers at a frequency of 33 Hz or less. A typical excitation current was 10^{-8} A in the high T regime and 10^{-9}–10^{-10} A in the low T regime. Care was taken to avoid electron heating for $T < 100$ mK. The outputs of the lock-in amplifiers were digitized through two digitizing voltmeters and the results were fed into a computer system for data analysis. The derivatives of ρ_{xx} and ρ_{xy} were obtained digitally. We calculated σ_{xy} and σ_{xx} from ρ_{xy} and ρ_{xx} by the formula $\sigma_{\alpha\beta} = \rho_{\alpha\beta}/({\rho_{xx}}^2 + {\rho_{xy}}^2)$, where $\alpha\beta = xx$ or xy.

3 Semi–classical regime

We plot in Fig.2 $(d\rho_{xy}/dB)^{max}$, which is the maximum of $d\rho_{xy}/dB$, and the peak value of σ_{xx} (denoted by σ_{xx}^{max}) in between the $i = 1$ and $i = 2$ plateaus, i.e. in the Landau level index $N = 0 \downarrow$ level for $0.1\,\mathrm{K} < T < 50\,\mathrm{K}$. When $T > 30\,\mathrm{K}$, $d\rho_{xy}/dB \sim 1900\Omega/\mathrm{T}$, which equals the classical value of $1/ne$. This result indicates that the transport can be described classically for $T > 30\,\mathrm{K}$. When T is less than 30 K, $(d\rho_{xy}/dB)^{max}$ starts to deviate from the

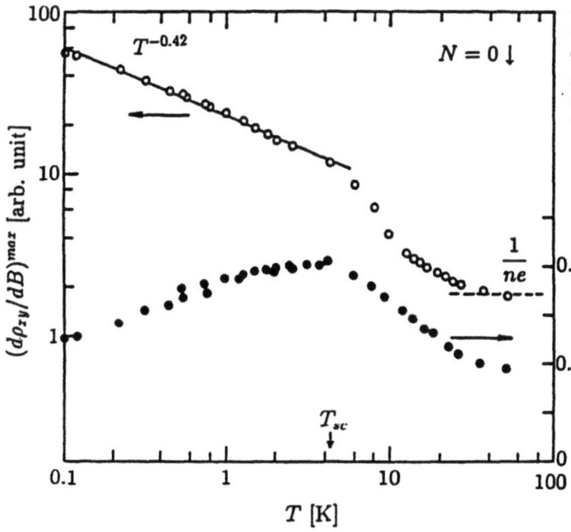

Fig.2 $d\rho_{xy}/dB(B^*)$ (the upper curve) and σ_{xx}^{max} (the lower curve) as a function of T for Landau level $N = 0 \downarrow$.

classical value and σ_{xx}^{max} increases with decreasing T. This increase can be attributed to the T dependence of the Fermi–Dirac distribution function, f. Within the framework of Ando's self consistent Born approximation (SCBA) [15], the density of states for each Landau level has a semielliptic form with a width $\Gamma^2 = (2/\pi)\hbar\omega_c(\hbar/\tau)$ for short range scatterers, where ω_c is the cyclotron frequency, and the relaxation time τ is related to m^* and μ via $m^*\mu/e$. At the center E_0 of the $N = 0$ Landau level, the σ_{xx} can be calculated through

$$\sigma_{xx}(T) \sim \int (-\frac{\partial f}{\partial E}) \left[1 - \left[\frac{E - E_0}{\Gamma}\right]^2\right] dE \tag{2}$$

Using m^* as a fitting parameter, we fit the σ_{xx}^{max} data in Fig.2 to this equation for $4.2\,\mathrm{K} < T < 30\,\mathrm{K}$. The result is plotted in the inset of Fig.1. The solid line is the best fitted curve with $m^* = 0.045 \pm 0.03$, which agrees with the cyclotron mass obtained from cyclotron resonance experiments [13]. This result indicates that the transport in this T range can be described satisfactorily within SCBA. On the other hand, σ_{xx}^{max} in this model is monotonously increasing with decreasing T and eventually saturates at a finite value in the $T = 0$ limit; whereas σ_{xx}^{max} in Fig.2 starts to decrease with decreasing T when $T < 4\,\mathrm{K}$, indicative of an incipient transport mechanism different from that within the semi–classical SCBA.

4 Localization to delocalization transition and scaling

When $T \leq T_{sc} \sim 4\,\mathrm{K}$, both $(d\rho_{xy}/dB)^{max}$ and σ_{xx}^{max} change their T dependent behavior, namely $(d\rho_{xy}/dB)^{max}$ follows a power law of $T^{-0.42}$ and σ_{xx}^{max} decreases with decreasing T. The transport is due to genuine scaling [16]. We

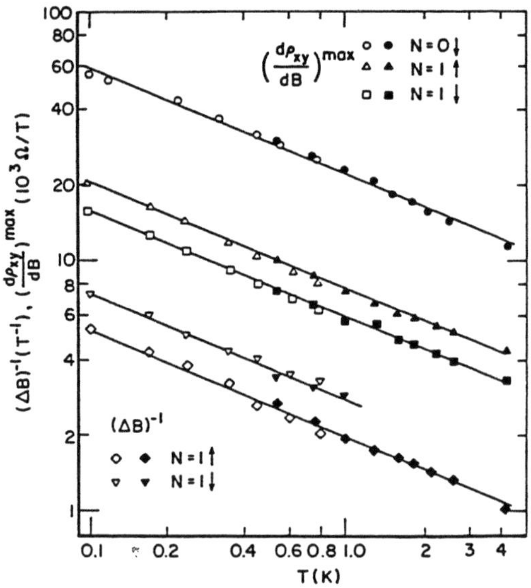

Fig.3 The upper portion shows the T dependence of $(d\rho_{xy}/dB)^{max}$ for Landau levels $N = 0\downarrow, 1\uparrow$, and $1\downarrow$; the lower portion shows that for $(\Delta B)^{-1}$ for $N = 1\uparrow$ and $1\downarrow$ levels.

plot the T dependence of $(d\rho_{xy}/dB)^{max}$ for the $N = 0\downarrow, 1\uparrow$, and $1\downarrow$ Landau levels on a log–log scale in the upper three curves of Fig.3. The bottom two curves are the T dependence of $(\Delta B)^{-1}$ in which ΔB is the charateristic width of ρ_{xx} and is operationally defined as $\Delta B \equiv B_2 - B_1$, with B_1 and B_2 being the positions of B at which $d\rho_{xx}/dB$ has extrema. All our data show the same linear dependence for all the quantities we studied. In another words, both $(d\rho_{xy}/dB)^{max}$ and $(\Delta B)^{-1} \sim T^{-\kappa}$, and $\kappa = 0.42 \pm 0.04$, independent of Landau levels. This identical power law dependence on T for both resistance components ρ_{xy} and ρ_{xx} in all three Landau levels indicate that the electron localization to delocalization transition at the center of each Landau level can be uniquely described by the critical exponent κ [11].

In Fig.4, we plot the T dependence of $d^2\rho_{xx}/dB^2(B^*)$ for the $N = 0\downarrow$, $1\uparrow$ and $1\downarrow$ Landau levels. On this log-log scale, the data for all three Landau levels show strikingly the same linear dependence on T in the range $0.04\,\mathrm{K} < T < 4.2\,\mathrm{K}$ for the $N = 0\downarrow$ level and $0.02\,\mathrm{K} < T < 4.2\,\mathrm{K}$ for the $N = 1\downarrow, 1\uparrow$ levels. The solid lines plotted in Fig.4 are best fitted to the data points. We, then, obtain $d^2\rho_{xx}/dB^2(B^*) \sim T^{-0.85}$, independent of the Landau levels. We have similarly studied the T dependence of $d^3\rho_{xy}/dB^3(B^*)$ and obtain $d^3\rho_{xy}/dB^3(B^*) \sim T^{-1.15}$. The uncertainty in both exponents is $\pm 10\%$. Within our experimenatl error, we find that $d\rho_{xy}/dB(B^*)$, $d^2\rho_{xx}/dB^2(B^*)$, and $d^3\rho_{xy}/dB^3(B^*)$ have a power law dependence on T with exponents $-\kappa$, -2κ and -3κ respectively.

These power law dependences are direct consequence of scaling [11]. They are manifestations of a divergent localization length ξ when E_F approaches E_c as given in Eq.1. They show that close to the singular magnetic field values B^*, the transport coefficients are differentiable functions of only one single parameter, $T^{-\kappa}(B - B^*)$. We can write the inelastic scattering length,

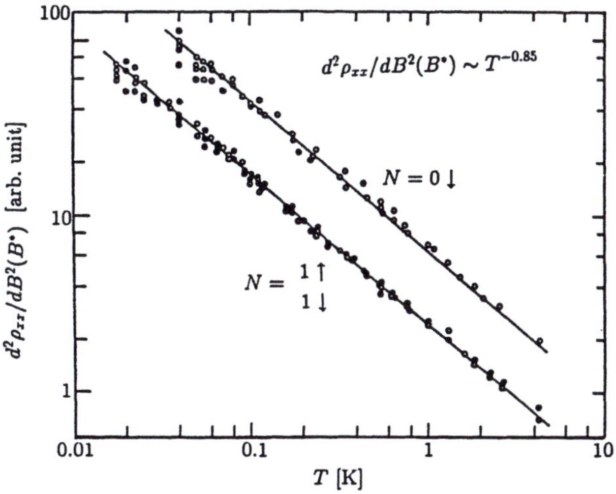

Fig.4 $d^2\rho_{xx}/dB^2(B^*)$ as a function of T for Landau levels $N = 0\downarrow$, $1\downarrow$, and $1\uparrow$.

within which quantum phase coherence is retained, as $l_{in} \sim T^{-p/2}$. Since $(B - B^*) \sim (E - E_c) \sim \xi^{-1/\nu}$, it is apparent that this single parameter is $(l_{in}/\xi)^{1/\nu}$ and $\kappa = p/2\nu$. Thus the experiment demonstrates explicitly and quantitatively that the electron transport in the IQHE shows scaling.

5 Discussion

If we draw an analogy with the theory of ordinary metalic conductors in two dimensions [17,18] and assume $p \sim 1$, then $\nu \sim 1.2$. This value is close to 4/3 predicted in the percolation picture [19], which neglects the quantum interference effect. A recent numerical finite–size scaling calculation found that $\nu = 2.34 \pm 0.04$ for the lowest Landau level [20]. If we use their calculated ν and our experimentally measured κ, $p \sim 2$. At present, however, there is no independent experimental value of p in high B to allow the determination of ν. Nevertheless, the fact that κ is independent of Landau levels is in disagreement with the calculations of Aoki and Ando [21], which predicts a ν dependent on the Landau level index.

Some comments about the characteristic temperature T_{sc} are in order. T_{sc} represents a crossover of the electron transport from the semi–classical regime to the quantum regime. In the semi–classical regime, the transport is described by the SCBA and the effect of decreasing T is the sharpening of the Fermi–Dirac distribution function. In the quantum regime, scaling sets in and coherent quantum interference dominates the phenomenon. The effect of T is its change of the inelastic scattering length, within which the quantum phase coherence is retained. The value of T_{sc} depends very much on the microscopic details of the disorder in the sample, in particular, whether short or long range potential fluctuations dominate. It can be as low as 0.06 K for the 2DEG in GaAs/Al_xGa_{1-x}As heterostructures, wherein long range potential fluctuations are known to dominate, compared to $\sim$ 4 K in the InGaAs/InP samples studied in these experiments.

Finally, we would like to mention that similar work has been extended to the fractional quantum Hall effect (FQHE) [22], in which the correlation between electrons is essential [23]. It is found that $(d\rho_{xy}/dB)^{max}$ and $(\Delta B)^{-1} \sim T^{-\kappa}$ in between the 1/3 and 2/5 plateaus and $\kappa = 0.43 \pm 0.02$ [24]. More work is currently in progress studying the transition regions between the IQHE and the FQHE plateaus.

6 Acknowledgement

The authors would like to thank Dr. M. Paalanen at AT&T Bell Laboratory and Dr. R. Webb at IBM Thomas J. Watson research center for their generocity. We are indebted to Dr. M. Razeghi for the InGaAs/InP samples used in this work.

References

[1] For a review, see P.A. Lee and T.V. Ramakrishnan, Rev. Mod. Phys. **57**, 287 (1985).

[2] D.J. Bishop, D.C. Tsui, and R.C. Dynes, Phys. Rev. Lett. **44**, 1153 (1980).

[3] D.J. Thouless, Phys. Rev. Lett. **39**, 1167 (1977).

[4] E. Abrahams, P.W. Anderson, D.C. Licciardello, and T.V. Ramakrishnan, Phys. Rev. Lett. **42**, 673 (1979).

[5] K. von Klitzing, G. Dorda and M. Pepper, Phys. Rev. Lett. **45**, 494 (1980).

[6] D.C. Tsui, and A.C. Gossard, Appl. Phys. Lett. **38**, 550 (1981).

[7] A. Briggs, Y. Guldner, J.P. Vieren, M. Voos, J.P. Hirtz, and M. Razeghi, Phys. Rev. **B 27**, 6549 (1983).

[8] M.A. Paalanen, D.C. Tsui, and A.C. Gossard, Phys. Rev. **B25**, 5566 (1982).

[9] A.M.M. Pruisken, Phys. Rev. **B 32**, 2636 (1985); and references therein.

[10] H.P. Wei, D.C. Tsui, M. Paalanen, and A.M.M. Pruisken, Phys. Rev. Lett. **61**, 1294 (1988).

[11] A.M.M. Pruisken, Phys. Rev. Lett. **61**, 1297 (1988).

[12] H.P. Wei, D.C. Tsui, and M. Razeghi, Appl. Phys. Lett. **45**, 666 (1984).

[13] Y. Guldner, J.P. Vieren, P. Voisin, M. Voos, M. Razeghi, M.A. Poisson, Appl. Phys. Lett. **40**, 877 (1982).

[14] G. Bastard, Appl. Phys. Lett. **43**, 591 (1983).

[15] T. Ando and Y. Uemura, J. Phys. Soc. Jpn. **36**, 959 (1974).

[16] H.P. Wei, D.C. Tsui, and A.M.M. Pruisken, Phys. Rev. **B33**, 1488 (1986).

[17] E. Abrahams, P.W. Anderson, P.A. Lee, and T.V. Ramakrishnan, Phys. Rev. **B 24**, 6783 (1981).

[18] B.L. Altshuler, A.G. Aronov, and D.E. Khmelnitsky, J. Phys. **C 15**, 7367 (1982).

[19] S.A. Trugman, Phys. Rev. **B 27**, 7539 (1983).

[20] B. Huckestein and B. Kramer, (preprint).

[21] H. Aoki, and T. Ando, Phys. Rev. Lett. **54**, 831 (1985).

[22] D.C. Tsui, H.L. Störmer, and A.C. Gossard, Phys. Rev. Lett. **48**, 1599 (1982).

[23] R.B. Laughlin, Phys. Rev. Lett. **50**, 1395 (1983).

[24] L. Engel, H.P. Wei, D.C. Tsui, M. Shayegan, workbook of EP2DS–8, Grenoble, France, Sep. 1989, p.692.

The Spatial Potential Distribution in $GaAs/Al_xGa_{1-x}As$ Heterostructures Under Quantum Hall Conditions Studied with the Linear Electro-Optic Effect

P.F. Fontein[1]*, J.A. Kleinen*[1]*, P. Hendriks*[1]*, F.A.P. Blom*[1]*, J.H. Wolter*[1]*, H.G.M. Lochs*[2]*, F.A.J.M. Driessen*[2]*, and L.J. Giling*[2]

[1]Eindhoven University of Technology, 5600 MB Eindhoven, The Netherlands
[2]University of Nijmegen, Toernooiveld, 6525 ED Nijmegen, The Netherlands

Abstract. We applied the linear electro-optic effect (Pockels effect) to investigate the spatial potential distribution in $GaAs/Al_xGa_{1-x}As$ heterostructures under quantum Hall conditions. With this method, which avoids electrical contacts and does not disturb the current distribution, we probed the electrostatic potential of the two-dimensional electron gas (2 DEG) locally. We observed a steep change of the Hall potential at the edges of the 2 DEG over a distance of about 200 μm. This change at the edges accounts for more than 80 % of the total Hall voltage. The remainder of the Hall voltage is distributed in the interior of the sample, where the potential shows a linear dependence on the position.

1. Introduction

Until recently, experimental access to the problem of the current and potential distribution in two-dimensional electron gases (2 DEGs) under quantum Hall conditions was possible only by attaching electrical contacts[1,2] to the interior of the 2 DEG. These electrical contacts, however, disturb the system to be investigated.

Our method to determine the spatial potential distribution makes use of the linear electro-optic effect[3] (Pockels effect). The method is not uncommon in the field of GaAs chip testing[4], but has until recently[5] never been applied under QH conditions. Since it is a contactless method we avoid the disturbing influences of the electrical contacts.

2. Experimental technique

GaAs becomes birefringent when an electric field is applied. We can take advantage of this effect to determine the potential difference between the 2 DEG and the back-gate of a $GaAs/Al_xGa_{1-x}As$ heterostructure. This has been demonstrated in Ref. 6.

Our samples consist of a 400 μm GaAs substrate with on one side the 80 Å Au-layer kept at ground potential. On the other side a 4 μm GaAs buffer layer, a 200 Å $Al_xGa_{1-x}As$ spacer layer, a 400 Å $Al_xGa_{1-x}As$ Si-doped ($n = 2\cdot 10^{24}$ m^{-3}) layer (both with $x = 0.3$) and a 180 Å GaAs cap layer are grown. The electron concentration of the sample is $5.0\cdot 10^{15}$ m^{-2} and the mobility 20 m^2/Vs.

Springer Series in Solid-State Sciences, Vol. 97
Localization and Confinement of Electrons in Semiconductors
Editors: F. Kuchar · H. Heinrich · G. Bauer

The potential difference between the 2 DEG and the Au-layer results in a change of the polarization of the transmitted light. This change in the polarization is analyzed by a quarter wave plate and a polarizer. Since the potential differences to be detected are fairly small we apply an alternating current (235 Hz) through the 2 DEG and thus modulate the transmitted light intensity. The detector output can hence be measured with a lock-in technique. We carefully checked that the measured signals had neither an out of phase nor a double frequency component. In order to determine the local potential in the 2 DEG we first perform a calibration measurement. To this end an alternating voltage of 5.6 V_{pp} is applied between the 2 DEG and the Au-layer and the resulting detector signal is measured. Next, an alternating current of known amplitude is sent through the 2 DEG (with one current contact also at ground potential) and again the lock-in signal is measured. Both measurements are taken at the same position of the laser beam. The ratio of the detected intensities in these two measurements yields the unknown potential at the position of the laser beam for the case of the alternating current flowing through the 2 DEG. Subsequently, the laser beam is scanned across the surface of the sample step by step. At each spot the calibration procedure is repeated. We checked that the results did not depend on the amplitude of the voltage applied in the calibration measurement.

We performed our measurements on a sample with Hall-bar geometry of 5.4 mm length and 2 mm width. No side arms were applied which conventionally are used as voltage probes. Current contacts (In) were alloyed into the 2 DEG at both ends (5.4 mm apart). Prior to our experiments we checked the homogeneity of our samples with a laser-scan technique and selected a sample which did not show interruptions of the 2 DEG[7] or other major defects. The temperature in all experiments was 1.5 K with the sample submerged in superfluid ^4He (in order to avoid both disturbing influences of boiling ^4He in our optical experiments and unwanted heating effects). The electrical current used was 5 μA_{eff}. We checked that heating effects did not show up for this current.

3. Results

In the QH plateaus we made scans of the potential across the width of the Hall-bar. These scans were made in the middle between the current contacts. The results of two scans at different positions within the same plateau are plotted in Fig. 1. The edges of the Hall-bar are at 0 mm and 2 mm.

A striking result of these measurements is the observation of a steep increase of the Hall potential at the edges of the 2 DEG. The presence of such a steep increase could be deduced only indirectly in Ref. 5. Although the width of the regions of steeply increasing potential has the same order of magnitude as the focal diameter of the spot of light (70 μm, diffraction limited) the difference seems to be significant. In the interior of the Hall-bar

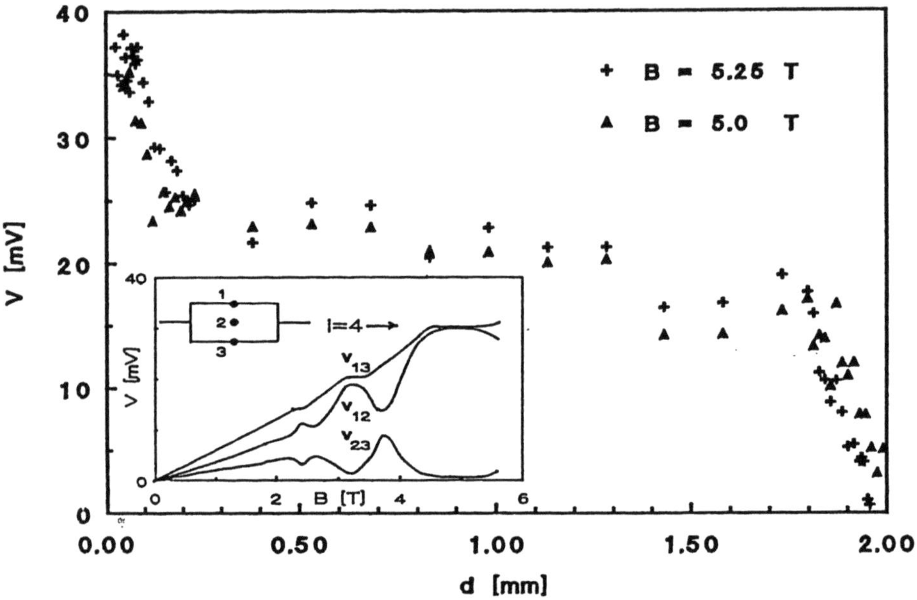

Fig. 1. *Plotted are the results of line scans of the potential as a function of position made at magnetic fields of 5.0 T and 5.25 T. The scans are made across the width of the structure (2 mm) in the middle between the current contacts. The edges are at 0 mm and 2 mm. It is evident that the potential increases rapidly at the edges of the structure. Away from the edges an almost linear behaviour is observed. No significant dependence of the potential distribution on magnetic field occurs in the entire plateau region. The inset shows the result of an electrical transport experiment on the same sample after an In-contact was alloyed into the interior of the Hall-bar. The measured potential differences strongly depend on the magnitude of the magnetic field.*

the potential shows a small linear increase. Furthermore, it is clear from Fig. 1 that within experimental error the potential distribution is the same in the whole plateau region. We also observe no significant difference in the potential distribution between measurements taken inside or outside a plateau region, the edge effects remain present.

In principle, the observed steep increase of the potential could be due to fringing fields at the edges, which might have a large component of the electric field parallel to the 2 DEG. However, experiments at room temperature show that in our experiments the influence of fringing effects is negligibly small. A further indication of the absence of any influence of fringing fields is the fact that at high currents (where the QHE disappears) the potential becomes a linear function of position across the width of the structure, see Fig 2. We also investigated whether the application of

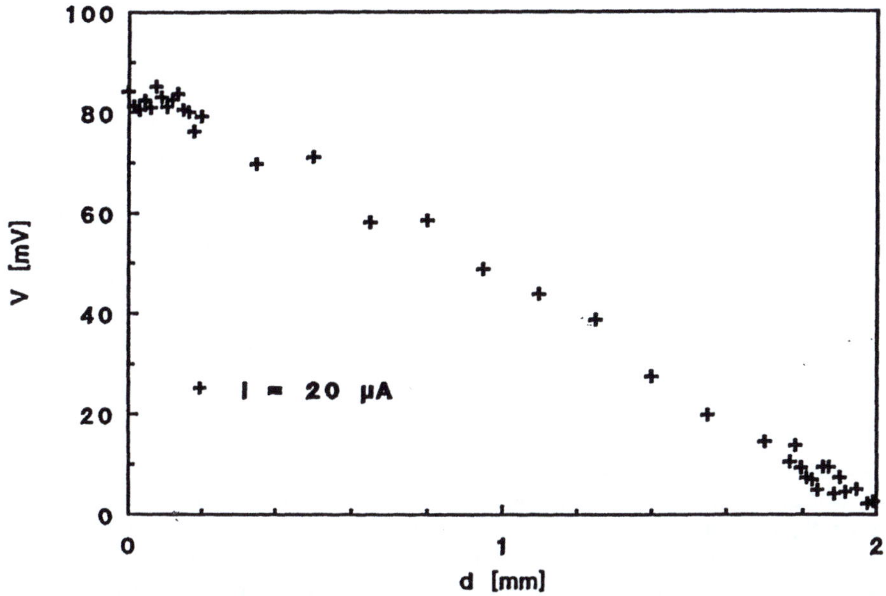

Fig. 2. Plotted is a line scan at 3.5 T (i=6) with a current of 20 µA. At this current some heating can be observed. It is clear that the edge effects are absent now and hence the current is homogeneously distributed.

alternating currents leading to current reversal in the sample may cause some problems. To this end we applied a DC offset current density to obtain a modulated current density which is not reversed. However, from both types of experiments we obtain the same results. If there is a spatial switching of current paths on time scales smaller than our measuring time, we may also expect a drawback on our experiments. To obtain a high resolution we had to average every single measurement for more than one minute. If switching occurs on a time scale much smaller than one minute these will be averaged and eventually cancelled. Results from measurements performed during the same cool-down cycle reproduce within experimental error. We believe that switching effects on these long time scales must be ruled out.

We now derive the current distribution from the voltage distribution. If one assumes the equation $j = \sigma E$ to hold under QH conditions (with j the current density, σ the local conductivity tensor and E the in plane electric field) and if one assumes that $\sigma_{xx} \ll \sigma_{xy}$ (as will be true in sufficiently high magnetic field) it then follows that $j_y = -\sigma_{xy}E_x$. Thus we see from Fig. 1 that more than 80 % of the current flows along the edges, the remainder in the interior. This behaviour might explain experiments like the asymmetries often observed in the Shubnikov-de Haas effect[8], which do not scale with sample width or current density, but which depend on the total current.

How do our experimental results compare to the theoretical models? Models which do not include edge effects obviously seem to be questionable. At first sight it also seems that the Büttiker model[9] does not apply, because there the edge currents are oppositely directed to each other (classically the so-called skipping orbits move in opposite directions at the opposite edges). In our experiments, however, the edge currents turn out to have the same direction. Note that due to the lock-in technique we measure exclusively the edge currents as induced by the *external* current. Models which assume a gradient in the electron concentration to be the origin of the QHE can only apply in the case that this gradient changes direction over distances much less than 70 μm (which is about our resolution). The same is true for the related percolation models, here current paths must be separated by distances smaller than 70 μm. Note also that the width of the edge currents as measured is much larger than the cyclotron radius to be associated with the skipping orbits.

It should be noted that with the linear electro-optic effect the electrostatic potential is measured and with electrical contacts the chemical potential. Thus even apart from disturbances introduced by electrical contacts, we do not expect a priori that the two experiments yield the same results. To illustrate this difference we alloyed electrical contacts into the interior of the 2 DEG and performed a Hall experiment. The results of these experiments are presented in the inset of Fig. 1 and look very similar to those presented in Refs 1,2. A current distribution will be derived from these measurements which is totally different from the one obtained with the linear electro-optic effect on the same sample.

We conclude that the Pockels experiments under quantum Hall conditions reveal the presence of edge currents. The width of these edge currents is less than 200 μm. We think it is probably just the classical Hall effect in the two-dimensional electron gas which is the origin of the potential distribution we measured[10].

Acknowledgments

The authors like to thank M. Leys and W. v.d. Vleuten for the growth of the heterostructures, P.A.M. Nouwens for the preparation of the sample and S.M. Olsthoorn for his assistance during the experiments. The work described is part of the research programme of the Dutch Foundation for Fundamental Research on Matter, which is financially supported by the Dutch Organization for the Advancement of Research.

References

[1]G. Ebert, K. von Klitzing, and G. Weimann, J. Phys. C **18**, L257 (1985).
[2]H.Z. Zheng, D.C. Tsui, and A.M. Chang, Phys. Rev. B **32**, 5506 (1985).

[3]A. Yariv, Quantum Electronics (Wiley, New York, 1967).
[4]B.H. Kolner, and D.M. Bloom, IEEE J. Quantum Electron. **22**, 79 (1986).
[5]P.F. Fontein, P. Hendriks, and J.H. Wolter, proc. EP2DS8, accepted for publication in Surf. Sc..
[6]P. Hendriks, F.J.M. Schnitzeler, J.E.M. Haverkort, J.H. Wolter, Kees de Kort, and G. Weimann, Appl. Phys. Lett. **54**, 1763 (1989).
[7]P.F. Fontein, P. Hendriks, J.H. Wolter, R. Peat, D.E. Williams, and J.-P André, J. Appl. Phys. **64**, 3085 (1988).
[8]B.E. Kane, D.C. Tsui, and G. Weimann, Phys. Rev. Lett. **59**, 1353 (1987).
[9]M. Büttiker, Phys. Rev. B **38**, 9375 (1988).
[10]R.R. Gerhardts, and V. Gudmundsson, Solid State Commun. **67**, 845 (1988).

Recent Experimental Studies of the Integer and Fractional Quantum Hall Effects in Steady and Pulsed Magnetic Fields

R.G. Clark

Clarendon Laboratory, University of Oxford, Oxford OX1 3PU, UK

Abstract. Results from four experimental programmes are each briefly reviewed:

1. Identification of a universal dissipative conductance $\sigma_{xx}^c = e^2/h$ from activation studies of the integer and fractional QHE, with e the electron charge $-e$ or quasiparticle fractional charge $e^* = \pm e/q$ $(q = 3,5,7...)$.
2. Identification of the fundamental importance of spin reversal in the FQHE by observation of phase transitions in tilted-field studies, from unpolarised states at $\nu = 2/3$ and $4/3$ to polarised configurations.
3. Pulsed field measurements of the FQHE at mK temperatures.
4. A definitive, optical detection of the integer and fractional QHE in GaAs by magneto-luminescence measurements of an ultra-high mobility, single heterojunction at 120 mK.

1. Universal Dissipative Conductance in the Integer and Fractional QHE

The quantum Hall effect [1] is remarkable for the sample independence of the Hall conductance σ_{xy}. In this section a universal behaviour (to a much lower precision) of the dissipative conductance σ_{xx} is described. We first address the IQHE regime. Activation data for even Landau level filling factors $\nu = 2$–20 in sample G160 are shown in Fig. 1 (characteristics of all samples referred to in this paper are given in Table 1). Additional results for $\nu = 1,2,4,6$ in sample G635 and $\nu = 2,4,6$ in sample G139 are presented in Refs. 2 and 3. The σ_{xx} data is obtained from ρ_{xx}, ρ_{xy} measurements in the Hall bar samples. Fits to the linear region of the Arrhenius plots are extrapolated to $1/T = 0$. The $\nu = 2,4,6$ data in all three samples lead to an intercept $\sigma_{xx}^c = \sigma_{xx}(1/T=0) = e^2/h \pm 10\%$. There are two sources of small departure from this behaviour. As ν is increased from 8 to 20 in G160, σ_{xx}^c falls systematically to $0.75e^2/h$, due to deviations from the Arrhenius form (see below). Also, the G635 $\nu = 1$ results give $\sigma_{xx}^c = 1.25e^2/h$, as do further studies of $\nu = 3$ in G160 and $\nu = 1$ at dark concentration $(n = 1.4 \times 10^{11}\, cm^{-2})$. Odd-$\nu$ experiments are problematic since the spin-splitting energy $g^* \mu_B B$ is small compared to the cyclotron energy that is the dominant gap at even ν. In short, there is substantial evidence for a constant σ_{xx}^c close to e^2/h for even integer ν and indications of the same behaviour at odd integer ν. The constancy of σ_{xx}^c is in striking contrast to predictions of RPA transport theory [4], in which σ_{xx}^c increases with ν, but is in accord with scaling calculations [5].

The linear extrapolations in Fig. 1 require qualification. The correct starting point is the relation $\sigma_{xx}(T) = -\int dE\ \sigma_E\ \partial f/\partial E$, where $f = \{exp[(E-E_F)/T] + 1\}^{-1}$. The simplest model applicable to high quality samples ('sharp' Landau levels) is to assume σ_E is a constant, σ_0, in a band $\pm\Gamma$ around the Landau level centre and $\sigma_E = 0$ elsewhere. At integer filling, this gives

$$\sigma_{xx}(T) = \sigma_{xx}^c\ [f(\Delta-\Gamma) - f(\Delta+\Gamma)] \qquad (1)$$

where 2Δ is the (ideal) Landau level splitting and $\sigma_{xx}^c = 2\sigma_0$ (the factor of 2 comes in because both the nearly full and nearly empty Landau levels make equal contributions to $\sigma_{xx}(T)$). At low T, where $e^{(\Delta-\Gamma)/T} \gg 1$ and f reduces to Boltzmann statistics, Eq. (1) simplifies to $\sigma_{xx}(T) = \sigma_{xx}^c e^{-(\Delta-\Gamma)/T}[1 - e^{-2\Gamma/T}]$ and for $e^{-2\Gamma/T} \ll 1$ the 'linear' form

Springer Series in Solid-State Sciences, Vol. 97
Localization and Confinement of Electrons in Semiconductors
Editors: F. Kuchar · H. Heinrich · G. Bauer

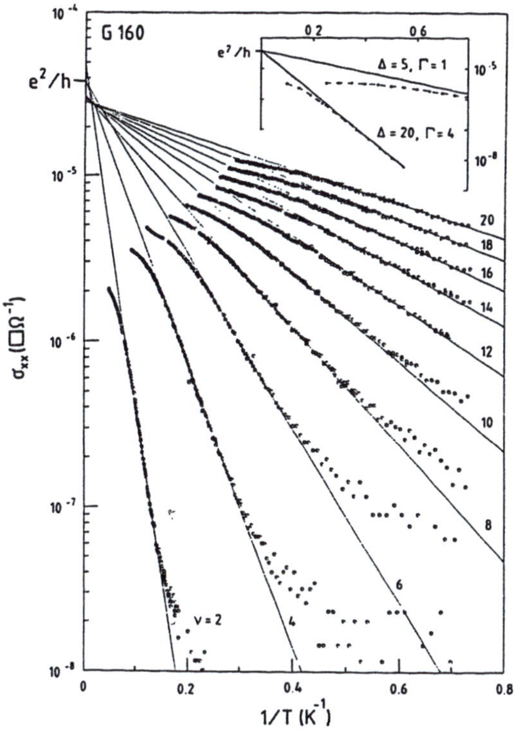

Fig. 1 *IQHE activation data*

Table 1. *GaAs-GaAlAs heterojunction characteristics*

Sample	Spacer (Å)	n_{sat} ($\times 10^{11}$cm^{-2}) *dark values	μ_{sat} ($\times 10^{6}$cm^2/Vs)
G641	1600	0.93 , 0.36*	9.0 , 3.7*
G139	1600	0.95	1.0
G635	800	1.45	7.2
G156	800	1.60	1.6
G160	400	2.20	2.0

$$\sigma_{xx}(T) = \sigma_{xx}^{c} e^{-(\Delta-\Gamma)/T} \tag{2}$$

is obtained. In general, σ_E will be more complicated and Γ will be both T and B-dependent [2, 6]. However, for a given ν (fixed B) and over a temperature range for which the linear form is valid, straight-line extrapolation of the raw σ_{xx} data in *high quality* samples would be expected to provide a good estimate of σ_{xx}^{c}. This is verified by the G139, G160 and G635 $\nu = 2,4,6$ data. The deviation from linearity in each of the Fig. 1 plots at high T (downward curvature) is due to a breakdown of assumptions leading to Eq. (2) and a full Fermi-Dirac analysis is required in this regime. Deviations at very low T can be due to two mechanisms; the dominance of hopping over activated conduction [7] and non-ideal contacts [8]. A meaningful analysis of this region is difficult as in general both effects will be present. The systematic fall-off in σ_{xx}^{c} at high ν is also due to a breakdown of the Arrhenius form. This is shown in the Fig. 1. inset, where Eqs. (1) and (2) with $\sigma_{xx}^{c} \equiv e^2/h$ are

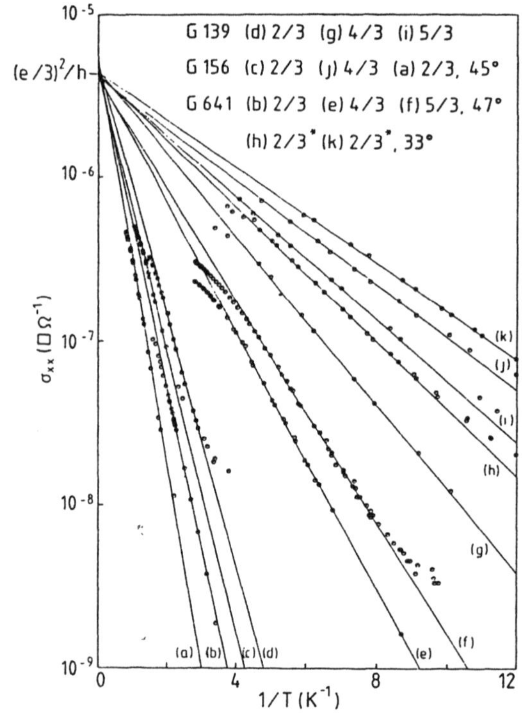

Fig. 2 *FQHE activation data*

compared for $\Delta = 20K$, $\Gamma = 4K$ and $\Delta = 5K$, $\Gamma = 1K$. In the experimental temperature range, the Arrhenius form is close to the simulated data for the first but not the second case. Consequently, for the (second) situation applicable to high ν, linear extrapolation of raw data leads to an artificially low estimate of σ^c_{xx}. It does not contradict the existence of a universal σ^c_{xx} value, which is masked by limitations of the extrapolation technique. It is difficult to progress beyond a linear approach however, as this requires an independent knowledge of $\sigma_E(E)$, $\Gamma(B,T)$ and Δ to obtain σ^c_{xx} in a rigorous fit. In low quality samples, where $\Gamma \sim \Delta$, the universality of σ^c_{xx} as judged by a linear fit is obscured at *low* ν by violation of $e^{(\Delta-\Gamma)/T} \gg 1$ [2].

The universal result $\sigma^c_{xx} \simeq e^2/h$ at integer filling is of fundamental importance when compared to activation studies of the fractional QHE. In Fig. 2 we show FQHE data ($\nu = p/q$) for 'parent' $q = 3$ states in three high quality samples. The data set includes results for dark(*) and saturated concentrations and different orientations ϑ of the B-field to the sample plane-normal. Linear extrapolation leads to $\sigma^c_{xx} = (e/3)^2/h \pm 10\%$. The only deviations from this result are for G156, G641 5/3 states ($\vartheta = 0°$), for which $\sigma^c_{xx} \sim 0.8(e/3)^2/h$ (not shown); and studies in the critical angular range where phase transitions are observed for 2/3, 4/3 states (see section 2). The Fig. 1 and 2 data provide striking confirmation of the correspondence of electrons in the IQHE and quasiparticles of charge $\pm e/q$ in the FQHE [9]. It is further substantiated by our earlier report [10] for G139 FQHE states with $q = 3,5,7$ and 9 for which $\sigma^c_{xx} = 0.91(e/q)^2/h$ on average (Sachrajda *et al.* [11] find impressive agreement with this result; see also Refs. 12, 13 for noise, tunnelling studies of e^*). Our measured $\sigma^c_{xx} = 2\sigma_0 \simeq e^2/h$ at integer ν agrees with the simulation result $\sigma_0 = 0.46e^2/h$ [14], further supporting the conclusion that σ^c_{xx} is a universal number close to e^2/h.

2. Spin Configurations of Fractional QHE Ground States in the N = 0 Landau Level

Recent tilted-field activation energy studies of the FQHE [15, 16] and finite size calculations [17-19] have revealed the importance of reversed spins in the ground states of the quantum electron liquid. In particular we have demonstrated that for the region $1<\nu<2$, the 4/3 state is unpolarised (↑↓) at low B (by observation of a 4/3 phase transition) whereas the 5/3 state is fully polarised (↑↑) (maximum possible polarisation) and is relatively unaffected on sample rotation [15]. A phase transition has also been reported for the $\nu = 8/5$ state [16], consistent with an ↑↓ assignment at low B. The 4/3 transition, first identified in sample G156, is shown with increased definition in the high quality sample G641 [3] in which the 8/5 transition is also observed. The angular evolution of G641 FQHE structure for $1<\nu<2$, consistent with low-field spin configurations 4/3↑↓ and 8/5↑↓, obtained on rotating the sample (at saturated concentration) *in-situ* at $T = 55mK$, is mapped out in Figs. 3 and 4 where ϑ is the angle of the sample plane-normal to the B-field. The initial collapse of the unpolarised 4/3, 8/5 states caused by the increased spin Zeeman energy with tilt angle (at fixed Coulomb energy) is identified in Fig. 3, and is followed by the emergence of re-entrant 4/3, 8/5 states signifying a field-induced phase transition to polarised configurations, shown in Fig. 4. The 8/5 transition takes place first (with respect to angular range) and is followed by the 4/3 transition. Explicitly, at $\vartheta = 0°$ strong 4/3 and 8/5 ρ_{xx} minima are observed in Fig. 3a. The 8/5 minimum is systematically destroyed over the range $\vartheta = 25°$ (not shown) to 52° (Fig. 3b-c) and is absent at $\vartheta = 58°$ (Fig. 3d). At $\vartheta = 43°$ (Fig. 3b) the steady destruction of the 4/3 ρ_{xx} minimum just begins and is complete by $\vartheta = 62°$ (Figs. 3c-d, 4a), at which point the 8/5 state has re-emerged (Fig. 4a). Re-entrant structure at $\nu = 4/3$ is first observed at $\vartheta = 64.7°$ (Fig. 4b) and a deep 4/3 ρ_{xx} minimum is recovered by $\vartheta = 67.4°$ (Fig. 4d). The 8/5 ρ_{xx} minimum is similarly well-defined at this high angle. The ρ_{xx} development is supported by ρ_{xy} data (not shown here); in the phase transition the 4/3, 8/5 ρ_{xy} plateaux are destroyed during the collapse and are recovered in the re-entrant phase. In contrast to the 4/3, 8/5 evolution, 5/3 structure is largely unaffected over the entire angular range (Figs. 3a-d, 4a-d) consistent with its polarised assignment. The behaviour of the 7/5 and 10/7 structure during the re-emergence is interesting. As the 4/3 re-entrant ρ_{xx} minimum strengthens, the 7/5 state is destroyed (Figs. 4b-d) as observed in G156 [15] whereas the 10/7 state remains largely unaffected.

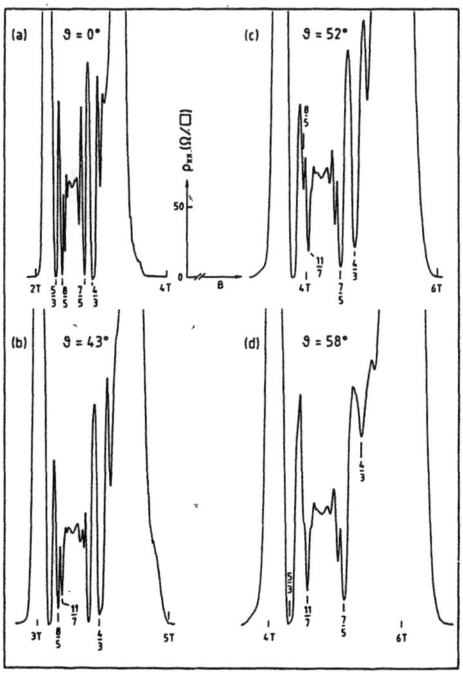

Fig.3 *Angular evolution of G641* $1<\nu<2$ *structure at* $55mK$; $\nu=4/3$ *collapse*

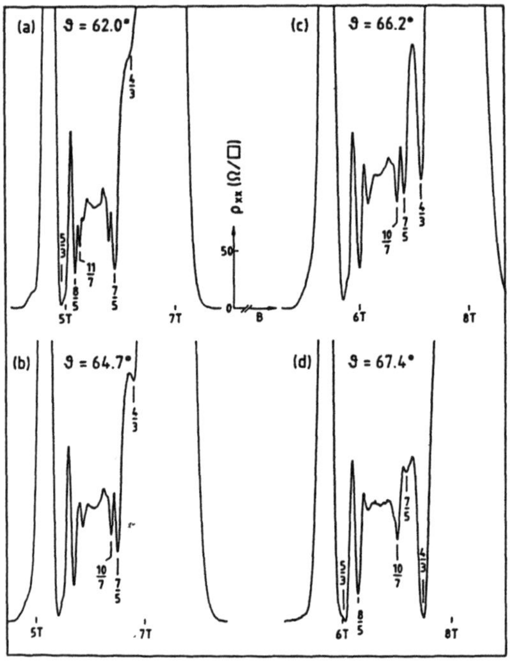

Fig. 4 *As for Fig.3 ; ν=4/3 re-emergence*

The situation is quantified by activation studies to determine the characteristic gap Δ defined by the relation $\rho_{xx} = \rho^c_{xx} e^{-\Delta/kT}$, and σ^c_{xx} from the ρ^c_{xx} and ρ_{xy} data. The destruction of the $4/3\uparrow\downarrow$ state in G641 is characterised by a linear collapse in $\Delta_{4/3}$ from $0.90K$ (at $\vartheta = 0°$) to zero over the field range $B = B_\perp/\cos\vartheta = 2.9$ to $5.4T$, shown in Fig. 5. The 4/3 state is absent (zero gap) over the range $B = 5.5-6.5T$. Beyond $6.5T$ the 4/3 state re-emerges, and $\Delta_{4/3}$ for the re-entrant state $\vartheta = 67.4°$ ($B = 8.1T$) is $0.26K$. At $\vartheta = 0°$ the extrapolated intercept σ^c_{xx} from the 4/3 activation data is $0.97(e/3)^2/h$ (Fig. 2e). This falls towards $(e/5)^2/h$ during the collapse of the unpolarised state. The re-entrant state at $B = 8.1T$ has an intercept of $0.8\,(e/5)^2/h$. Studies at very high angles to determine whether $\Delta_{4/3}$ and σ^c_{xx} remain at this value or increase have proved difficult, due to rising ρ_{xx} background problems (which may be associated with the large parallel B-field component) and an apparent instability of the 4/3 re-entrant state at and beyond the 4/3 data point circled at $7.5T$. The most reproducible data from several repeat activation studies are plotted. The general development of the G641 4/3 results closely resembles the behaviour reported for the lower quality, higher density sample G156, for which the linear collapse of the (weaker) gap $\Delta_{4/3}$ with B occurred at higher absolute field values. On defining $B_c(4/3)$ as the extrapolated field at which $\Delta_{4/3}$ collapses to zero ($B_c(4/3) = 5.4T$ in G641) we find that all reported $B_c(4/3)$ [3, 11, 15, 20] are consistent with the relation $B_c(4/3) = An^{1/2}$ with $A = 5.51$ (B_c in Tesla, n in units of $10^{11}cm^{-2}$). A change in σ^c_{xx} from $(e/3)^2/h$ to $(e/5)^2/h$ in the re-entrant phase implies a change in e^* for the 4/3 state from $e/3$ to $e/5$. Within the hierarchical model [21] this would provide an explanation for the observed destruction of the 7/5 state at the 4/3 transition (see open circle $\Delta_{7/5}$ data in Fig. 5), as $e/3$-charged quasiparticles form the fundamental basis of the 7/5 state.

With the inclusion of spin (but not Landau level mixing) we asserted [15] that the correct symmetry for FQHE states is between ν and $2-\nu$ (not $1-\nu$ as expected in a fully polarised framework, explaining previous suggestions of ν, $1-\nu$ asymmetry [22]) and that projected low-B assignments for the important parent states in the region $\nu<1$ are $2/3\uparrow\downarrow$, $1/3\uparrow\uparrow$ in agreement with calculations by Maksym [19]. An early indication of the existence of a $2/3\uparrow\downarrow$ state was reported in Ref. 3. A quantitative, tilted-field study of sample G641 at dark concentration [23] is shown in Figs. 6-9,

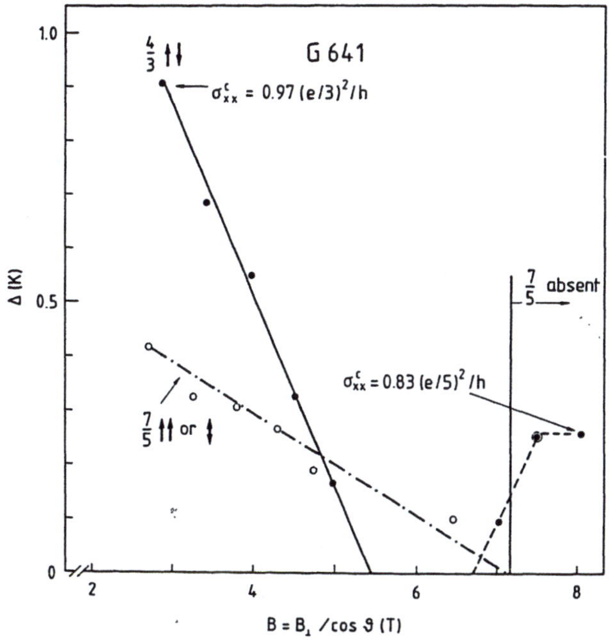

Fig. 5 *ν= 4/3 (and 7/5) gap vs. total field B in G641 at saturated concentration.*

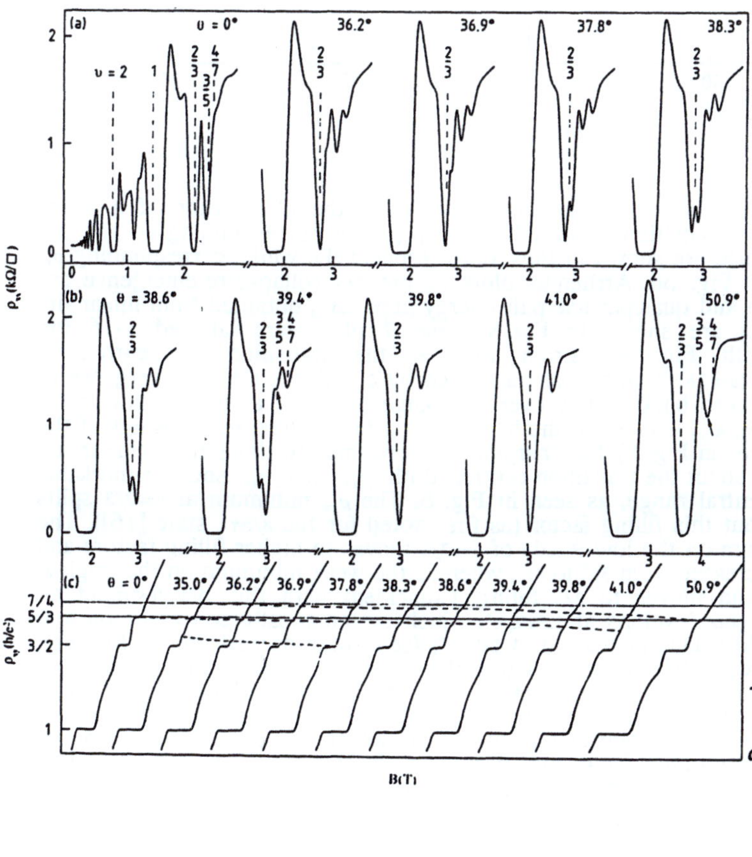

Fig. 6

Angular evolution of ν<1 structure at 45mK in G641 at dark concentration

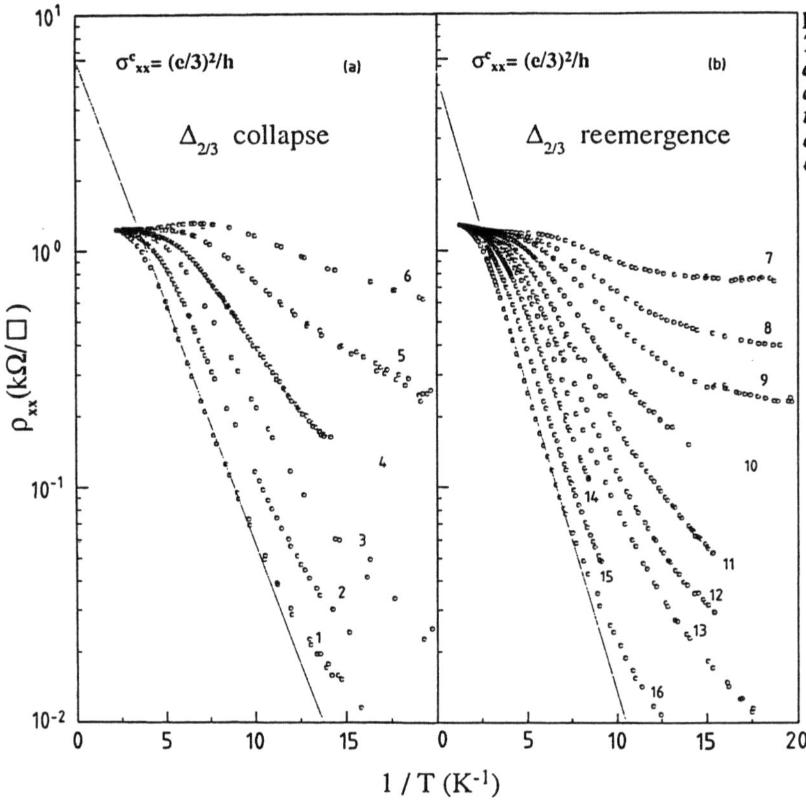

Fig.7 *Tilted-field activation data at $\nu = 2/3$ in G641 at dark concentration*

which confirms the low-B $2/3\uparrow\downarrow$ assignment through the observation of a sharp $2/3\uparrow\downarrow \rightarrow 2/3\uparrow\uparrow$ phase transition. Selected G641 ρ_{xx} data corresponding to the 2/3 collapse/re-emergence are shown in Figs. 6a/b respectively, whilst corresponding ρ_{xy} data are shown in Fig. 6c. Arrhenius plots of the 2/3 collapse/re-emergence are shown in Fig. 7a/b and quasiparticle-pair energy gaps $2\Delta_{2/3}$ obtained from linear fits to the Fig. 7 data are shown in Fig. 8 (Fig. 7 plots are labelled 1-16 for cross-reference with Fig. 8). From Fig. 8, the transition is centred at $B = B_{\perp}/cos\vartheta = 2.87T$ ($\vartheta = 38.3°$), and is characterised by a collapse and re-emergence of $2\Delta_{2/3}$ from $\sim 1K$ to zero back to $1.3K$ over the range $\vartheta = 0°–59°$ at two distinct rates $R_{2/3} = (2/\mu_B)|d\Delta_{2/3}/dB|$; $R_{2/3}$ is close to the bare GaAs g-factor (0.44) in the wings of the transition (solid lines in Fig. 8), but sharpens to more than 10 times this value in a narrow field range about the transition centre. Both ρ_{xx} and ρ_{xy} show remarkable structure in the central range, as seen in Fig. 6. The ρ_{xx} minimum at $\nu = 2/3$ splits asymmetrically about this filling factor (as first noted for the $8/5\uparrow\downarrow$ state [16]). The (strong) ρ_{xx} minimum on the low-B side of $\nu = 2/3$ moves to higher filling factors and weakens as ϑ is increased from 36° to 41°, whereas the weak minimum on the high-B side strengthens and moves to 2/3 filling. Coincident with this development an anomalous ρ_{xy} plateau develops at $1.56(h/e^2)$ which systematically converges on and merges with the dominant plateau at $(3/2)(h/e^2)$ (see lower dashed line in Fig. 6c). Open circles/squares in Fig. 8 correspond to the lower/upper minimum of the 2/3 ρ_{xx} doublet structure. Points 6–8 are tentative due to the form of the Fig. 7 data. Correlation of the 2/3 splitting with unprecedented effects in the daughter 3/5 and 4/7 ρ_{xx}, ρ_{xy} evolution is also identified. In particular, strong 3/5 ρ_{xx}, ρ_{xy} structure moves smoothly to a higher filling factor and disappears in the central transition region, whereas weak 4/7 ρ_{xx}, ρ_{xy} structure moves smoothly to 3/5 filling and strengthens,

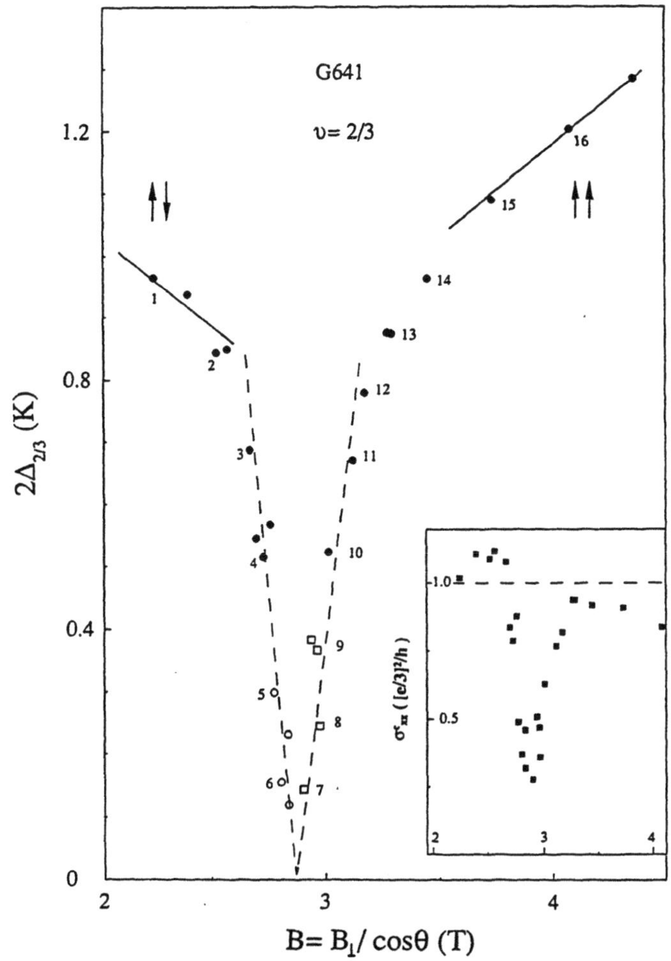

Fig. 8 *$\nu = 2/3$ quasiparticle-pair energy gap ($2\Delta_{2/3}$) vs. total field B in G641 at dark concentration*

effectively *becoming* the 3/5 FQHE state. Whilst this can be seen from close inspection of Fig. 6, filling factors of ρ_{xx} and ρ_{xy} structure obtained from this figure are plotted against ϑ in Fig. 9, where the trends are seen more clearly. Arrows at 39.4°, 50.9° in Fig. 9 relate to those in Fig. 6b to provide a cross-reference point. The Fig. 9 data cannot be explained by sample inhomogeneity and we interpret the results as evidence for the formation of an intermediate state in the critical region of the 2/3 phase transition which, unlike any FQHE state yet observed, is characterised by ρ_{xy} structure at non-rational fractional multiples of h/e^2. On the low and high-B side of the 2/3 transition the Fig. 7 intercepts σ^c_{xx} are close to $(e/3)^2/h$ as shown, whilst in the transition region σ^c_{xx} falls towards $(e/5)^2/h$ as shown in the Fig. 8 inset.

Overall, the 2/3 data are consistent with a phase transition from an ↑↓ to ↑↑ configuration, with an intermediate state exhibiting new features which remain to be explained in the central transition region. Whilst the 4/3 re-entrant state is complex and shows properties similar to the 2/3 state in the intermediate regime, the observation of phase transitions at $\nu=2/3$ and 4/3, in contrast to the absence of transitions at $\nu = 1/3$ and 5/3, provides strong evidence for ν, $2-\nu$ symmetry. Ideal

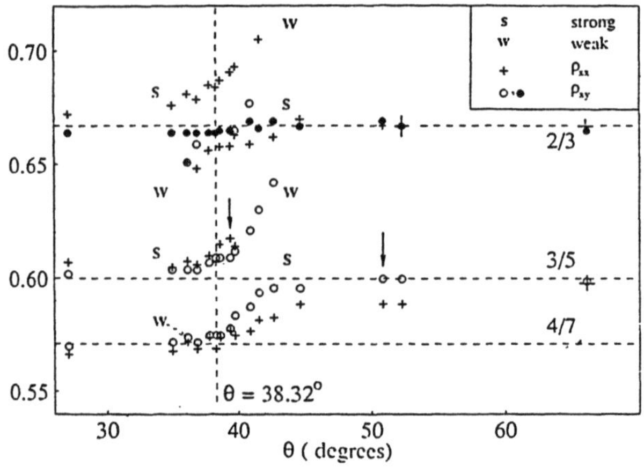

Fig. 9 *Filling factor* ν *from Fig. 6 angular evolution of* ρ_{xx} , ρ_{xy} *FQHE structure*

symmetry will be perturbed by Landau level mixing, which is important at low B. Finally, we note that the finite extent (in field) of the 'zero gap' region characteristic of the 4/3 transition, evident in Fig. 5, reduces with sample density (e.g. from $3.14T$ in G156 [15] to $1.24T$ in G641) and our observation of an extremely sharp 2/3 transition at very low n does not necessarily indicate a difference with the 4/3 data.

3. Measurements of the Fractional QHE in Pulsed Magnetic Fields

Transport measurements of the fractional QHE in the extreme quantum limit have been made by combining a Clarendon toploading ^3He system and $40T$ pulsed coil [24] at the Leuven pulsed field facility: $500kJ$ ($40mF$, $5kV$) capacitor bank with optically-triggered thyristor switching [25]. Using fast electronics, this unique combination has allowed observation of well-resolved FQHE structure at temperatures $\sim 400mK$ with pulse durations ~ 10 ms. The ^3He system, shown schematically in Fig. 10, is mounted on airsprings and has four concentric long glass tails inserted in the 20 mm bore pulsed magnet; precise tail alignment in this critical assembly is checked using a portable X-ray machine. To achieve a 5 mm sample space in this confined geometry, the tail set incorporates an OVC bellows arrangement and is surrounded by a split liquid nitrogen dewar which serves both as the cryostat liquid nitrogen shield and as the magnet coolant. The split dewar design facilitates lowering of the cryostat into the coil. The sample, a red LED, and Ruthenium Oxide thermometer are mounted as shown in Fig. 11 on a replaceable t/l probe tail which, along with shrapnel plates, safety valves and special ^3He dump arrangements, allows for possible explosion of the pulsed coil. No significant heating of the sample is observed in the pulsed field (on the timescale of the pulse) with this arrangement. Using Cu-Nb and specially-developed wires, prototype 20 mm bore $60T$ pulsed coils are being manufactured both at Oxford and Leuven for this project. To date ν=1/3, 2/3, 1/5, 2/5, 3/5, 3/7, 4/7, and 4/9 FQHE structure has been resolved in shots up to $35T$ in samples G641 and G635. Preliminary pulsed field data for G635 are shown in Figs. 12 and 13. Shots to beyond $40T$ will examine the critical region $\nu<1/5$ where the onset of an electron solid has been reported in low density samples. The main thrust of this programme is to determine how far the electron liquid state persists in the extreme quantum limit under conditions of intense magnetic fields in samples of the highest quality. To quantify the measurements, FQHE activation energy studies in pulsed fields are in progress and provide a rigorous test of the data. Capacitive effects in this experimental regime are found to shift the absolute value of ρ_{xx} minima and produce differences in traces for increasing/decreasing field. The problems can be overcome by careful analysis of two and four terminal data and a comparison of pulsed/steady field FQHE activation measurements will be reported elsewhere [26].

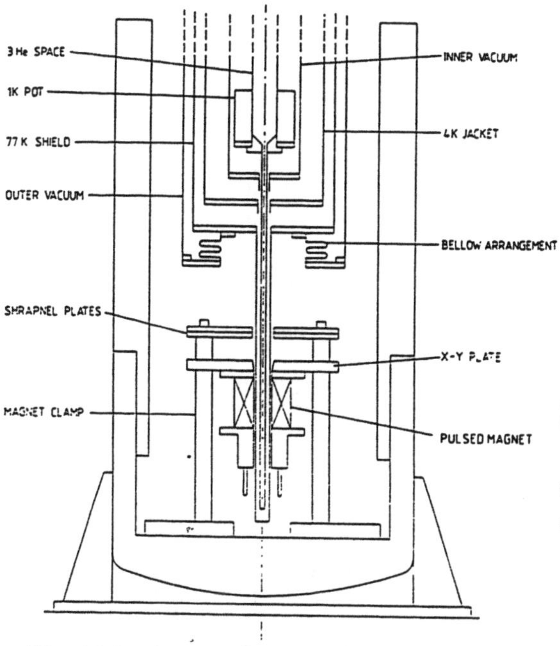

Fig. 10 *Toploading ^3He system/40T coil*

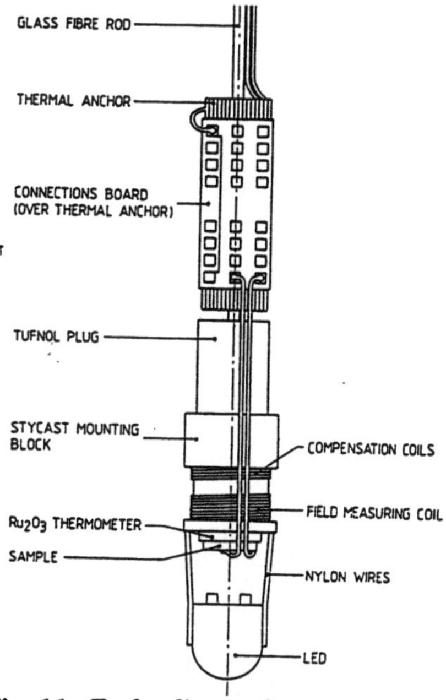

Fig. 11 *Toploading probe arrangement*

Figs. 12, 13
G635 FQHE data; Pulsed field shots to 20 , 34*T*

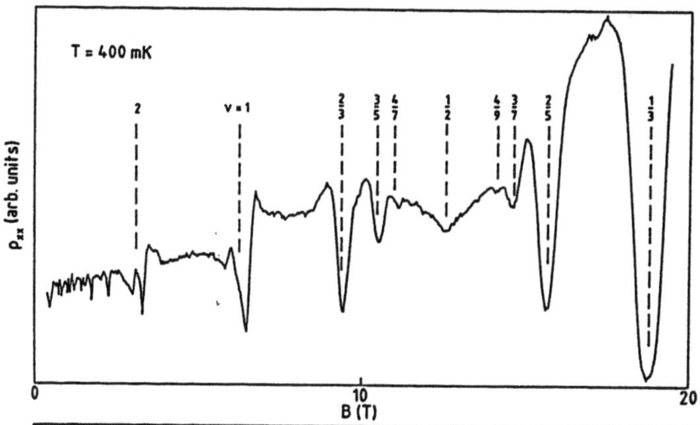

Fig. 12

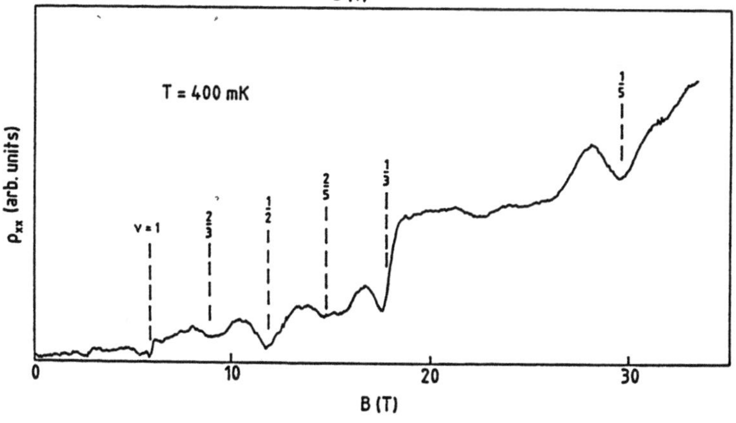

Fig. 13

4. Optical Detection of the Integer and Fractional QHE in GaAs

Whilst spectroscopic measurements of the FQHE quasiparticle gap at $\nu=7/3$ and $8/3$ in a Si metal-insulator-semiconductor transistor at $1.5K$ have been reported [27], optical experiments on GaAs have remained a challenge. A variation in the intensity of electron-hole Landau level luminescence has been reported for a GaAs/GaAlAs modulation-doped multiple quantum well (MQW) at $1.8K$, in the IQHE regime [28], but there was no correlation with ρ_{xx} minima. Optical anomalies near $\nu=2/3$ and $\nu=1$ have been reported in magneto-luminescence measurements of GaAs MQW at $T\approx0.4K$ [29], but an energy splitting of the luminescence line could not be assigned to quasiparticle gaps. We have recently reported a definitive, optical detection of the integer and fractional QHE in GaAs by a comprehensive study of the $\nu=2/3$ hierarchy out to the $6/11$ daughter state, and integer studies from $\nu=1$ to 10 in an optimised, single heterojunction at $120\ mK$ [30].

Our experiment measures bandgap photoluminescence from electrons in $n=0$ and $n=1$ subbands of the two-dimensional electron gas (2DEG) confined at a single $GaAs/Ga_{0.68}Al_{0.32}As$ heterojunction, with an electron density $n=9.7\times10^{10}cm^{-2}$ and mobility $\mu=9\times10^{6}cm^2/Vs$. The sample structure is shown schematically in the Fig. 14 inset. Ordinarily, band bending in GaAs heterojunctions would be expected to give rise to broad photoluminescence. In this structure, the GaAs layer is very weakly n-type, producing flat bands that give sharp luminescence lines. The incorporation of a GaAs/GaAlAs superlattice (SL) buffer acts to prevent carbon acceptors reaching the 2DEG. The sample was mounted in the dilute phase of a dilution refrigerator, and a sample temperature $\sim40\ mK$ was achieved with the laser off, rising to $120\ mK$ under continuous illumination. The laser wavelength was $740nm$ with an intensity $<10^{-4}W/cm^2$. The luminescence spectrum obtained at $120\ mK$ is shown in Fig. 14. It shows two lines, E_0 and E_1, whose zero-field energies are 1.509 and $1.511\ eV$ respectively. The much higher intensity of E_1 compared to E_0 is due to the greater E_1 electron wavefunction penetration into the GaAs layer where there is overlap with the photoexcited hole wavefunctions. On increasing the magnetic field, both lines shift to higher energy as expected for electron-hole $l_e=0\rightarrow l_h=0$ inter-Landau level transitions. The intensity of E_1 decreases with increasing field and that of E_0 increases. For $B\geq2.5T$ the intensities of the lines are reversed from the zero-field

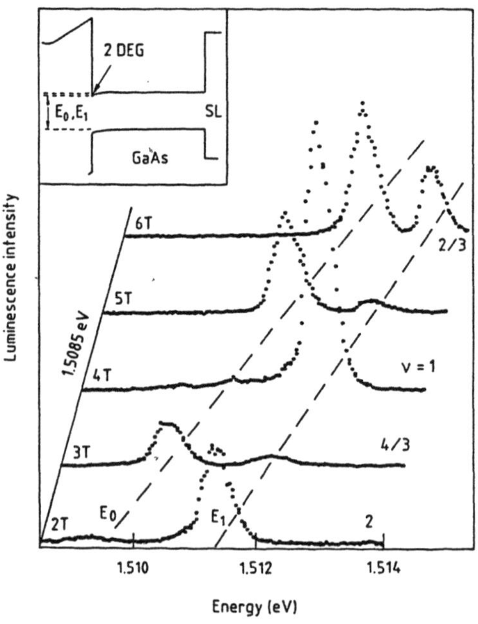

Fig. 14 *Photoluminescence spectra at* 120*mK*

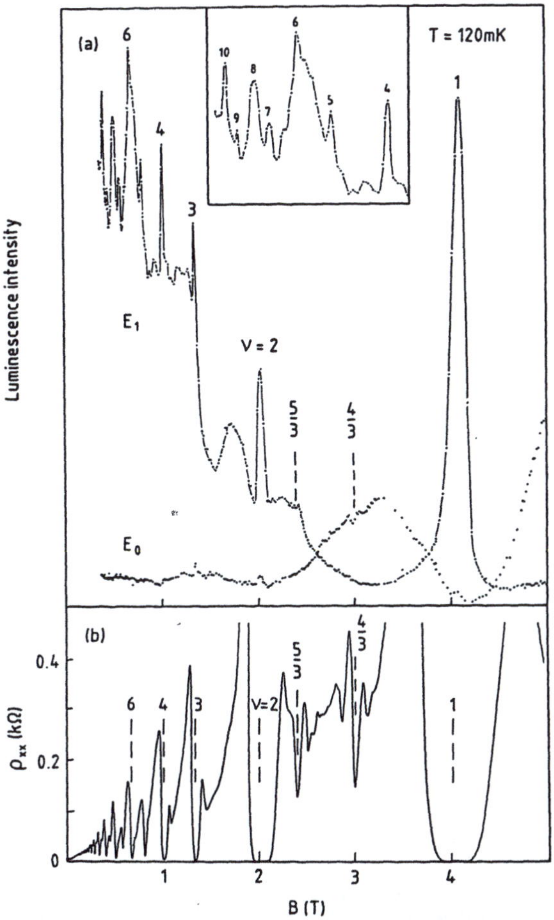

Fig.15 *E_0, E_1 intensities and ρ_{xx} data at 120mK for $\nu \geq 1$*

situation, with E_0 now being the more intense. An exception to this occurs near 4T ($\nu=1$) where E_1 shows a dramatic enhancement, whereas E_0 decreases substantially. When mapped out in detail, the remarkable modulation of the intensities of both lines correlates with changes in the many-electron ground state.

The low field ($\nu \geq 1$) and high field ($\nu < 1$) correlations are shown in Figs. 15 and 16 respectively, where peak intensities of E_0 and E_1 at 120 *mK* (Figs. 15, 16a) are compared with ρ_{xx} data taken simultaneously (Figs. 15, 16b). From Fig. 15 it is seen that E_1 has sharp maxima at integer ν. Strong enhancement occurs at ν even and weaker enhancement at ν odd (with the exception of $\nu=1$). Additionally, the intensity of E_1 shows a broad maximum near $\nu=7/3$, and a distinct shoulder at $\nu=5/3$. For fields up to $\nu=2$ the intensity of E_0 shows little structure. For $\nu<2$, E_0 shows a weak minimum at 4/3, and is suppressed to a broad minimum at $\nu=1$. In contrast to this behaviour E_1 undergoes a dramatic enhancement by a factor ~35 at $\nu=1$; the peak occurs at 4.08*T*, whereas the field for $\nu=1$ filling estimated from ρ_{xx} FQHE structure is 4.0*T*. The situation for $\nu<1$ is presented in Fig. 16, where transport data show well-resolved FQHE states at $\nu=2/3$, 7/11, 3/5, 4/7, 5/9, and 6/11. The ρ_{xx} resolution and structure are mirrored in detail in the E_0 intensity (even for the weak 7/11 state), and the similarity with the transport data is striking. This comparison provides *definitive* evidence for optical detection of the FQHE. The E_1 luminescence line also responds strongly to changes in the electronic ground state. Whereas the E_0

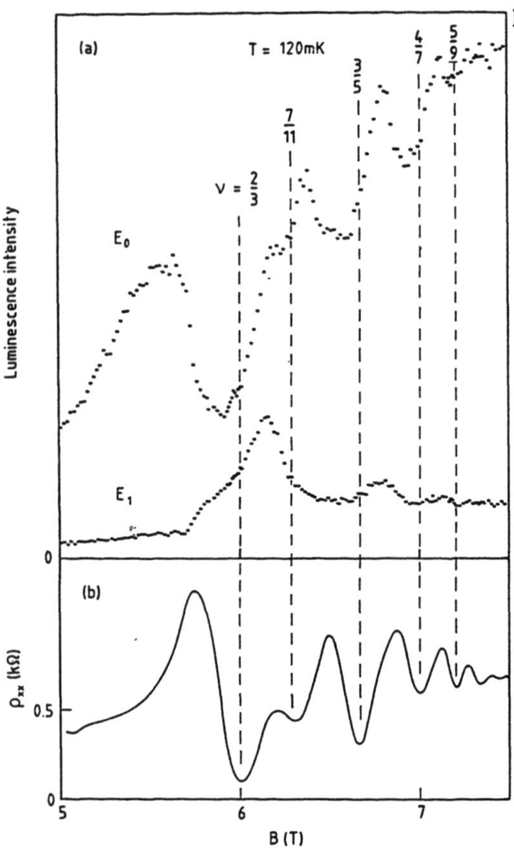

Fig. 16 *E_0, E_1 intensities and ρ_{xx} data at 120mK for $\nu < 1$*

intensity shows minima at the above fractions, the E_1 intensity develops maxima. In addition, the strengths of the E_0 minima and E_1 maxima decrease as the filling factor is swept through higher-order fractions in the 2/3 hierarchical sequence, similar to ρ_{xx}. The Fig. 16 data reveal that the optical and transport structure are not in precise alignment (see dashed lines in Fig. 16). The E_1 maxima occur at fields above exact $\nu = p/q$ filling whereas E_0 minima occur at fields below those required for exact filling. Correlation with the B-extent of FQHE ρ_{xx} minima is discussed in Ref. 30.

The spectra also reveal that the energy of the E_0 line changes dramatically as the field is varied through the $\nu < 1$ fractional states. Fig. 17 shows this behaviour for the $\nu = 2/3$ and 3/5 regions. The effect around $\nu = 2/3$ is striking: at $B = 6.25T$ ($\nu = 0.64$) a doublet is resolved in E_0, shown in the inset to Fig. 17 where arrows mark the energy corresponding to the linear B-dependence (dashed line) extrapolated from lower field. The lower component of the doublet lies on this line, while the upper component is shifted up in energy by 0.16 *meV* (1.8*K*). With decreasing field the higher-energy component becomes the more intense, and the relative strength of the lower component rapidly diminishes. The upward energy shift of the peak luminescence then remains constant for a range of fields spanning $\nu = 2/3$, before gradually decreasing to zero at $B = 5.75T$ ($\nu = 0.69$), although an upper (shifted) component is still discernible at this and lower fields (see inset spectra for position a). In the neighbourhood of $\nu = 3/5$ a similar trend in the E_0 energy is observed, although it is less distinct. In contrast, E_1 remains unshifted from the normal linear B-dependent behaviour throughout this entire range of magnetic field. A detailed interpretation of the data presents a considerable theoretical challenge, and these

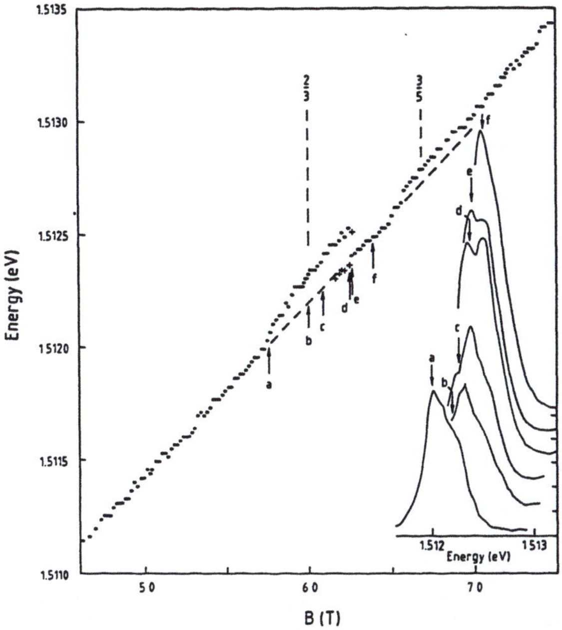

Fig. 17
B–dependence of E_0 energy (+ shows weak component of resolved doublet)

results open the way for further optical spectroscopy of the many-body electron liquid states and the Wigner solid in the extreme quantum limit. The reader is referred to Ref. 30 for a full discussion of this recent work.

The span of this work has involved a number of collaborations. I am indebted to my research group Simon Haynes, Jennifer Mallett, Andrew Suckling, Jane Branch, Philippa Wright, Philip Oswald and Richard Ford; Jeffrey Harris and Tom Foxon of Philips Research Laboratories, UK; Andrew Turberfield, John Ryan and Harry Jones of the Clarendon Laboratory; Fritz Herlach, Maarten van der Burgt and Guido Heremans of the Katholieke Universiteit Leuven; Chris Andrikidis and Stephen Collocott of the CSIRO National Measurement Laboratory, Australia; and to Peter Maksym and John Chalker of the Universities of Leicester and Southampton.

References

[1] K. v. Klitzing, G. Dorda and M. Pepper, *Phys. Rev. Lett.* **45**, 494 (1980).
[2] J.V. Branch, R.G. Clark, C. Andrikidis, S.J. Collocott, G.J. Griffiths, J.J. Harris, C.T. Foxon and J.T. Chalker, submitted to *Phys. Rev. Lett.*
[3] R.G. Clark, S.R. Haynes, J.V. Branch, A.M. Suckling, P.A. Wright, P.M.W. Oswald, J.J. Harris and C.T. Foxon, *Proc. 8th Int. Conf. on Electronic Properties of 2D Systems*, Grenoble 1989; *Surface Science* , to be published.
[4] T. Ando and Y. Uemura, *J. Phys. Soc. Jpn.* **36**, 959 (1974).
[5] M.P.A. Fisher, G. Grinstein and S.M. Girvin, *Phys. Rev. Lett.* **64**, 587 (1990).
[6] H.P. Wei, D.C. Tsui, M.A. Paalanen and A.M.M. Pruisken, *Phys. Rev. Lett.* **61**, 1294 (1988).
[7] Y. Ono, *J. Phys. Soc. Jpn* **51**, 237 (1982).
[8] S. Komiyama, H. Hirai, S. Sasa and T. Fujii; to be published.
[9] R.B. Laughlin, *"The Quantum Hall effect "* Ch. 7, editors R.E. Prange and S.M. Girvin, Springer-Verlag (1987).
[10] R.G. Clark, J.R. Mallett, S.R. Haynes, J.J. Harris and C.T. Foxon, *Phys. Rev. Lett.* **60**, 1747 (1988).

[11] A. Sachrajda, R. Boulet, Z. Wasilewski, P. Coleridge and F.Guillon; to be published.
[12] J.A. Simmons, H.P. Wei, L.W. Engel, D.C. Tsui and M. Shayegan, *Phys. Rev. Lett.* **63**, 1731 (1989).
[13] A.M. Chang and J.E. Cunningham, *Sol. Stat. Comm.* **72**, 651 (1989).
[14] J.T. Chalker and G.J. Daniell, *Phys. Rev. Lett.* **61**, 593 (1988).
[15] R.G. Clark, S.R. Haynes, A.M. Suckling, J.R. Mallett, P.A. Wright, J.J. Harris and C.T. Foxon, *Phys. Rev. Lett.* **62**, 1536 (1989); *Proc. 19th Int. Conf. Physics of Semiconductors*, Warsaw 1988 Vol. 1, p.101.
[16] J.P. Eisenstein, H.L. Störmer, L. Pfeiffer and K.W. West; **a.** *Phys. Rev. Lett.* **62**, 1540 (1989); **b.** *Proc. 8th Int. Conf. on Electronic Properties of 2D Systems*, Grenoble 1989; *Surface Science*, to be published.
[17] B.I. Halperin, *Helv. Phys. Acta* **56**, 75 (1983).
[18] T. Chakraborty and F.C. Zhang, *Phys. Rev.* **B 29**, 7032 (1984); **30**, 7320 (1984); **34**, 7076 (1986).
[19] P.A. Maksym, *J. Phys. Condens. Matt.* **1**, 6299 (1989); see also R.G. Clark and P.A. Maksym, *Physics World*, Sept. 1989, p.39.
[20] J.E. Furneaux, D.A. Syphers and A.G. Swanson, *Phys. Rev. Lett.* **63**, 1098 (1989).
[21] F.D.M. Haldane, *Phys. Rev. Lett.* **51**, 605 (1983).
[22] R.J. Haug, K. v. Klitzing, R.J. Nicholas, J.C. Maan and G. Weimann, *Phys. Rev.* **B 36**, 4528 (1987).
[23] S.R. Haynes, J.V. Branch, R.A. Ford, R.G. Clark, J.J. Harris, C.T. Foxon and J.T. Chalker, submitted to *Phys. Rev. Lett.*.
[24] H. Jones, F. Herlach, J.A. Lee, H.M. Whitworth, R.G. Day, D. Jeffrey, D. Dew-Hughes and G. Sherratt, *IEEE Transactions on Magnetics*, **25**, 1055 (1988); *Proc. 11th Int. Conf. on Magnet Technology*, Tsukuba 1989.
[25] F. Herlach, L. Van Bockstal, M. van der Burgt and G. Heremans, *Proc. 2nd Int. Symposium on High Field Magnetism*, Leuven 1989, Elsevier Publishers, Ch. 2, p. 61.
[26] Invited paper, *Application of High Magnetic Fields in Semiconductor Physics*, Würzburg 1990, to be published in Springer Series on Solid State Physics.
[27] I.V. Kukushkin and V.B. Timofeev, *Pis'ma Zh. Eksp. Teor. Fiz.* **44**, 179 (1986).
[28] C. H. Perry, J.M. Worlock, M.C. Smith and A. Petrou, in *High Magnetic Fields in Semiconductor Physics*, Springer-Verlag, New York, 1987, p. 202.
[29] D. Heiman, B.B. Goldberg, A. Pinczuk, C.W. Tu, A.C. Gossard and J.H. English, *Phys. Rev. Lett.* **61**, 605 (1988).
[30] A.J. Turberfield, S.R. Haynes, P.A. Wright, R.A. Ford, R.G. Clark and J.F. Ryan, submitted to *Phys. Rev. Lett.*.

New Condensed States in the Fractional Quantum Hall Effect

J.P. Eisenstein

AT&T Bell Laboratories, Murray Hill, NJ 07974, USA

Experiments on the fractional quantum Hall effect which have provided evidence for the importance of the electron spin are reviewed. These include tilted-field experiments suggesting the presence of an unpolarized ground state for the even-denominator FQHE at $\nu=5/2$ and the evidence for a ground state phase transition at $\nu=8/5$. Recent findings of anomalies in the $\nu=2/3$ FQHE will also be discussed.

1. Introduction

The fractional quantum Hall effect (FQHE) is the result of the condensation, at high magnetic field, of a two-dimensional electron system into novel incompressible quantum liquid states. These states occur only at special rational fractional values of the Landau level filling factor $\nu=p/q$. In magneto-transport they exhibit vanishing diagonal resistivity ρ_{xx} and a Hall plateau quantized to $\rho_{xy}=qh/pe^2$. Extensive effort, in both theory and experiment, has been expended over the last several years in elucidating this phenomenon[1]. For the most prominent fractional state, that at $\nu=1/3$, a detailed understanding, based upon the seminal work of R. B. Laughlin[2], has been obtained. For the 1/3-state experimental determinations of the energy gap Δ are in reasonable quantitative agreement with the Laughlin theory. On the other hand, for the so-called hierarchical states at $\nu=p/q$ with $p>1$, the situation is not nearly so clear. Recent experimental findings[3-11] suggest an essential role for the electron spin in the structure of these ground states and their excitations. In the original Laughlin theory the spin was irrelevant, being assumed fully polarized by the magnetic field. In this paper we review our experiments which bear on this issue. This review will include the tilted-field experiments on the even-denominator FQHE[3,4,8] and the evidence for phase transitions[6,9,11] between distinct states occupying the same filling factor. We will conclude by mentioning some aspects of the data which are not as yet understood.

Springer Series in Solid-State Sciences, Vol. 97
Localization and Confinement of Electrons in Semiconductors
Editors: F. Kuchar · H. Heinrich · G. Bauer

2. Even-Denominator State

The first experimental indication that the spin structure of an FQHE ground state might be non-polarized came with the discovery[3] of the even-denominator state at $\nu=5/2$ filling. This observation was very surprising given the large number of distinct odd-denominator states that have been found since the original work of Tsui, Stormer, and Gossard[12]. Furthermore, the apparently successful theory of Laughlin and others[1] in describing the $\nu=1/3$ state and its hierarchical descendants contains no provision for even-denominator fractions. For the Laughlin variational wavefunctions at $\nu=1/m$, the odd-denominator rule stems from the requirement of exchange antisymmetry for many-fermion states. Since the spins are all taken to be polarized, this requirement implies that m must be an odd integer. An obvious suggestion is that the $\nu=5/2$ FQHE, really a 1/2-state in the second Landau level, might be a Laughlin-like state, but one with a non-polarized spin configuration. In fact, Haldane and Rezayi[13] have proposed an elegant model system for which an unpolarized FQHE ground state exists at 1/2-filling. They have further argued that such a state is most probable in the second, rather than the lowest, Landau level. On the other hand, recent numerical work[14,15] shows the Haldane-Rezayi state is *not* the ground state for a system with real coulomb interactions. Experimental tests of the spin-reversal idea are clearly essential.

To accomplish this, we have performed tilted-field experiments on the 5/2-state[4,8]. With this technique the spin-flip energy may be enhanced while the coulomb and exchange energies are held fixed. This follows because the spin-flip energy depends on the total magnetic field B_{tot} while the coulomb and exchange energies, as well as the filling factor ν, depend solely upon the perpendicular field $B_{\perp}$. When the sample is tilted by an angle θ relative to the applied magnetic field, $\theta=0$ corresponding to the field perpendicular to the 2D plane, a larger B_{tot} must be applied in order to achieve the same filling factor ν. If the 5/2-state contains a significant number of reversed spins, then increasing the spin-flip energy should suppress the state.

Figure 1 shows representative resistivity ρ_{xx} data for the filling factor range $3>\nu>2$ with the field perpendicular to the 2D plane. This data was obtained with a GaAs heterostructure sample of density $N_s=2.3\times10^{11}cm^{-2}$ and mobility $\mu\sim7\times10^{6}cm^{2}/Vs$. The prominent minimum at 3.73T is the $\nu=5/2$ state, it is the strongest feature in this filling factor range. Although not shown in the figure, this deep minimum is accompanied by a strong plateau in the Hall resistance at $\rho_{xy}=2h/5e^2$. While our earliest tilted-field experiments[4] showed a qualitative collapse of the 5/2-state as θ was increased, a quantitative study required higher quality samples and

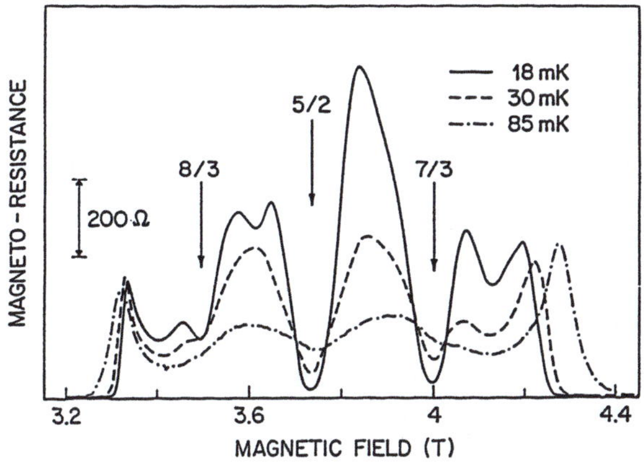

Figure 1. Magneto-resistance ρ_{xx} in filling factor range $3 > \nu > 2$.

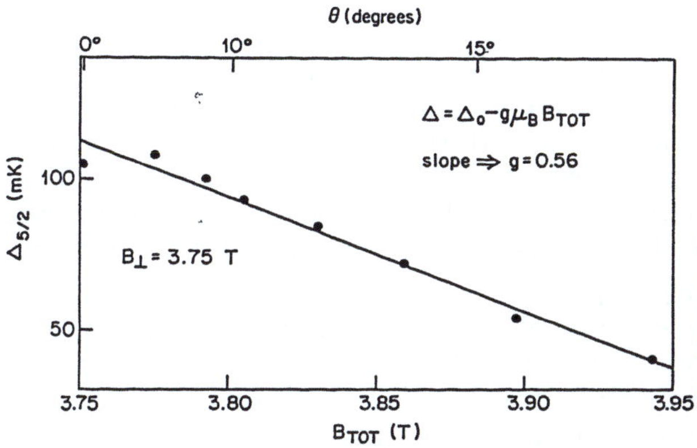

Figure 2. Energy gap Δ for $\nu=5/2$ FQHE *vs.* total magnetic field, B_{tot}. Straight line gives g-factor $g=0.56$.

systematic measurements of the activation energy for the ρ_{xx} minimum[8].

In Fig. 2 the energy gap Δ for the 5/2-state, defined via $\rho_{xx}= const. \times \exp(-\Delta/2T)$, is plotted vs. B_{tot}. This data was obtained with the same sample used for Fig. 1. While the absolute value of Δ is hard to interpret, the linear decline with B_{tot} is certainly suggestive of a spin Zeeman origin for the angular dependence. With the plausible assumption of spin-1/2 quasiparticles above an *unpolarized* ground state, we expect an energy gap of the form $\Delta = \Delta_0(B_\perp) - g\mu_B B_{tot}$. From the slope $d\Delta/dB_{tot}$, a g-factor can be extracted; we obtain $g=0.56$, not far from the bare g-factor ($g\sim 0.4$) for electrons in GaAs[16]. Thus the data suggest a spin-unpolarized ground state for the $\nu=5/2$ FQHE.

How strong is the conclusion that these data imply a unpolarized ground state rather than spin-flip excited states above a polarized condensate? Such excitations have been predicted[17], and apparently observed[6]. The relevant point is that for a polarized ground state, excitation of spin-reversed quasiparticles decreases the *net* z-component of the spin, S_z, of the system. This will give a positive contribution to the energy gap, so that one expects $\Delta = \Delta_0(B_\perp) + g\mu_B B_{tot}$. This Zeeman contribution is of opposite sign to that shown in Fig. 2. The negative Zeeman component we observe implies S_z has increased, which is inconsistent with a polarized ground state. It is, however, expected if the condensate is unpolarized.

3. Phase Transitions in the FQHE

For the 5/2-state there is, so far, no evidence of any *polarized* ground state appearing at high total magnetic fields. The possibility that some FQHE states can exist in either polarized or unpolarized configurations was actually suggested several years ago[18]. Numerical calculations[19] have established that, ignoring the Zeeman energy, the $\nu=2/5$ state should be unpolarized. In a real sample, whether this is actually the case depends on the relative size of the Zeeman energy and the difference in coulomb/exchange energies between the unpolarized and polarized ground states. The possibility arises of driving a phase transition between the two configurations. As we shall see, this can be done either by tilting or by changing the 2D density, N_s.

The strongest evidence for such a phase transition[6] has been obtained at $\nu=8/5$. Figure 3 shows tilted-field energy gap data for the $\nu=8/5$ FQHE. The sample is the same as that used for the 5/2-state experiment just described. The 8/5-state first weakens under tilt, its energy gap falling linearly with B_{tot}. At about $\theta=30$ deg. this decay stops and further increase in the angle leads to the gap *increasing* linearly with B_{tot}. Interestingly, near the low point of this re-entrant behavior the ρ_{xx} resistivity minimum at 8/5 is seen to *split*

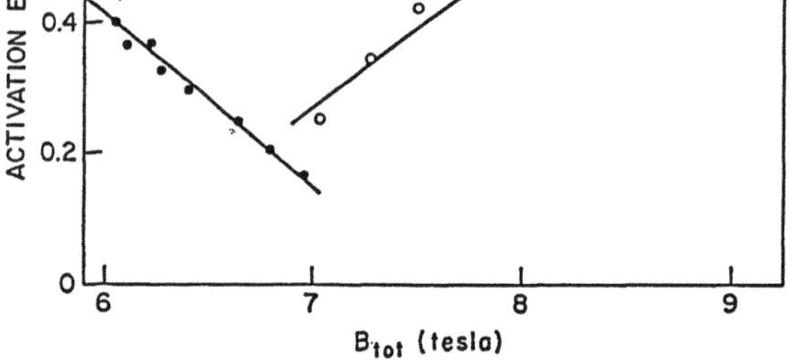

Figure 3. Energy gap Δ for $\nu=8/5$ FQHE *vs.* B_{tot}.

into two components. Away from the low point, however, a single ρ_{xx} minimum is obtained and the Arrhenius plots are linear for almost two decades in resistivity.

This behavior has been cited as evidence[6] for a spin transition in the FQHE ground state at $\nu=8/5$. In the low-angle phase, the energy gap data suggests an unpolarized ground state for the same reasons as discussed above for the case of $\nu=5/2$. The slope $d\Delta/dB_{tot}$ now gives a g-factor of $g=0.39$, assuming a net spin change $\Delta S_z=+1$. In the high-angle phase, for $B_{tot}>7T$, the slope is about the same magnitude, but with opposite sign. In this region the data point to a polarized ground state with $\Delta S_z=-1$ spin-flip excitations. Although not yet observed, there should be a transition to polarized quasiparticles at still higher magnetic fields. In the vicinity of the gap minimum at $B_{tot}=7T$, the splitting of the ρ_{xx} minimum raises the possibility of phase separation[6] in the sample.

The existence of an unpolarized-to-polarized phase transition at $\nu=8/5$ would be expected assuming particle-hole symmetry and the known possibility of such an effect at $\nu=2/5$. Neglecting Landau level mixing, but allowing for possible spin flips, particle-hole symmetry connects states at filling factor ν and $2-\nu$. Recent numerical work[20] has predicted a critical magnetic field of about $B_c=5T$ for the expected transition at $\nu=2/5$ and has supported our identification of the quasiparticle spin states for both phases.

The second example[11] of a possible FQHE phase transition is the $\nu=2/3$ state. This state, the second FQHE state to be observed, is often assumed a particle-hole conjugate to the canonical $\nu=1/3$ state. This connection relies on the symmetry between filling fraction ν and $1-\nu$ expected at fields high enough so that both Landau level and spin mixing can be ignored. As we have seen, the latter assumption is often not valid, leaving only the symmetry between ν and $2-\nu$. Interestingly, recent experiments have suggested a spin transition in the $\nu=4/3$ FQHE state[7]. Since 4/3 is the conjugate of 2/3, it is worthwhile studying the latter fraction at the lowest possible fields.

Figure 4 displays energy gap data[11] on the 2/3-state in the field range around 3T. For this sample, applying a back-gate bias allowed the density, N_s, to be varied from about 2.6 to $7\times10^{10}cm^{-2}$ while the mobility, μ changed from about 2.2 to $6\times10^6 cm^2/Vs$. The solid dots in the figure show the magnetic field dependence of the 2/3-state gap at *zero tilt*. These data are obtained by changing the N_s, and thus the field position of $\nu=2/3$. Initially rising slowly with field, the gap begins to drop sharply around 2.9T, reaching a minimum value at about 3.3T. Further increase in the density yields a rapid rise in the gap which persists to beyond 4.3T.

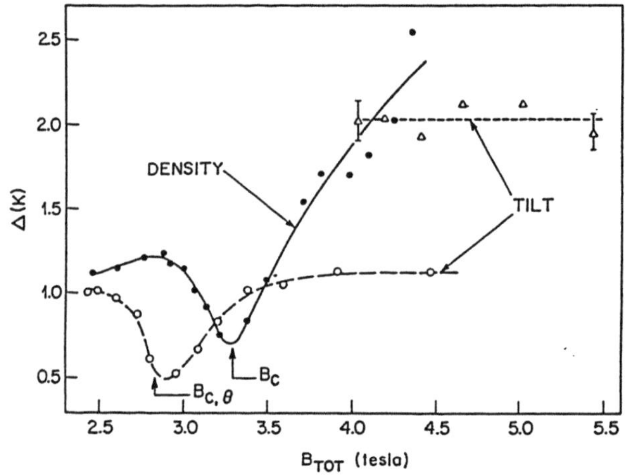

Figure 4. Energy gap Δ for $\nu=2/3$ FQHE. Both zero-tilt density shift and fixed density tilt data are shown.

A similar re-entrance is observed on tilting the sample at fixed density, so long as the density N_s is less than about $N_{s,c} \sim 5.3\times10^{10}cm^{-2}$. This density, for which $\nu=2/3$ occurs at $B_c=3.3\,T$, exhibits the minimum gap in the zero-tilt experiment. The open dots in Fig. 4 show the tilt dependence of the 2/3-state gap for $N_s=3.9\times10^{10}cm^{-2}$, corresponding to a perpendicular field of $B_\perp=2.43\,T$. Now the minimum gap occurs at $B_{tot}=2.9\,T$, well below the $B_c=3.3\,T$ seen in the zero-tilt experiment. At higher angles, with B_{tot} well above the minimum, the observed gap becomes independent of field.

If the density is raised, via the back gate, to beyond $N_{s,c}$, tilting shows no re-entrant gap structure. For the open triangles in Fig. 4, $N_s=6.5\times10^{10}cm^{-2}$ and the observed gap is essentially independent of angle.

These findings[11] are suggestive of a ground state spin transition at $\nu=2/3$, although the case is not as strong as that for $\nu=8/5$. For $\nu=8/5$ the transition was traversed only via tilting. Now, at $\nu=2/3$, this can be done either by tilting or by changing the density. It is apparent from the data that the same transition is being probed with these two methods. This supports a Zeeman origin for the effect. Possible effects of Landau level mixing or inter-subband coupling for example, will depend very differently on whether the sample is tilted or has its density changed.

For a transition from an unpolarized ground state to a polarized one, it is clear that the critical field measured by changing density (B_c) will exceed that seen by tilting ($B_{c,\theta}$). Raising the density increases the both the coulomb energies and the Zeeman energy, the transition occurring because the former scales as $B^{1/2}$ and the latter as B. Tilting raises only the Zeeman term and thus closes the energy difference between the two competing ground states at a lower total field. A simple model[11] for the ground state energy in each phase yields a compact relation between the critical fields:

$$B_{c,\theta}=(B_{\perp}B_c)^{1/2}. \tag{1}$$

In this equation, $B_{\perp}$ is the perpendicular field component held fixed in the tilt experiment. Assuming the critical fields display the minimum energy gap, we can compare Eq. 1 to experiment. Taking $B_c=3.3\,T$, and, for the open dot tilt data in Fig.4 $B_{\perp}=2.43\,T$, Eq. 1 yields $B_{c,\theta}=2.83\,T$. The arrows in the figure show these fields, and reasonably good agreement with the data is apparent. As already noted, if the density is such that $B_{\perp}$ exceeds $B_c\sim 3.3\,T$, tilting does not show a re-entrant gap. Within the model this is obvious, the system begins in the polarized state and tilting only further stablizes it.

Unlike the transition[6] seen at $\nu=8/5$, this $\nu=2/3$ effect does not yield a simple picture of the quasiparticle excitation processes taking place. For the zero-tilt data this is not surprising since changing the density changes the coulomb and Zeeman energies simultaneously and also alters the sample mobility. All affect the energy gap. But even for the tilt experiments, where only the Zeeman energy is being varied, the situation is unclear. There are no simple linear dependences like those seen at $\nu=8/5$ that can be used to calculate a g-factor. Near the critical field the variation of Δ with B_{tot} is considerably sharper than expected from simple spin-flip processes. Only at the highest fields can a simple picture be formed. There, where the tilted-field gap becomes independent of B_{tot}, $\Delta S=0$ spin-preserving transitions may be operative. This, at least, is to be expected.

4. Conclusion

The experiments described above give ample evidence for the importance of spin in the FQHE ground states and their excitations. Relatively simple models suffice to interpret the bulk of the observations. There are, however, aspects of the data which are not understood and require further study. Lack of space precludes discussion of these points here, but most of the details have been published elsewhere[6,8]. Perhaps the most prominent of these puzzles are the splitting[6] of the $\nu=8/5$ resistivity minimum near the transition point and the lack of clear evidence[8] for spin-reversed quasiparticles at $\nu=5/3$, the supposed parent of the 8/5-state. While the splitting effect may signal a coexistence of the unpolarized and polarized phases near the transition point, the data on the 5/3-state suggests our understanding of the parent-daughter relationship within the hierarchical model may require revision.

It is clear that the inclusion of the spin degree of freedom considerably enriches the spectrum of phenomena within the FQHE. Presumably future experiments will reveal more states with zero or partial spin polarization. Further theoretical work is required to generalize the hierarchical model to include spin. This should provide a better understanding of the ground state phase transitions, spin-flip excitations, and hopefully the even-denominator FQHE.

It is a pleasure to thank my collaborators H. L. Stormer, R. Willett, L. N. Pfeiffer, and K. W. West for their essential contributions to the work described here.

References

1. *The Quantum Hall Effect*, edited by R. E. Prange and S. M. Girvin (Springer-Verlag, New York) and *The Fractional Quantum Hall Effect* by T. Chakraborty and P. Pietilainen, vol. 85, Springer Series on Solid State Sciences (Springer-Verlag, Berlin 1988).
2. R. B. Laughlin, Phys. Rev. Lett. 50, 1395 (1983).
3. R. L. Willett, J. P. Eisenstein, H. L. Stormer, D. C. Tsui, A. C. Gossard and J. H. English, Phys. Rev. Lett. 59, 1776 (1987).
4. J. P. Eisenstein, R. L. Willett, H. L. Stormer, D. C. Tsui, A. C. Gossard and J. H. English, Phys. Rev. Lett. 61, 997 (1988).
5. R. G. Clark, J. R. Mallett, S. R. Haynes, P. A. Maksym, J. J. Harris and C. T. Foxon in *High Magnetic Fields in Semiconductor Physics II*, edited by G. Landwehr, vol. 87, Springer Series on Solid State Sciences (Springer-Verlag, Berlin 1989).
6. J. P. Eisenstein, H. L. Stormer, L. Pfeiffer and K. W. West, Phys. Rev. Lett. 62, 1540 (1989).
7. R. G. Clark, S. R. Haynes, A. M. Suckling, J. R. Mallett, P. A. Wright, J. J. Harris and C. T. Foxon, Phys. Rev. Lett. 62, 1536 (1989).
8. J. P. Eisenstein, R. L. Willett, H. L. Stormer, L. N. Pfeiffer and K. W. West, Proc. Eighth Intl. Conf. on Electronic Properties of Two-Dimensional Systems, Grenoble 1989.
9. J. P. Eisenstein, H. L. Stormer, L. N. Pfeiffer and K. W. West, *ibid.*
10. R. G. Clark, S. R. Haynes, J. V. Branch, J. R. Mallett, A. M. Suckling, P. A. Wright, P. M. W. Oswald, J. J. Harris and C. T. Foxon, *ibid.*
11. J. P. Eisenstein, H. L. Stormer, L. N. Pfeiffer and K. W. West, to be published.
12. D. C. Tsui, H. L. Stormer and A. C. Gossard, Phys. Rev. Lett. 48, 1559 (1982).
13. F. D. M. Haldane and E. H. Rezayi, Phys. Rev. Lett. 60, 956 (1988).
14. A. H. MacDonald, D. Yoshioka and S. M. Girvin, Phys. Rev. B 39, 8044 (1989).

15. T. Chakraborty and P. Pietilainen, Phys. Rev. B 38, 10097 (1988).

16. M. Dobers, K.v. Klitzing and G. Weimann, Phys. Rev. B 38, 5453 (1988).

17. T. Chakraborty, P. Pietilainen and F. C. Zhang, Phys. Rev. Lett. 57, 130 (1986).

18. B. I. Halperin, Helv. Phys. Acta 56, 75 (1983).

19. T. Chakraborty and F. C. Zhang, Phys. Rev. B 29, 7032 (1984).

20. X. C. Xie, Yin Guo and F. C. Zhang, Phys. Rev. B 40, 3487 (1989).

Experiments on the Nature of the Extreme Quantum Regime of a 2-D Electron System

F.I.B. Williams[1], *E.Y. Andrei*[2], *R.G. Clark*[3], *G. Deville*[1], *B. Etienne*[4], *C.T. Foxon*[5], *D.C. Glattli*[1], *J.J. Harris*[5], *E. Paris*[4], *and P.A. Wright*[3]

[1]Laboratoire de Physique du Solide et de Résonance Magnétique, CEN de Saclay, F-91191 Gif-sur-Yvette Cedex, France
[2]Department of Physics, Rutgers University, Piscataway, NJ 08855, USA
[3]Clarendon Laboratory, Parks Road, Oxford OX1 3PU, UK
[4]Laboratoire de Microstructures et Microélectronique, 196 Avenue Henri Ravera, F-92220 Bagneux, France
[5]Philips Research Laboratory, Redhill, Surrey RH1 5HA, UK

1. Introduction

Electrons in a uniform 2-D space interacting through (3-D) Coulomb forces are expected to present a phase transition driven by quantum effects: at zero temperature, the ground state of the dilute system is a classical 2-D crystal with small zero-point vibrational energy; as the density is increased, this energy grows faster than the Coulomb correlation energy until at some critical density, n_W, the system transforms into a correlated quantum liquid /1/. The 2-D electron system (2DES) we study here is well into this liquid regime with a density some 200 times greater than the critical Wigner density. It is however possible to restore the importance of the Coulomb interaction by localizing the electrons with a magnetic field to orbits of size $l_c = (\hbar c/eB)^{1/2}$; when $\sqrt{2}l_c < a = (\pi n_s)^{-1/2}$, the mean distance between electrons at surface density n_s, the independent particle kinetic energy becomes identical for all the electrons (degeneracy of fundamental Landau state exceeds number of particles) and the Coulomb interaction benefits from the multiplicity of choice to minimise its energy by introducing correlations. The first spectacular effect of this is the specially favoured correlated liquid states /2/ which occur at rational values p/q, q = odd integer, of the filling factor $\nu = 2l_c^2/a^2$. At yet higher fields a new state showing long range correlation between orbit centres is expected to occur: a state one might call by extension a magnetically induced Wigner solid (MIWS). This has been variously estimated to occur for zero temperature at some critical filling factor ν_c between 1/3 and 1/10 /3/.

Ideally we would like to study the transition to this MIWS state experimentally; first in its simplest context of $T \longrightarrow 0$ and no perturbing potential, then to study the effects of temperature and finally to determine its stability under application of a controllable random field.

To do this we start by choosing what seems to be the best physical realisation found so far, the electron layer at the interface (heterojunction) between ultra pure *GaAs* and modulation doped *GaAlAs* fashioned by molecular beam epitaxy. In such systems, the effective electron mass $m \approx 0.07$ (free elec-

Springer Series in Solid-State Sciences, Vol. 97
Localization and Confinement of Electrons in Semiconductors
Editors: F. Kuchar · H. Heinrich · G. Bauer

tron mass) and the dielectric screening constant $\varepsilon \approx 13$, making the estimated Wigner density $n_W \approx 5.10^8\text{cm}^{-2}$ for a Bohr radius $a_B \approx 100\text{Å}$. We work with samples whose density ranges over $3 < n < 12\ 10^{10}\text{cm}^{-2}$ and whose "quality" is at least hinted at by zero field mobilities $1 < \mu < 8 \times 10^6\text{cm}^2V^{-1}s^{-1}$. But even at this level of quality the random field still produces notable effects and we have to start our programme from the more complicated end.

1.1 Pure System

In zero random field, zero temperature and neglecting Coulomb admixtures to higher Landau states ($l_c \ll a_B$), the ground manifold energies scale like $(e^2/\varepsilon l_c)\, g(\nu)$ or equivalently $(e^2/\varepsilon a)\, h(\nu)$ and the thermodynamic state of the system is determined by ν alone; if we work at finite temperature, the free energy scales like $F = (e^2/\varepsilon a)\, f(\nu, t)$, where $t = T/\left(e^2/\varepsilon a\right)$. This tidy scaling behaviour is broken upon introducing an outside potential into the problem, particularly if that potential varies in a random way from sample to sample. Nonetheless, if one does find such scaling despite the non-ideality, one can have some confidence that the underlying physics is determined by Coulomb interaction and Landau degeneracy.

1.2 Random Field Effects

Real heterojunctions have at least three sources of disorder: the nearly random distribution of ionised donors, the residual acceptor impurities in the *GaAs* and the *Ga/Al* disorder introduced by the alloying. The effect of the first of these is much reduced by withdrawing the donors by a large spacer distance d (100nm and 300nm in our double delta doped samples /4/ and 160nm in the volume doped samples /5/); the second source of disorder is reduced by the introduction of a superlattice trap in the *GaAs* buffer region /5/; the third source of disorder is not thought to be very important but could in principle be reduced by a few monolayers of pure *AℓAs* at the interface /6/, but in practice this has not made a significant improvement.

Whatever the cause, it is helpful to have in mind a picture like Fig.1 of how one might expect the 2DES solid to react to a random field. We imagine the random field to be in competition with the Coulomb correlation effects at zero temperature and high field, where the pure 2DES would be in a crystalline configuration. Our speculative picture is inspired by studies /7/ of classical systems with the idea that the dominant role of the quantum effects for this purpose is to soften the crystal.

We consider first the strong pinning scenario which might result from a low concentration n_p of strong pinning centres of strength V_p : the lattice will be strongly perturbed at well-localized points of average separation $l_p \approx n_p^{-1/2}$, between which it is strained, thereby fixing the correlation length: there will be a threshold depinning field for displacing the lattice given by $E_{dp} \approx (n_p/en_s)\, V_p(2\pi/a)$ defined to precision $\sim (a/l_p)^{-1}$; associated with the pinning length will be a characteristic frequency v/l_p (v is a sound velocity) unrelated

Configuration	PERFECT CRYSTAL	STRAINED MONOCRYSTAL	DOMAINED POLYCRYSTAL	GLASS
Schematic representation	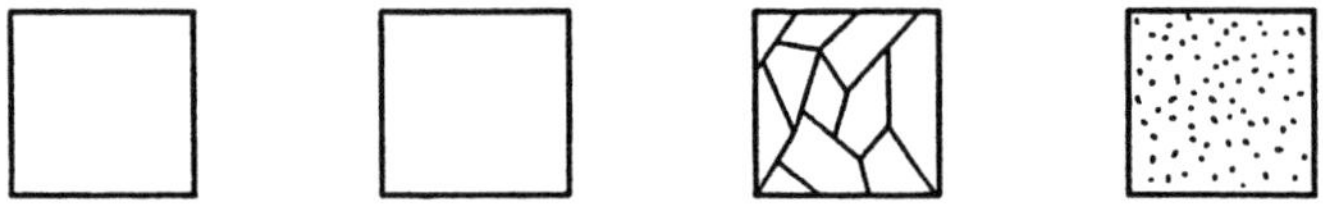			
Correlations	Long range order	Power law decay	Exp. decay $\sim L_D$	Random $L_D \rightarrow a$
Depinning field E_T	0	0	$\sim 1/L_D$	Wide distribution
Pinning frequency	0	0	$\sim 1/L_D$	Wide distribution
Melting temperature	$T_m^{(0)}$	$<T_m^{(0)}$	$<T_m^{(0)}$ Rounding	Suppressed (gel)
$\langle V_R\rangle / \langle V_{COUL}\rangle$	0	⟶	⟶	∞

Figure 1: Schematic representation of the effects of increasing random field V_R compared with Coulomb interaction V_c. The threshold depinning field E_T, the pinning frequency f_p and the average domain size are discussed briefly in the text.

to the strength of the pinning potential. The condition for this situation is that the cost in elastic energy be small compared with the pinning potential. As the concentration of such pinning centres is increased, either the elastic term becomes important because of the shorter length scale and one goes over into the weak pinning regime, or one goes all the way to a glass, where in the limit of more pinning centres than electrons, each electron is individually bound to a randomly situated pinning centre; this limit has no correlation and if the V_p are supposed to have a statistical distribution, there is no averaging effect and both the pinning field and the pinning frequency, now associated with V_p, reflect directly the distribution of the random field.

The weak pinning situation arises from more frequent, weaker, potential fluctuations. As the random field is turned on, the lattice first strains to reduce its energy in the potential fluctuation; this introduces a power law decay in correlation, no finite depinning field and zero characteristic pinning frequency in the large N limit. As the random field becomes stronger, it becomes more profitable for the lattice to shift more or less rigidly over a certain domain to take advantage of a local potential fluctuation and to adjust its elastic strain smoothly to that of a neighbouring domain which has itself shifted into a fluctuation minimum; the mean domain size $L_D \approx a^2K/V_R$ or $a^2K/V_p n_p^{1/2}$ where K is an elastic constant ($\approx e^2/\varepsilon a^3$), V_R is energy fluctuation $(\text{area})^{-1/2}$; the correlation length $\xi \sim L_D$; there is a depinning field $E_{dp} \approx \left(\pi^2 a^3 K/eL_D^2\right)$ related to a characteristic pinning frequency $f_p \approx v/L_D$ by $f_p^2 \approx eE_{dp}/4\pi ma$; both are defined to a precision $\sim (a/L_D)^{-1}$. As V_R becomes stronger, the domain size reduces until eventually one again goes over into a glass state with the same properties as mentioned before.

The random field also has an effect on the melting of a 2D solid /8/: typically, limited disorder introduces power law decay of correlations independently of the temperature contribution and depresses the melting temperature; above some value of disorder field the phase transition is suppressed. For finite domain size, the thermodynamic limit cannot be taken and the sharpness of a phase transition is lost.

2. Experiments

We have employed three different types of experiment: (a) contactless, short length scale ($\lesssim$ 10 microns), capacitance measurements of screening by the 2DES to estimate the importance of pinning and to give a thermodynamic signature of any phase transition; (b) contactless radio-frequency absorption measurements, again on short length scales (using a well-defined set of wavenumbers) and short time scales (frequencies continuously swept $0.01 < f < 4GHz$); (c) contacted, low frequency, low current AC transport on millimeter length scales to show both linear and nonlinear response.

All except the very last can be expressed in terms of a linear response function $\sigma(q,\omega) = \sigma' + i\sigma''$ defined by

$j_q = \sigma(\mathrm{q},\omega)E_q\exp{(i\omega t)}\exp{(i\mathrm{q}.\mathrm{r})}$,

bearing in mind that four point contacted measurements give the relationship between the current density and the screened field at the electron plane whereas the contactless measurements give the relationship to the unscreened field applied by the electrodes. Figure 2 illustrates what a small portion of the (ω, q) plane relevant to our 2DES we can explore: the small rectangle shown enlarged in the inset occupies about $10^{-2} \times 10^{-2}$ of the "Debye" area. The usual "DC" transport measurements are situated within a $10^{-8} \times 10^{-3}$ portion around the origin of this. Our capacitance and radio-frequency (rf) measurements impose a set of wavevectors all at once: the rf absorption is sensitive to a sum $\sum_{p=0}^{\sim 10} \sigma'_{xx}(pq_0,\omega)\, c_p$ where p is discrete ($p = 0, 1, ..., 10$ and the c_p are weighting factors fixed by the geometry while ω is a continuous variable which sweeps out the comb shown on the inset. Capacitance measurements give a sum $\sum_{p=1}^{\sim 10} \sigma''(pq_0,\omega)\, c_p$ where the ω is limited to fixed values below about $10^5 Hz$ and $p = 0$ is excluded by design. $q_0 \approx 2\pi/16\mu m$ is fixed by electrode geometry.

2.1 Capacitance

Provided thermodynamic equilibrium can be maintained, capacitance measures directly the variation of the chemical potential with density /9/. In practice this is often masked by the charge transport problems which tend to arise in just the more interesting regions of filling factor (around integers and fractions and in the high field domain, in each of which $\sigma \longrightarrow 0$). The problem is illustrated by the measurement scheme of Fig.3 where one measures the capacitance between

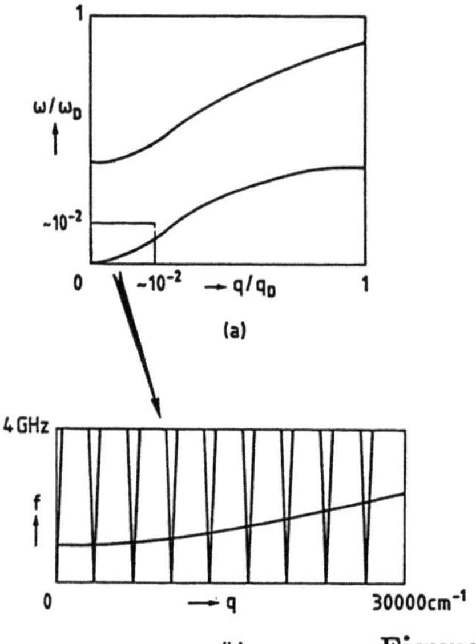

Figure 2

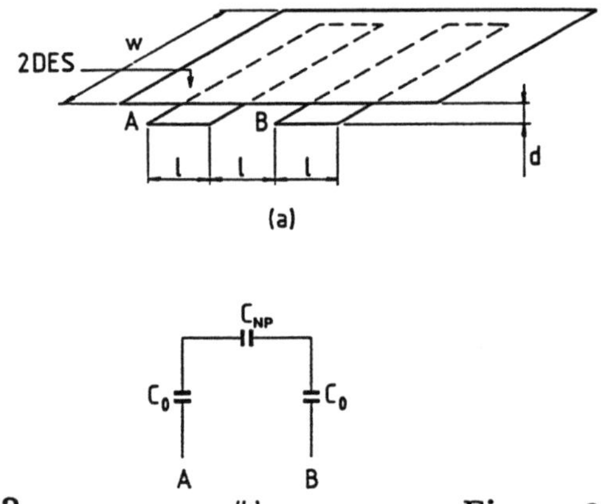

Figure 3

Figure 2: (a) Schematic dispersion relation normalized to Debye frequency and wavelength for elementary excitations of an electron solid in magnetic field. The (lower) shear branch is centred on zero frequency for a liquid state and this basic difference in excitation spectrum is exploited to distinguish liquid from solid. (b) The lower left hand rectangle occupying an area $\sim 10^{-2} \times 10^{-2}$ of Figure 2a is magnified to show crossing of the multibranch electromagnetic probe field and the lower branch of the electron solid excitations. The radiofrequency absorption $\sim \sigma'(q,\omega)$ where (q,ω) are defined by the comb swept out experimentally by sweeping the frequency; the electron excitation branch is a line of singularities in the electron response, giving rise to a resonant absorption from the probe field at the crossing points with the comb. This lower hybrid branch of electron excitations is gapless in an infinite crystal, but can have a finite gap at $q = 0$ for finite domain size (pinning frequency). Capacitance ($\sim \sigma''(q,\omega)$) is measured at fixed frequency, also using a spatially periodic probe field as represented by the series of points. The customary "DC transport" measurements give $\sigma'_{ij}(q,\omega)$ very close to the origin ($q < 10\text{cm}^{-1}$, $\omega/2\pi < 10\text{Hz}$) on this figure.

Figure 3: (a) The capacitance measured between strips A and B reflects the screening behaviour of the 2D electron system by which they are capacitively coupled. In reality, the 2DES is placed on a periodic interdigital array of $\ell \approx 8\mu m$ fingers. (b) The equivalent circuit of Figure 3a. $C_0 \sim \ell/h$ is coupling capacitance to the 2DES, itself represented by a perfect conductor in series with a capacitor which is infinite for perfect screening. If there is a non-zero stiffness $\frac{\partial\psi}{\partial n}\frac{1}{C_{AB}} \sim \frac{1}{\ell}(2h + \frac{1}{e^2}\frac{\partial\psi}{\partial n})$, allowing access to the derivative of the non-electrostatic component ψ of the chemical potential.

strips A and B each coupled capacitively to the 2DES a distance d removed. The characteristic time associated with charge transport across the distance ℓ is $\tau \leq C_{AE}/G_{AB} \sim \ell^2$ where G_{AB} is the conductance of the 2DES between capacitances C_{AE} and C_{BE} from the strips of crosswise dimension ℓ. Reducing ℓ from the typical 3mm to 10 microns decreases τ by 10^5, so that even for σ^{-1} = 100 Mohms per square, thermodynamic equilibrium is still assured at frequencies $\lesssim 100kHz$.

The equilibrium condition in the 2DES is that $\nabla\mu = 0$ where $\mu = \phi + \Psi$ is the electrochemical potential decomposed into a purely electrostatic (Poisson) part ϕ and a non-Poisson part Ψ which reflects the rest of the physics like correlation, pinning, etc. If Ψ = constant there is perfect electrical screening; deviations from perfect screening result when $\nabla\Psi \neq 0$. In equilibrium $\nabla\phi = \nabla\Psi$ and we measure electrically $\phi_{AB} = -\int_A^B \nabla\Psi \, dx$. For example, if Ψ is a local function of filling factor $\Psi = \Psi(\nu_0) + \frac{\partial\Psi}{\partial\nu}\delta\nu = \Psi(n_0)\,|_B + \frac{\partial\Psi}{\partial n}\big|_B \delta n$ so that $\nabla\Psi = \frac{\partial\Psi}{\partial n}\nabla n$. Application of a potential v_{AB} across AB gives rise to a charge transfer $\delta Q(\lesssim v_{AB}C_{AE}/2)$ which, because of $\frac{\partial\Psi}{\partial n}$ sets up a counter potential in the charge sheet of $eV = \frac{\partial\Psi}{\partial n}(n_A - n_B) = \frac{\partial\Psi}{\partial n}\delta Q/elw$. This relation is reproduced by the equivalent circuit of Fig.3 where the capacitance in series with the perfect conductor is given by $1/C_{NP} = (1/wl)(1/e^2)\frac{\partial\Psi}{\partial n}$.

$(1/e^2)\frac{\partial\Psi}{\partial n}$ is a stiffness constant or inverse compressibility of the 2DES of non-electrostatic origin. Its behaviour is often a precious indication of the nature of a new phase; e.g. at fractional filling factors the stiffness is thought to diverge (incompressibility), at a first order phase transition $\frac{\partial\Psi}{\partial n}$ undergoes a jump, etc. It is also sometimes referred to as the inverse of the thermodynamic density of states to underline its connexion with an independent (quasi)particle picture.

The actual experiments are done with a periodic interdigital array on which the sample is placed "face down" (2DES on the array side of the sample) in its middle in such a way that it only sees the periodically modulated field to be sure to eliminate any longer length scale phenomena arising from border effects. It is then more convenient to work in Fourier space and to compare the capacitance $C_q = n_q/V_q$ associated with wave vector q with that obtained for perfect screening $C_q(PS)$:

$$C_q(PS)/C_q(2DES) = 1 + \omega_q^2(\text{stiffness})/\omega_q^2(\text{electrostatics})$$

where ω_q (electrostatics) is the plasmon frequency in the continuum limit and ω_q (stiffness) is the frequency associated with the non-Poissonian stiffness of the system ($\omega_q^2 = (1/n_s m)\frac{\partial\Psi}{\partial n}q^2$). If the 2DES is pinned to the host, that too inhibits screening and if there is a single frequency ω_p associated with that pinning, ω_q^2 (stiffness) $\longrightarrow \omega_p^2 + \omega_q^2$ (stiffness). Deviations from the perfectly screening electrons capacitance towards the unpolarisable electrons capacitance reflects the $(\omega_p^2 + \omega_q^2$ (stiffness)$)/\omega_q^2$ (electrostatics) term and gives a quantitative measure of the importance of incompressibility and pinning. Figure 4 compares two different samples, one of which has been maltreated by current pulsing to reduce its density, but which has also introduced additional disorder. One sees on the same curves the relative incompressibility of the $\nu = 1/3$ and

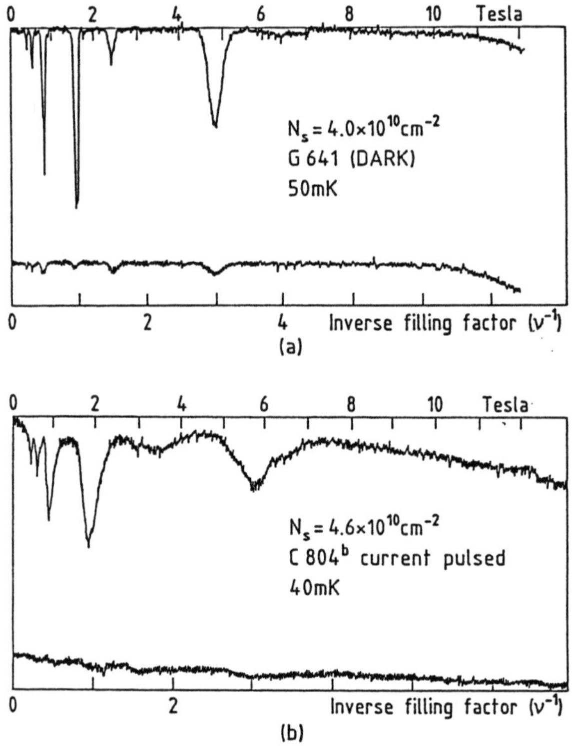

Figure 4a,b: Interdigital array capacitance measurement on 2 different 2DES samples showing: the relative incompressibility at integral and fractional filling factors and the still relatively good screening by the 2DES in the high field regime. Sample G641 is in its as cooled state, sample C804 has been maltreated, in situ, by a series of current pulses to reduce its density by a factor of 2, which treatment also increases disorder. The out of phase component is also shown to indicate the transport contribution. The key to equilibrium is the short wavelength (32 microns) of the array.

2/3 and integer QH states; the non-zero compressibility reflects a non-zero density of states in the gap due also to the disorder. The signal in quadrature to that giving capacitance is also shown, on the same scale, to confirm that the dissipative transport contribution is small, in contrast to the case of long length scale measurements where the latter swamps the contribution of the thermodynamic potential.

As an illustration, on the better sample the high field deviation of capacitance is some 10% of the peak variation due to the $\nu = 1$ integer which we can use to put a bound on the non-polarisable limit in order to establish an upper bound for the pinning frequency of $\omega_p^2 < 10^{-1}\omega_q^2(ES) \approx (40GHz)^2$.

2.2 Radiofrequency Spectroscopy

This technique allows us to explore the behaviour of $\sigma_{xx}(pq_0, \omega)$ over the comb defined by $p = 0, 1, ..., 10$ and $0.01 < \omega < 4GHz$ depicted by Fig.2. The radio-frequency excitation is applied to the 2DES in the form of a spatially modulated longitudinal electric field created by a meander transmission line /10/ on which the sample is placed face down in such a way that the 2DES is at most some 300nm away, allowing the electrons to feel up to about the tenth harmonic of the meander periodicity (see Fig.5 for experimental configuration). The presence of an electronic excitation of wavevector k and frequency ω_k leads to a resonant response in $\sigma(q, \omega)$ around the point (k, ω) (represented by a pole in the generalised susceptibility); if q is fixed and ω is swept continuously there will be a resonant absorption as ω is swept through ω_q. As there are a series of spatial Fourier components in the coupling field, there will be a resonant response at each ω_q thereby allowing one to sample a number of points in the dispersion relation of the branch of excitations (see Fig.2).

The experiment has been designed to reveal the presence of any low-lying excitation branch in the 2DES with the idea that a liquid should have only damped modes of zero frequency resulting from purely viscous reaction to shear. A free solid, on the other hand, is expected to have a gapless finite frequency excitation branch arising from elastic reaction to shear. But if the solid is pinned to the host and this pinning is characterised by a well-defined pinning frequency ω_p, we expect the lower hybrid branch frequency in a magnetic field to become

$$\omega_-(q) \approx \left(\omega_p^2 + (\mu/\mathrm{mn}_s)\, q^2\right)^{1/2} \left(\omega_p^2 + (2\pi e^2 n_s/m)\, q\right)^{1/2} /\omega_c$$

where $q = pq_0$, $p = 0, 1, ..., 10$ and $q_0 = 2\pi/16\mu m$ for our meander line, $\mu \approx (e^2/\varepsilon m a^3)\, f(\nu, t)$, m is the effective mass of an electron, ω_c is its cyclotron frequency. A distribution of ω_p leads to a corresponding spread in ω_- : i.e. a line width in excess of the line width expected from momentum damping and

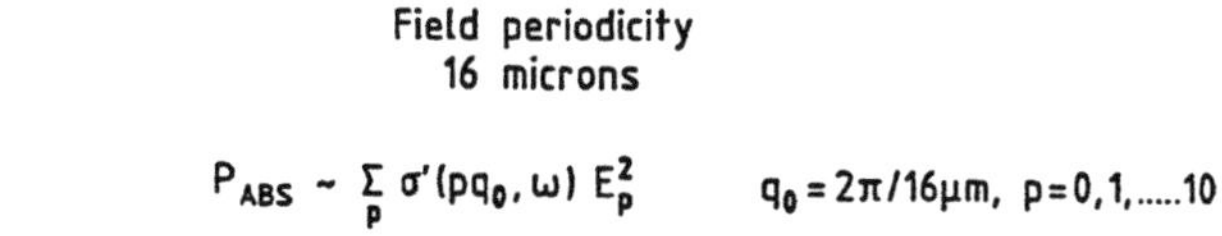

Figure 5: Schematic view of rf spectrometer. The role of the back gate modulation and subsequent synchronous detection is to eliminate background absorption from other sources.

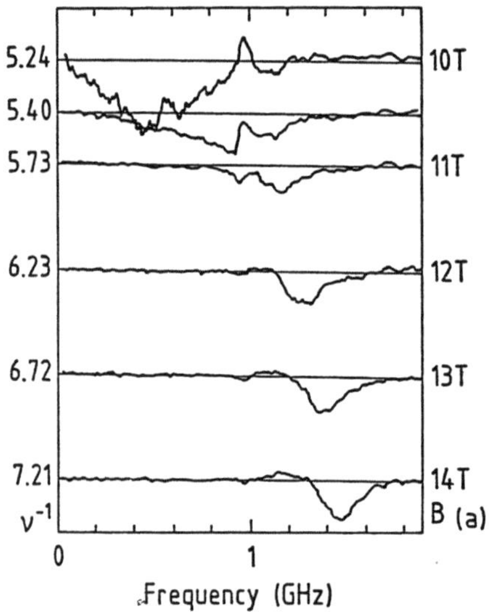

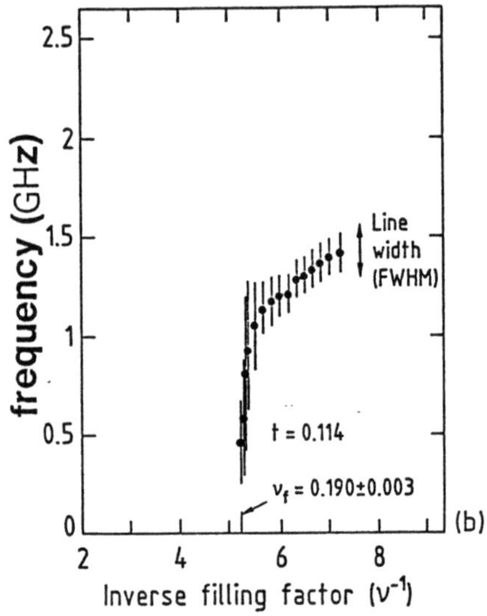

Figure 6: (a) The rf absorption spectrum (derivative with respect to a perpendicular electric field modulation of $\sim 30V\mathrm{cm}^{-1}$) for several values of magnetic field. The inverse filling factor is shown on the left hand scale. Sample G641 non-illuminated at 45mK; $n_S = 4.65\ 10^{10}\mathrm{cm}^{-2}$, $\mu \approx 4\ 10^6\mathrm{cm}^2V^{-1}\mathrm{sec}^{-1}$ at $B = 0$. (b) The central frequency of the absorption plotted against inverse filling factor to show how phase boundary filling factor ν_f is determined. The vertical bars represent line width, not estimated errors.

anharmonic phonon interaction. To give a feeling for orders of magnitude, if one takes the $\nu \longrightarrow 0$ limit for the shear modulus at density $n_s \approx 7\ 10^{10}\mathrm{cm}^{-2}$ one finds $\omega_- \approx 25GHz \times 400GHz/10000GHz \sim 1GHz$ at $B = 25T$ ($\nu \approx 0.12$) for the tenth harmonic $p = 10$ supposing $\omega_p = 0$.

The radio-frequency absorption spectrum does indeed show very distinctive behaviour /11/. At low temperatures ($t \lesssim 0.1$, $T \lesssim 50$mK typically) there is a broad, weak, frequency response for all ν above a well-defined value ν_f, below which the spectrum undergoes a qualitative change: the weak broad response turns into a strong resonant response which sharpens in width and whose basic frequency moves rapidly up from zero as the field is increased to reach a plateau value at the highest fields and lowest temperatures available to us. This plateau value varies from sample to sample: in samples which give a single broadish resonance of relative width $\Delta f/f_0 \sim 1/10$, this plateau value lies in the range 1.5 to $3GHz$ (Fig.6); in the sample which gives a multiplet structure each of whose components has a relative width $\sim 1/20$ the structure lies between 0.6 and $1.3GHz$ /11/. The resonant form broadens and disappears as t is increased at fixed ν : the locus of points (ν_f, t_f) at which the resonant spectral form appears defines a low t, low ν corner of the (ν, t) plane in which the behaviour of the spectrum is monotonic in ν and t, exhibiting none of the Shubnikov-de Haas-type oscillations characteristic of the higher filling factors.

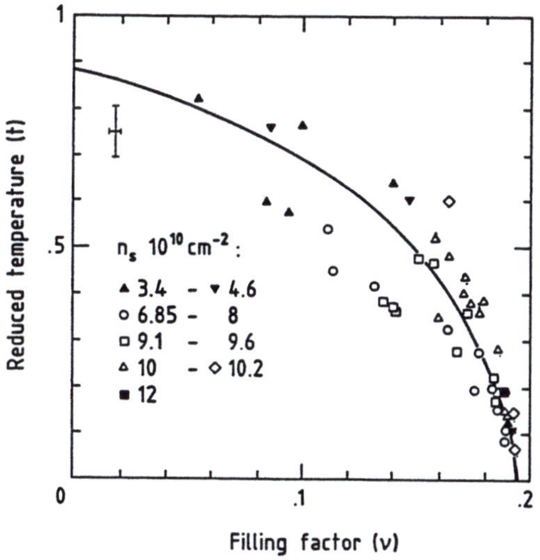

Figure 7: Points (ν_f, t_f) such as that determined from Figure 6 plotted in the reduced variable plane (t, ν) where t is the temperature normalized by the classical melting temperature (~ Coulomb interaction) and ν is the filling factor. In the limit $\ell_c \ll a_B$, one expects phenomena depending only on Coulomb interaction and degeneracy to show universal behaviour in such coordinates (independence with respect to density).

A plot of the points at which the spectrum changes its form is given in Fig.7 in the reduced variables (ν, t) for a large range of different samples (six differently grown samples from two different laboratories with some samples recycled to alter their density). Despite the fact that the total range of densities varies over a factor of about 3.5, the plot shows a remarkably common locus in these reduced coordinates. The most striking feature is the precision to which the $t \longrightarrow 0$ limit of ν_f is defined, indifferent to sample (disorder) and density: $\nu_f(t \longrightarrow 0) \equiv \nu_c = 0.192 \pm 0.003$. The next most salient feature is the independence of the plot with respect to sample density; the upper existence (reduced) temperature shows scatter, but this is not correlated with density. Thirdly the $\nu \longrightarrow 0$ extrapolation gives $t = 0.85 \pm 0.15$, quite close to the classical point charge melting temperature that one expects in this limit. That the phase diagram so obtained scales with ν and t indicates that the phenomenon is governed by magnetic degeneracy and Coulomb interaction.

But is it a real phase transition? The behaviour of the frequency of the resonances upon approaching ν_c at low t illustrated on Fig.6b is very suggestive. The much softer disappearance with t is less convincing in this sense.

The conclusion from this data is: (i) there is a distinct high field, low temperature phase associated with magnetic degeneracy and Coulomb interaction, (ii) the resonance line widths put a lower bound on the correlation length of $L_D/a > f/\delta f$ which varies from 20 to 10 according to the sample, (iii) there is scatter in the upper existence temperature which it is tempting to interpret as a depression of a critical temperature due to a random field, (iv) the resonance structure can be interpreted as arising from a combination of pinning and shear elasticity, but as it stands it cannot be used to unambiguously measure the finite shear modulus due to the possible importance of the pinning frequency.

It is clearly desirable to have precise, independent, information on the random field. One approach to that is the screening behaviour as seen by the capacitance measurement described earlier. Another is to measure threshold field effects and this is described in the following section.

2.3 Low Frequency AC Contacted Transport

In all our experiments, sample density and quality are checked by contactless Shubnikov-de Haas oscillations in the fixed frequency rf absorption, the capacitance or the contacted low current AC resistivity or, sometimes, all three together.

Here, however, we describe some low frequency ($\sim 7Hz$) contacted transport results on a sample not mounted on an electrode structure for rf or capacitance so that we could illuminate it uniformly and thereby considerably improve its "quality". The sample ($G641$ /5/), with $n_s = 4.5\ 10^{10}\mathrm{cm}^{-2}$ and $\mu \sim 4\ 10^6\mathrm{cm}^2V^{-1}s^{-1}$ in the dark at $4K$, was illuminated briefly with a red light emitting diode at $\sim$ 200mK to attain saturation in density at $n_s = 9.5\ 10^{10}\mathrm{cm}^{-2}$, upon which its mobility increases to $8.10^6\mathrm{cm}^2V^{-1}s^{-1}$. A first indication of the improvement in its quality is given by the longitudinal (ρ_{xx}) resistance ratio of the minimum on the $\nu = 1/5$ oscillation to the maximum on the low field shoulder which goes from about 1.5 to about 75.

But perhaps the most interesting behaviour lies on the low filling factor side of $\nu = 1/5$: as usual, the apparent linear longitudinal resistance in this region becomes very high ($\sim$ 100Mohms per square) /12/ and for this reason we use a broad and short geometry (length to width ratio $\sim 10^{-1}$). The new features are that it shows clear minima at filling factors $\nu = 2/11$ and $\nu = 1/7$, illustrated in Fig.8. As the temperature is raised above $\sim$ 50mK ($t = 0.09$) these minima disappear. The other important feature is the non-linearity at low currents accompanied by large phase shifts. This is a consequence of the appearance of a threshold field for conduction as discussed below.

The measuring configuration is composed of a high impedance (10^{14}ohm) low frequency (period $T_0 = 150ms$) square wave current source which is shunted by a $330pF$ capacitance C before feeding the current contact of the sample (whose other current contact is at earth return) Fig.9. The longitudinal potential drop across two contacts approximately 0.5mm apart is measured with a high impedance (100 Mohm shunted by $1.8nF$) differential input phase sensitive synchronous detector operating in the modulus/phase mode. At low currents ($\lesssim 1nA$) the voltage signal suddenly dephases with respect to the current source as the magnetic field is swept up through a very well-defined filling factor ν_p. This phase lag, θ, is inversely proportional to the current and a simple interpretation can be made in terms of an $I - V$ characteristic with

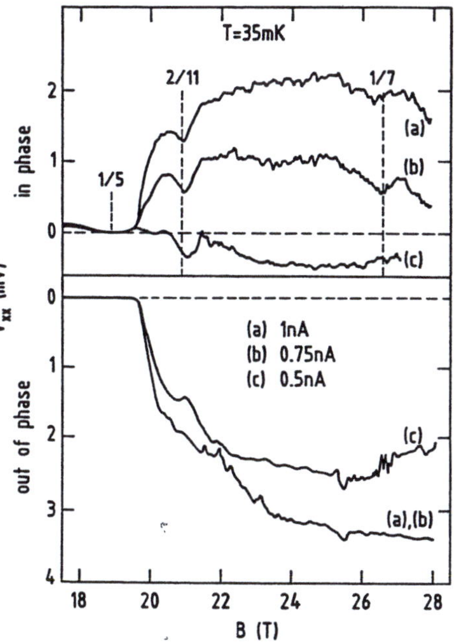

Figure 8: In phase and out of phase components of the longitudinal potential drop at 3 different AC(6.66Hz) current levels through sample G641 illuminated at low temperature ($n_S = 9.2\ 10^{10}\text{cm}^{-2}$, $\mu_0 = 8\ 10^6\text{cm}^2V^{-1}\text{s}^{-1}$). Both $\nu = 2/11$ and $1/7$ features and non-linearity are clearly seen.

a threshold voltage V_T of the sort depicted in Fig.9. While the sample is non-conducting the current charges the capacitor until the voltage reaches the threshold value, V_T, at which point it flows through the sample to give rise to a longitudinal potential drop. The phase shift is due to the time required to charge the capacitor to $V_T : \theta/2\pi = CV_T\ i_s T_0$. Formally we cannot preclude a contact effect, but the fact that at low temperature the filling factor at which this phase shift appears, $\nu_p = 0.194$, is just the value $\nu_c = 0.192 \pm 0.003$ at which the radio-frequency spectrum (on other samples) undergoes the qualitative change in form impels one to believe this to be another manifestation of the same basic physical phenomenon. This is reinforced by the displacement of ν_p with reduced temperature t, the locus fitting very well the phase diagram constructed from the rf absorption data. The jump and the phase shift in the high field region decrease with temperature and increase with field.

The threshold field that one derives from the phase shift has all the hallmarks of a depinning field for a charge density wave and of course this would be expected for a magnetically induced Wigner solid phase of finite domain size pinned by a random field. The lowest temperature data give a threshold field of $\sim 30mV\text{cm}^{-1}$ which, on a weak pinning model, would correspond to a pinning frequency of $\sim 7GHz$ and a domain size of about 5 microns if one uses the calculated infinite field shear elastic constant.

The new information from this experiment is twofold: (i) a threshold field for conduction appears as one crosses the frontier $\nu_f(t)$ in the (ν, t) plane at which the rf absorption shows the appearance of a new phase, (ii) in this new phase, the longitudinal resistivity shows clear minima at $\nu = 2/11$ and $1/7$

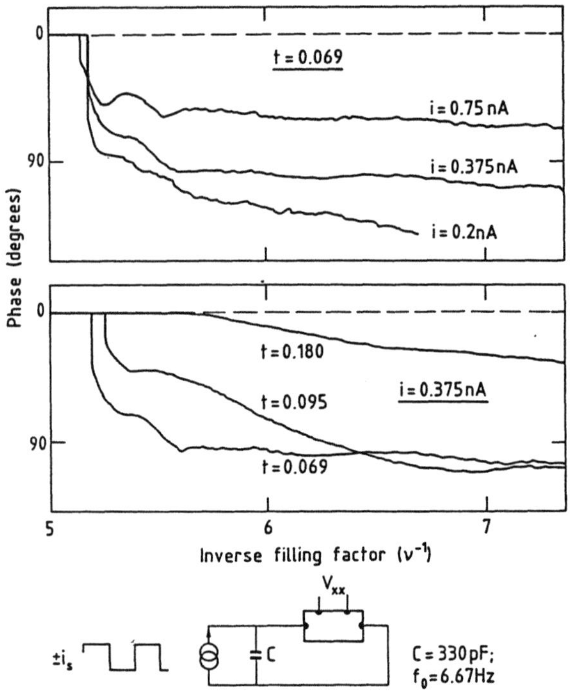

Figure 9: Phase lag of longitudinal potential drop across sample with respect to current. A threshold voltage for depinning, V_T, would give rise to a phase lag $\theta/2\pi = CV_T/i_S T_0$ where C is the capacitance shunting the current source and T_0 is the $\pm i_S$ square wave period. The sharpness of the phase lag onset is partially an instrumental effect.

which disappear at temperatures above $t \approx 0.1$ ($\sim$ 50mK) for this density of $n_s = 9.6\ 10^{10}\text{cm}^{-2}$.

3. Conclusions

There is a distinct phase in the low filling factor, low reduced temperature corner of the phase diagram of the 2DES which is associated with Coulomb interaction and magnetic degeneracy. It shows a radio-frequency absorption spectrum very different from that of the higher ν higher t portion and also displays threshold behaviour for conduction. The rf spectrum, which can be interpreted in terms of pinning and shear, indicates a sharpness which sets a lower bound on the domain size in excess of 10^2 particles for all the high quality samples we examined, indicating that the disorder is not sufficient to override the Coulomb interaction between electrons to form a glass. Depinning effects occur which should allow a quantitative comparison with the rf data and from this an unambiguous test for intrinsic shear elasticity. The general nature of the phase diagram in reduced variables (ν, t) is independent of density in the range we have investigated except for the appearance of features in the diagonal resistivity at fractional filling factors $\nu = 1/7$ and $\nu = 2/11$; at density $n_s = 4.5\ 10^{10}\mathrm{cm}^{-2}$, a $\nu = 1/7$ feature appears at *high* temperatures both above /13/ and extending some way into the new phase, while at $n_s = 9.6\ 10^{10}\mathrm{cm}^{-2}$ there are features at filling factors 2/11 and 1/7 which exist only at *low* temperatures $t < 0.1$. This inverted behaviour between the two different densities suggests that there is another relevant energy scale in competition with the Coulomb interaction.

The virtually inescapable conclusion is that this new phase is the magnetically induced Wigner solid with minimum crystallite size in excess of 10^2 electrons even in our least good samples.

4. Acknowledgments

It is a pleasure to thank V. Duburcq for her collaboration in part of this work, and to thank the staff of the CNRS-MPI high magnetic field laboratory - particularly J.C. Vallier, C. Warth, P. Sala and J.C. Picoche. Thanks are also due to A. Janossy and G. Mihaly for illuminating discussion of conduction and screening properties of charge density waves. C. Chaleil and J. Verrier made essential technical contributions. We also gratefully acknowledge a NATO travel grant which has considerably eased the problems of international collaboration.

References

[1] E.P. Wigner, Phys. Rev. **46**, 1002 (1934).

[2] D.C. Tsui, H.L. Stormer and A.C. Gossard, Phys. Rev. Lett. **48**, 1559 (1982); R.B. Laughlin, Phys. Rev. Lett. **50**, 1395 (1983).

[3] K. Maki and X. Zotos, Phys. Rev. **B28**, 4349 (1983);
P.K. Lam and S.M. Girvin, Phys. Rev. **B30**, 473 (1984);
D. Levesque, J.J. Weiss and A.H. MacDonald, Phys. Rev. **B30**, 1056 (1984);
Yu. E. Lozovik, V.M. Fartzdinov and B. Abdulaev, J. Phys. **C18**, L807 (1985);
S.T. Chui, T.M. Hakim and K.B. Ma, Phys. Rev. **B33**, 7710 (1986);
S. Kivelson, C. Kallin, D.P. Arovas and J.P. Schrieffer, Phys. Rev. **B36**, 1620 (1987).

[4] B. Etienne and E. Paris, J. Phys. (Paris) **48**, 2049 (1987).

[5] C.T. Foxon, J.J. Harris, D. Hilton, J. Hewett and C. Roberts, J. Semicon. Sci. Tech. **4**, 5.. (1989).

[6] M. Shayegan, V.J. Goldman, T. Sajito, M. Santos and C. Jiang, J. Crystal Growth **95**, 250 (1989).

[7] Y. Imry and S. Ma, Phys. Rev. Lett. **35**, 399 (1975);
H. Fukuyama and P.A. Lee, Phys. Rev. Lett. **B17**, 535 (1978);

P.A. Lee and H. Fukuyama, Phys. Rev. **B17**, 542 (1978);
P.A. Lee and T.M. Rice, Phys. Rev. **B19**, 3970 (1979).

[8] D.R. Nelson, Phys. Rev. **B27**, 2902 (1983);
S. Sachdev and D.R. Nelson, J. Phys. **C17**, 5473 (1984).

[9] F. Stern, Appl. Phys. Lett. **43**, 974 (1983);
T.P. Smith, B.B. Goldberg, P.J. Stiles, M. Heiblum, Phys. Rev. **B32**, 2695 (1985).

[10] G. Deville, A. Valdes, E.Y. Andrei and F.I.B. Williams, Phys. Rev. Lett. **53**, 588 (1984).

[11] E.Y. Andrei, G. Deville, D.C. Glattli and F.I.B. Williams, E. Paris and B. Etienne, Phys. Rev. Lett. **60**, 2765 (1988).
F.I.B. Williams, D.C. Glattli, G. Deville, B. Etienne, E. Paris and E.Y. Andrei, in "High Magnetic Fields in Semiconductor Physics II", Proc. of Int. Cond. Würzburg FRG, Aug. 1988, 157 (Springer-Verlag).
D.C. Glattli, G. Deville, V. Duburcq, F.I.B. Williams, E. Paris, B. Etienne and E.Y. Andrei, Proc. 8th Int. Conf. on Electronic Properties of 2-D Systems (EP2DS-8), Sept. 1989 to appear in Surface Science 1990.

[12] R.L. Willett, H.L. Störmer, D.C. Tsui, L.N. Pfeiffer, K.W. West and K.W. Baldwin, Phys. Rev. (Rapid Communication) **B38**, 7881 (1988).
E. Paris, Thesis Université de Paris 6 (1989) and the second of references [11].

[13] V.J. Goldman, M. Shayegan and D.C. Tsui, Phys. Rev. Lett. **61**, 881 (1988); Radio-frequency absorption measurements done in our laboratory also show a feature in σ at $\nu = 1/7$ and that it extends into the new phase, but disappears below $t \approx \frac{1}{2}t_f$.

Electron States of Nearly Free, Uniform-Density, Dilute Electron Systems in Wide Parabolic Wells

M. Shayegan

Department of Electrical Engineering, Princeton University,
Princeton, NJ 08544, USA

We report the fabrication, characterization, and physics of selectively-doped parabolic AlGaAs wells which contain low-disorder, nearly uniform-density, dilute electron systems. The electron layer in these structures has a thickness of ~1000Å, an effective three-dimensional density of ~ $1\times10^{16}cm^{-3}$, and a mobility of ~ $3\times10^5 cm^2/Vs$. These structures provide a rather unique system for the studies of the single-electron as well as many-body effects in space-charge layers in semiconductors. Examples of new physical phenomena observed in these systems are presented.

1. Introduction

In selectively doped semiconductor structures, the carriers are spatially separated from the dopant atoms to reduce the ionized-impurity scattering. Because of the reduced disorder and scattering, these structures provide a nearly ideal system for the study of many-body effects. The high-mobility two-dimensional electron system (2DES) at the interface of $GaAs/Al_xGa_{1-x}As$, for example, made the discovery of the fractional quantum Hall effect [1] possible and has been used to observe new quantum phenomena. Studies of the electron-electron interaction effects can be extended to three-dimensional (3D) semiconductor systems. In fact, a variety of collective phenomena such as charge-density waves, spin-density waves, and Wigner crystallization have been theoretically proposed to occur in dilute 3D *free*-electron systems under appropriate conditions (e.g., at very low temperatures and in intense applied magnetic fields). [2] Experimentally, dilute electron systems in *uniformly* doped degenerate semiconductors have been studied extensively; however, the electron-impurity interaction in these systems is strong and usually leads to the magnetic freeze-out of the carriers. Realization of a 3D dilute electron system with low disorder is, therefore, of considerable interest.

This paper deals with the fabrication, characterization, and the physics of a selectively-doped semiconductor structure which may be considered as a first step towards the realization of a 3D or quasi-3D free-electron system (Q3DES) [3-11]. The structures, which are grown by molecular beam epitaxy, consist of a wide undoped $Al_xGa_{1-x}As$ well bounded by undoped (spacer) and doped layers of $Al_yGa_{1-y}As$ ($y > x$) on two sides. The alloy composition (x) in the well is graded in such a way as to result in a quadratic dependence of

Springer Series in Solid-State Sciences, Vol. 97
Localization and Confinement of Electrons in Semiconductors
Editors: F. Kuchar · H. Heinrich · G. Bauer

the conduction band edge (E_C) on the distance from the well center. Owing to the self-consistent electrostatic potential, the electrons in the parabolic well screen the quadratic potential, and a system of nearly uniformly distributed electrons in a flat potential is obtained. In this system, the electron layer thickness increases with increasing areal density in the well so that the effective 3D density and the Fermi energy remain essentially constant.

Several aspects of these electron systems will be discussed here including: (a) the basic growth and structural parameters, (b) self-consistent quantum mechanical calculations of the electronic structure, (c) the experimental verification of the electronic structure via magnetotransport measurements, (d) the behaviour of the electron system (probed by magnetotransport and magnetooptics measurements) in the presence of an in-plane magnetic field, and (e) the preliminary studies of the electron-electron interactions in these systems.

2. Realization of Q3DES in wide parabolic wells: fabrication and characterization

The structures are grown on undoped GaAs substrates in a molecular beam epitaxy system whose details have been reported previously[3]. Here we give the structural parameters of a particular parabolic well (wafer M128). This structure consists of a 1600Å -wide well in which x is quadratically varied between x=0 at the well center to x=0.15 at the well edges. The well is bounded on both sides by 450Å of $Al_{0.4}Ga_{0.6}As$, and the doping consists of five δ-layers of Si (each with a density of $1.3 \times 10^{11} cm^{-2}$ and separated by 30Å).

In the absence of any space charge, and assuming that the conduction band offset (ΔE_C) for GaAs/$Al_xGa_{1-x}As$ is linearly proportional to x, the compositional grading in the well is expected to result in a parabolic potential well as shown in Fig. 1. In order to determine the subband structure of the electron system, we performed self-consistent calculations by solving Schrödinger and Poisson equations [5,8,10], and taking into account the exchange-correlation via local-density-functional approximation [12]. The results of these calculations are shown in Fig. 1 for several measured areal densities (n_s) in our structure. Note that as n_s increases, the width of the electron system (curve C) also increases so that the effective three-dimensional density as well as the Fermi energy remain essentially constant.

The experimental and quantitative evidence for the realization of the designed electron system is provided by our magnetotransport data. The transverse (ρ_{xx}) and Hall (ρ_{xy}) resistivities shown in Fig. 2 were measured in the conventional manner (**B** perpendicular to the sample plane) at low temperature ($T \simeq 0.4$ K) in the Hall bridge geometry. Each set of measurements was done following a combination of careful and controlled cooling of the sample as well as illumination by a red light-emitting diode. Particular care was exercised to obtain a homogeneous electron system. The data in Fig. 2 exhibit the integral and fractional quantum Hall effects, with the positions (in B) of ρ_{xx} minima and ρ_{xy} plateaus for $B \gtrsim 1$T being consistent with the total areal electron densities of (a) 1.10 x $10^{11}cm^{-2}$, (b) 1.87 x $10^{11}cm^{-2}$, and (c) 2.90 x $10^{11}cm^{-2}$. In the low-field range (B$\lesssim$1T), many Landau levels of each subband are occupied. The oscillatory magnetoresistance data in this range (Fig. 3) are periodic in 1/B and may contain more than one oscillation fre-

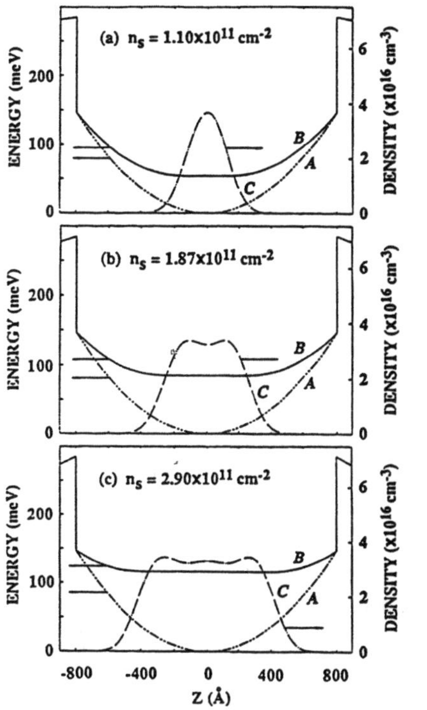

Fig. 1. The results of self-consistent calculations for structure M128 are shown for different n_s. Curve *A* for each density corresponds to E_C in the absence of any space charge. The conduction-band offset at the GaAs/$Al_xGa_{1-x}As$ heterointerface is assumed to be linearly proportional to x and equal to 980x (meV). Curve *B* shows the self-consistently calculated final potential of the well for each density. A constant effective mass of 0.067 m_e is assumed, with m_e being the free electron mass. The total charge distribution for each density, calculated by appropriately summing the wavefunctions for the individual subbands is shown as curve *C*. The number of occupied electric subbands are one, two, and three for the cases a, b, and c, respectively.

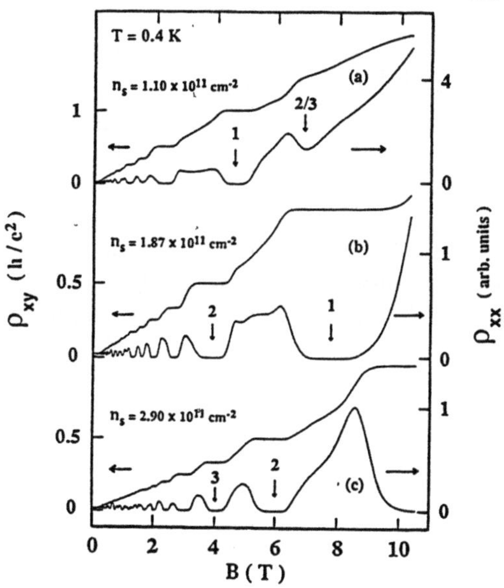

Fig. 2. Magnetotransport coefficients ρ_{xx} and ρ_{xy} for sample M128 at T$\simeq$0.4 K are shown for three different total areal electron densities. The vertical arrows indicate the Landau-level filling factors at which the integral or fractional quantum Hall effect is observed.

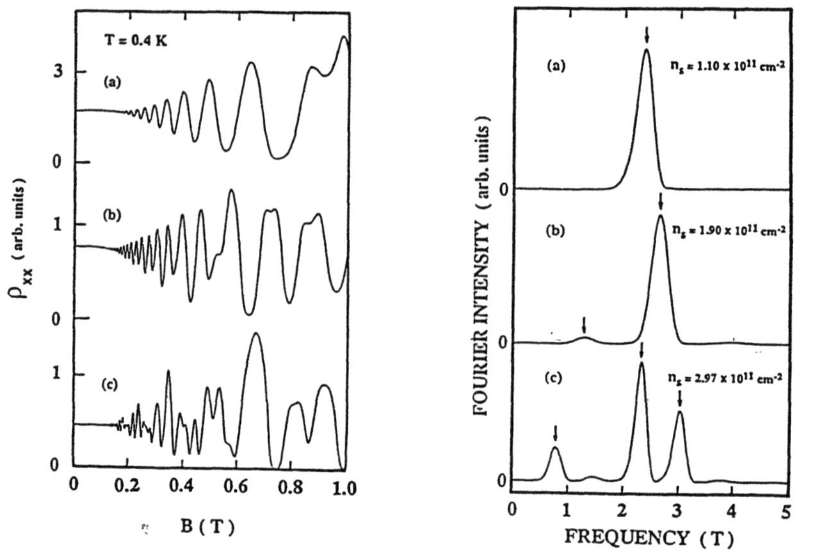

Fig. 3. On the left, the oscillatory magnetoresistance ρ_{xx} data of Fig. 2 are shown in detail at low magnetic fields. On the right, the Fourier transform power spectra of the oscillatory ρ_{xx} vs 1/B data of the left traces are shown.

quency. These frequencies can be determined from the Fourier transforms of the ρ_{xx} vs 1/B data. The positions of the peaks in the Fourier power spectrum (indicated by the vertical arrows in Fig. 3) give the areas of the Fermi-surface cross sections (circles in our case), and therefore the areal electron densities for the different occupied electric subbands (n_i for i=1,2...). The total areal electron densities deduced from the power spectra of Fig. 3 by summing the subband densities are also indicated in Fig. 3 and are in good agreement with the densities deduced from the high-field data (Fig. 2).

In Fig. 4, the measured subband densities obtained from each of the power spectra of Fig. 3 are plotted as a function of the corresponding measured n_s. For comparison, the results of our self-consistent calculations are also shown (as solid curves). The only fitting parameter used in these calculations is the curvature of the parabolic potential. This curvature is determined by the well-width, the value of x (the Al composition) at the well edge, and the dependence of ΔE_C on x. The experimental data are best fitted by assuming ΔE_C = 980x (in units of meV) as shown in Fig. 4. This value of ΔE_C is somewhat larger than the values reported in literature [13]. Considering the uncertainties in the growth parameters, however, the agreement between the measured and calculated subband densities is quite good.

The transport data presented so far, together with the results of the simulations, provide evidence for the realization of a high-quality electron system with a density which is nearly uniform over several hundred Å. Note that the low-temperature electron mobility (μ) for this structure, deduced

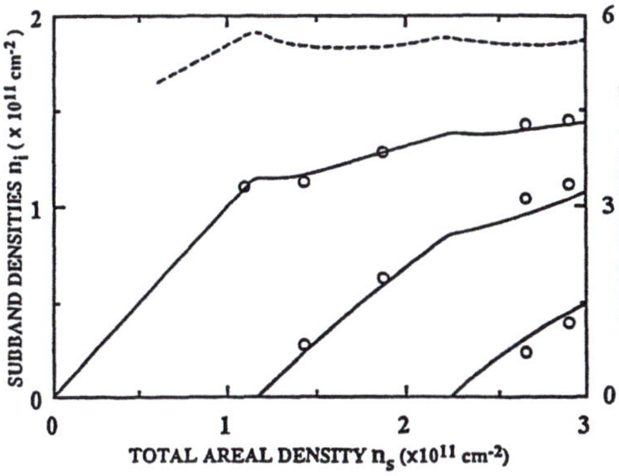

Fig. 4. The measured subband densities (circles) obtained from each of the Fourier transform power spectra of Fig. 3 are plotted as a function of the corresponding total measured areal electron density. The results of self-consistent calculations are shown as solid curves. The dashed curve shows the position of the calculated Fermi level (with respect to the bottom of the conduction band) for the system.

from the transport data, is about $3 x 10^5 cm^2/Vs$ -- two orders of magnitude larger than μ in n-doped bulk GaAs with a similar density of $\sim 3x10^{16} cm^{-3}$ (Fig. 1). Such selectively-doped Q3DES, therefore, provide a unique system to study certain novel physical phenomena. Here we present some examples.

3. Plasma-shifted cyclotron resonance in Q3DES

A very interesting property of the electron system in wide parabolic wells is demonstrated by the observation of plasma-shifted cyclotron resonance (CR) by Karrai et al. [11]. In a thick metallic film, CR is usually detected in magnetotransmission with normally incident light of wave vector **q**. In this Faraday geometry ($\mathbf{q} \| \mathbf{B}$) one observes a resonant absorption of the electromagnetic radiation at the frequency $\omega_c = eB/m^*$, corresponding to the single-particle cyclotron excitation. In the Voigt configuration ($\mathbf{q} \perp \mathbf{B}$), the magnetic field lies in the plane of the electron slab and the carriers in their cyclotron motion are coupled to one another by the macroscopic depolarization field leading to a collective or plasma-shifted cyclotron resonance (PSCR) [14]. The PSCR is excited by the electric component **E** of the incident electromagnetic radiation perpendicular to **B** at the frequency $\omega = (\omega_c^2 + \omega_p^2)^{1/2}$, where ω_p is the plasma frequency of the uniform 3D electron gas.

The PSCR is usually not observed in a quasi-2D electron system since cyclotron motion perpendicular to the confinement plane is not possible. In our Q3DES, however, even at relatively small magnetic fields ($B \gtrsim 1T$), the magnetic length l_o ($l_o^2 = \hbar/eB$) is smaller than the effective electron layer thickness and PSCR is possible. In Fig. 5 the CR data measured in both the Faraday and Voigt geometries are shown [11]. The data clearly demonstrates that the Voigt geometry CR is plasma-shifted. Analysis of the magnetic field dependence of the PSCR frequency leads to a value for the plasma frequency of $\omega_p = 51.5 cm^{-1}$. Relating the 3D electron density (n_o) to ω_p by $\omega_p^2 = n_o e^2/m^* \epsilon_o \epsilon$, we find $n_o \simeq 2.7 x 10^{16} cm^{-3}$, in reasonable agreement with the results of the magnetotransport measurements and the self-consistent calculations (Fig. 1). Recent theoretical work by Brey et al. [15] has demonstrated a spe-

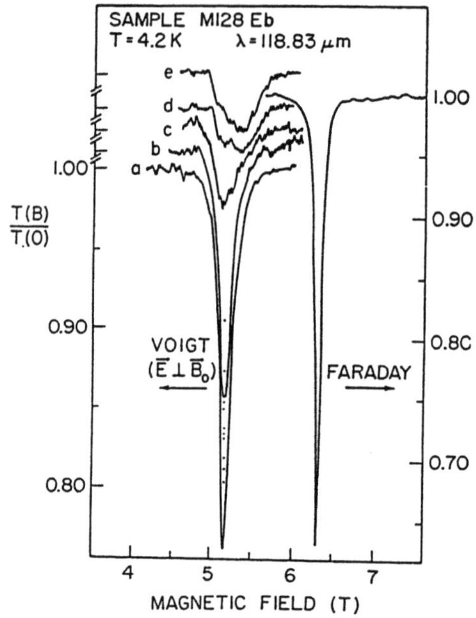

Fig. 5. Faraday CR (right) and Voigt PSCR (left) measured at the same laser frequency are shown. The Voigt resonance is shifted to lower field compared to the CR. The PSCR are measured under different filling conditions. The points are the positions of the minima of transmission for different areal electron densities.

cial property of the electron system in a parabolic well; namely, that in a magnetic field the absorption of light in this system is characterized by two harmonic frequencies ω_c and ω_o, where ω_o is the empty well harmonic oscillator frequency (ω_o is equal, by construction, to ω_p). The optical response of the system is then shown to be always at the frequencies of the mixed modes of these two oscillators, independent of the electron density or electron-electron interactions. The PSCR spectra taken for different values of n_s (Fig 5) clearly confirm this theorem: the resonance position of the PSCR is nearly independent of n_s.

4. Electron-electron interactions in Q3DES

The Q3DES provide a high-mobility (low-disorder) electron system in which, unlike the 2DES at GaAs/AlGaAs heterojunctions, the electron layer thickness can be substantially larger than the magnetic length (for $B \gtrsim 1T$). Therefore, new physical phenomena arising from the electron-electron interactions may be expected in these systems. Here we briefly discuss two new experimental observations we have made in the high-field magnetotransport data for these structures. These include: (1) the results of our study of the fractional quantum Hall effect (FQHE) in the Q3DES as a function of the electron layer thickness [16], and (2) certain anomalies in the magnetotransport data taken with a strong in-plane magnetic field [17].

Recently, we have been able to fabricate a high-quality Q3DES with very low density ($n_s < 1x10^{11}cm^{-2}$) and high mobility ($\mu \simeq 3.6x10^5 cm^2/Vs$). An example of the ρ_{xx} data for this structure is shown in Fig. 6. The high

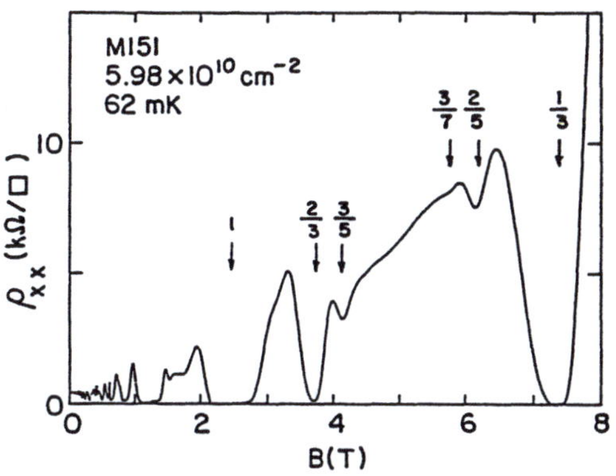

Fig. 6. The ρ_{xx} data is shown for structure M151 with n_s=5.98x10^{10}cm^{-2}. For this n_s, the analysis of the low-field data reveals that (at B=0), two electric subbands are occupied. The width of the electron layer, deduced from the self-consistent calculations, is ≃630Å .

quality of the electron system is evidenced by the observed higher order FQHE states at $\nu = 2/5$, $3/5$, and $3/7$ in addition to the fundamental 1/3 and 2/3 states. In this structure, n_s can be varied over the range 4x10^{10}cm^{-2} to 9x10^{10}cm^{-2}, corresponding to an electron layer thickness w_e (which we define as the full-width at half-maximum of the charge distribution) of about 400 to 1000 Å . We studied the temperature dependence of the ρ_{xx} minimum at $\nu =$ 1/3 and 2/3 in this structure and determined the energy gaps (Δ) [1,18] for the FQHE states at these filling factors for several n_s. We observe a dramatic decrease in the measured Δ as n_s (and therefore w_e) increases, in sharp contrast to the commonly observed increase of Δ with n_s for the 2DES at single GaAs/AlGaAs heterojunctions [18,19]. We believe that the reduction in Δ is dominated by the increase in w_e [16,18] .

The electron states of the Q3DES in the presence of an in-plane magnetic field are also intriguing. Recently, calculations for these systems have been reported suggesting the possibility of certain collective phenomena, such as a spin-density-wave instability, at low temperatures and in the presence of an in-plane magnetic field [20]. Such phenomena are expected to result in anomalies in the magnetotransport coefficients.

Here we briefly discuss some of our recent magnetotransport data taken with the magnetic field applied in the plane of the sample [9,17]. In Fig. 7a the data are shown for M128 for two different orientations of the in-plane magnetic field (**B** ||**I** and **B**⊥**I**, where **I** is the sample current). With B applied parallel to the quantum well, the potentials arising from both the magnetic field and the spatial confinement in the well must be considered. The result of such a mixed (magnetic and electric) potential is a nonlinear energy vs B fan diagram in which the energy levels at B=0 are determined by the well potential, while as B→∞, they approach the *bulk* Landau levels. This is schematically shown in Fig. 7b. At B=0, the energy levels E_n for our self-consistent potential (Fig. 1) are shown. Also shown are the bulk Landau levels (indexed by N). The solid and dotted curves *schematically* represent the energy vs B fan diagram and the position of E_F respectively [3,9]. Spin-splitting is not considered in Fig. 7b. As B is raised from zero, the degeneracy of the levels increases and magnetic depopulation of the levels occurs.

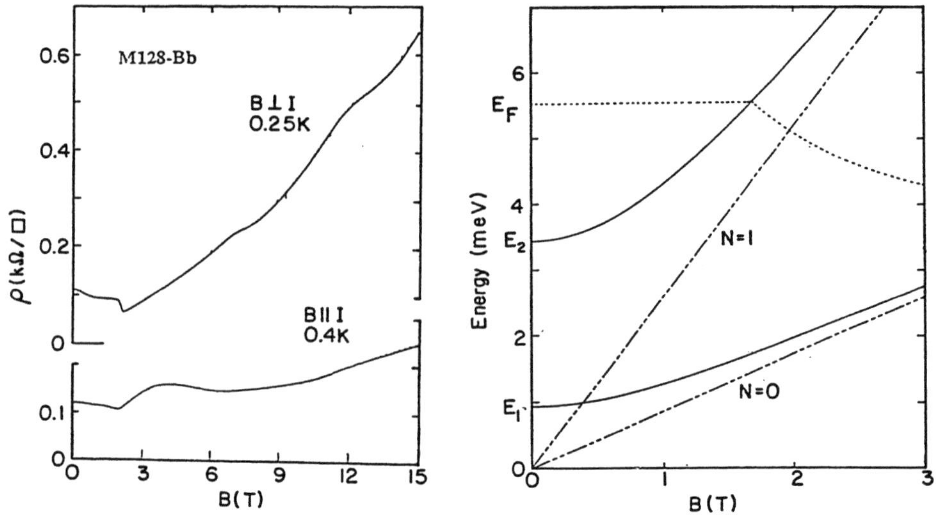

Fig. 7. On the left, the resistance of sample M128 as a function of an in-plane magnetic field is shown for two different orientations of sample current (I) with respect to the direction of magnetic field (B). For the upper and lower traces, n_s is $\simeq$ 2.0 and $\simeq 1.8\text{x}10^{11}\text{cm}^{-2}$, respectively. On the right, the energy vs B fan diagram is shown.

The strong oscillation in the resistance observed at B$\simeq$2T in Fig. 7a is a result of the depopulation of the E_2 level [3,9]. In the case of **B⊥I**, two additional oscillations at higher fields are observed. The oscillation at $\simeq$7.5T may be attributed to the depopulation of a spin-split E_1 level (not shown in Fig. 7b). The origin of the oscillation at B$\simeq$13.5T, however, is presently unknown. Another intriguing feature is the absence of similar oscillations in the high-field range for the **B**||**I** data of Fig. 7a. Further experimental work to elucidate these potentially interesting anomalies is planned.

5. Summary

In this paper, we have presented the basic fabrication and characterization of selectively-doped parabolic AlGaAs wells which contain low-disorder, nearly uniform-density, dilute electron systems. The electron layer thickness in these structures is large and, moreover, can be changed in a given sample (by changing n_s). Examples of new physical phenomena observed in these systems were also presented.

Acknowledgments. This presentation is based on collaborative work with J. Jo, T. Sajoto, M. Santos, K. Karrai, L. Engel, H.D. Drew, and V.J. Goldman. I also thank D.C. Tsui for useful discussions. Support of this work by National Science Foundation grant Nos. ECS-8553110 and DMR-8705002, the

Army Research Office grant No. DAAL03-89-K0036, grants by the New Jersey Commission on Science and Technology, Xerox Corporation, and GTE Laboratories, Inc., and an Alfred P. Sloan Fellowship grant is acknowledged.

References

1. For a review, see D.C. Tsui and H.L. Stormer, IEEE J. Quantum Electron. **QE-22**, 1711 (1986).
2. B. I. Halperin, Jpn. J. Appl. Phys. **26**, Suppl. 26-3, 1913 (1987).
3. M. Shayegan, T. Sajoto, M. Santos, and C. Silvestre, Appl. Phys. Lett. **53**, 791 (1988).
4. M. Sundaram, A.C. Gossard, J.H. English, and R.M. Westervelt, Superlattices and Microstructures **4**, 683 (1988).
5. T. Sajoto, J. Jo, H.P. Wei, M. Santos, and M. Shayegan, J. Vac. Sci. Technol. **B7**, 311 (1989).
6. K. Karrai, H.D. Drew, M.W. Lee, and M. Shayegan, Phys. Rev. B**39**, 1426 (1989).
7. E.G. Gwinn, R.M. Westervelt, P.F. Hopkins, A.J. Rimberg, M. Sundaram, and A.C. Gossard, Phys. Rev. B**39**, 6260 (1989).
8. T. Sajoto, J. Jo, L. Engel, M. Santos, and M. Shayegan, Phys. Rev. B**39**, 10464 (1989).
9. M. Shayegan, T. Sajoto, J. Jo, M. Santos, and H.D. Drew, Phys. Rev. B**40**, 3476 (1989); M. Shayegan, T. Sajoto, J. Jo, M. Santos, and L. Engel, Surf. Sci., in press.
10. T. Sajoto, J. Jo, M. Santos, and M. Shayegan, Appl. Phys. Lett. **55**, 1430 (1989).
11. K. Karrai, X. Ying, H.D. Drew, and M. Shayegan, Phys. Rev. B**40**, 12020 (1989).
12. F. Stern and S. Das Sarma, Phys. Rev. B**30**, 840 (1984).
13. R.C. Miller, A.C. Gossard, D.A. Kleinman, and O. Munteanu, [Phys. Rev. B**29**, 3740 (1984)] reported ΔE_C=750x (meV).
14. S. Iwasa, Y. Samada, E. Burstein, and E.D. Palik, J. Phys. Soc. Jpn. Suppl. **21**, 742 (1966).
15. L. Brey, N.F. Johnson, and B.I. Halperin, Phys. Rev. B**40**, 10647 (1989).
16. M. Shayegan, J. Jo, Y.W. Suen, M. Santos, and V.J. Goldman, submitted for publication.
17. J. Jo, M. Shayegan, V.J. Goldman, L. Engel, and M. Santos, unpublished.
18. For a recent review, see: T. Chakraborty and P. Pietlainen, *The Fractional Quantum Hall Effect* (Springer, Berlin, 1988).
19. G.S. Boebinger, H. L. Stormer, D.C. Tsui, A.M. Chang, J.C.M. Hwang, A.Y. Cho, C.W. Tu, and G. Weimann, Phys. Rev. B**36**, 7919 (1987).
20. L. Brey and B.I. Halperin, Phys. Rev. B**40**, 11634 (1989).

Density of States and Kinetic Coefficients of a Two-Dimensional Electron Gas in a Strong Magnetic Field

K.B. Efetov[1] *and V.G. Marikhin*[2]

[1]Max-Planck-Institut für Festkörperforschung, Heisenbergstr. 1, D-7000 Stuttgart 80, Fed. Rep. of Germany
[2]Moscow Physico-Technical Institute, Moscow, USSR

Abstract. A model of a two dimensional electron gas in a random potential of impurities is considered. It is assumed that a strong magnetic field perpendicular to the plane is applied. For high Landau levels we calculate the density of states, the density-density correlation function, the longitudinal and Hall conductivity. All calculations are carried out in the region of tails of the density of states using the method of integration over supermatrices. It is shown that density-density correlation function falls off exponentially, the static longitudinal conductivity is zero and the Hall conductivity is quantized. The height of the steps is proportional to the number of the corresponding Landau levels.

1. Introduction

Recently a lot of work was devoted to the study of a two dimensional electron gas in magnetic fields. This interest is mainly due to the discovery of the quantum Hall effect [1]. The usual point of view is that the integer quantum Hall effect can exist only if some amount of disorder is present in the system. The plateau regions for the Hall conductivity correspond to the regions of localized states in the spectrum of the system [2,3]. The quantization of the Hall conductivity can be proven for systems of non-trivial geometry [4]. However, usually one starts from rather strong assumptions such as the existence of a gap in the energy spectrum or the existence of an energy region where all states are localized.

A direct analytical calculation of kinetic quantities is rather difficult because it is necessary to consider localization effects. The phenomenon of localization exists only in the energy regions where the interaction of electrons with impurities is strong and perturbation theory methods are not applicable. Therefore one should use nonperturbative methods. We use below the method based on the use of integration over both commuting and anticommuting Grassmann variables. We work in the energy regions which are not very close to the points $E_n = \omega_c\left(n + \frac{1}{2}\right)$. In these regions the density of states is small and one can use the instanton approximation. Using this scheme one can calculate explicitly the density of states, and such kinetic quantities as the density-density correlation function the longitudinal and the Hall conductivity.

The kinetic quantities are calculated in the most interesting limit of vanishing external frequencies.

2. The model

Let us consider a two-dimensional system of non-interacting electrons in an impurity potential U and subjected to a perpendicular magnetic field B. The Hamiltonian H of this system has the form

$$\begin{aligned} H &= H_0 + U(r), \\ H_0 &= \frac{(-i\vec{\nabla} - \frac{e}{c}\vec{A})^2}{2m} \end{aligned} \tag{1}$$

where m is the electron mass, $\vec{A}$ is the vector potential.

Springer Series in Solid-State Sciences, Vol. 97
Localization and Confinement of Electrons in Semiconductors
Editors: F. Kuchar · H. Heinrich · G. Bauer

It is assumed that U(r) is a random Gaussian potential with the standard averages

$$< U(r) >= 0, < U(r)U(r') >= \gamma\delta(r-r'). \quad (2)$$

The density of states $\rho(\epsilon)$ in the system under consideration can be written in terms of the retarded G^R and the advanced G^A Green functions

$$\rho(\epsilon) = -\frac{1}{2\pi i}(G^R_\epsilon(r,r) - G^A_\epsilon(r,r)) = -\frac{1}{\pi} Im G^R_\epsilon(r,r) \quad (3)$$

where

$$G^{R,A}_\epsilon(r,r') = \sum_k \frac{\phi_k(r)\phi_k^*(r')}{\epsilon - \epsilon_k \pm i\delta} \quad , \quad (4)$$

$\phi_k(r)$ are eigenfunctions of the Hamiltonian H.

For studying kinetic quantities it is not enough to know the density of states. An important quantity characterizing the decay of wavefunctions is the correlation function $K(r,r',\epsilon,\omega)$,

$$K(r,r',\epsilon,\omega) =< G^R_{\epsilon+\omega}(r,r')G^A_\epsilon(r',r) > \quad . \quad (5)$$

The longitudinal and Hall conductivities can also be written in terms of the Green functions,

$$\begin{aligned}
\sigma_{\mu\nu}(\omega) &= \frac{1}{i\omega}(\chi^R_{\mu\nu}(-\omega) - \chi^R_{\mu\nu}(0)), \\
\chi^R_{\mu\nu}(\omega) &= \frac{e^2}{i\pi V}\int d\epsilon dr dr' n(\epsilon) < [v_\mu G^R_{\epsilon+\omega}(r,r')v'_\nu(G^R_\epsilon(r',r) - \\
&- G^A_\epsilon(r',r)) + v_\mu(G^R_\epsilon(r,r') - G^A_\epsilon(r,r'))v'_\nu G^A_{\epsilon-\omega}(r',r)] > \quad (6)
\end{aligned}$$

where $v = \frac{1}{m}(-i\nabla_\mu - \frac{eA_\mu}{c})$, V is the volume, n (ϵ) is the Fermi distribution $\mu,\nu = \{x,y\}$.

For the calculation of the average in eq. (3,5,6) it is convenient to represent the Green functions in the form of integrals over commuting and anticommuting variables [5,6]. This form gives the possibility to average over impurities before all other calculations. For example the average of the Green functions $G^{R,A}_\epsilon(r,r')$ takes the form

$$\begin{aligned}
< G^{R,A}_\epsilon(r,r') > &= \pm\int \Phi_\alpha(r)\Phi^+_\alpha(r')\exp\left[-L^\pm_1\{\Phi,\Phi^+\}\right] D\Phi \\
L^\pm_1\{\Phi,\Phi^+\} &= \int dr \left[\pm i\Phi^+(r)(H_0-\epsilon)\Phi(r) + \frac{\gamma}{2}(\Phi^+(r)\Phi(r))^2\right] \quad (7)
\end{aligned}$$

where $\Phi(r) = \begin{pmatrix} \chi(r) \\ s(r) \end{pmatrix}$, $\chi(r)$ are anticommuting variables, s(r) are commuting ones.

For the calculation of averages of two Green functions entering eqs. (5,6) one should double the number of the variables. Then the average product $< G^R_\epsilon G^A_{\epsilon-\omega} >$ takes the form

$$\begin{aligned}
< G^R_\epsilon(r,r')G^A_{\epsilon-\omega}(r',r) > &= \int \Phi^1_\alpha(r)\bar{\Phi}^1_\alpha(r')\Phi^2_\beta(r')\bar{\Phi}^2_\beta(r)\exp(-L_2)D\bar{\Phi}D\Phi \\
L_2 &= \int dr \left(-i\Phi H_0\Phi + \frac{\gamma}{2}(\bar{\Phi}\Phi)^2 - \frac{\omega+i\delta}{2}\bar{\Phi}\Lambda\Phi\right) \quad (8)
\end{aligned}$$

In eq. (8) $\Phi = \begin{pmatrix} \Phi_1 \\ \Phi_2 \end{pmatrix}$, $\bar{\Phi} = (\Lambda\Phi)^+$. Φ_1 and Φ_2 have the same structure as Φ in eq. (7) and Λ is a diagonal supermatrix with components $\Lambda^{11}_{\alpha\alpha} = -\Lambda^{22}_{\beta\beta} = 1$. Analogous formulae can be obtained for other averages.

In the next chapter we calculate the density of states. The calculation of this quantity is simpler than that of kinetic quantities but it demonstrates the applied method and helps to understand basic approximations.

3. Density of states

In order to perform the calculations in eq. (7) it is convenient to decouple the $(\Phi^+\Phi)^2$ term by integration over supermatrices P. Using the identity

$$\exp\left(-\frac{\gamma}{2}(\Phi^+\Phi)^2\right) = \int \exp\left[-(\Phi^+ P\Phi + \frac{1}{2\gamma} STrP^2)\right] DP \tag{9}$$

where $P = \binom{a\ \sigma^*}{\sigma\ ib}$, a, b are real numbers, σ and σ^* are anticommuting variables, and integrating over the supervectors Φ we represent eq. (7) for the averaged Green functions $G^{R,A}$ in the form

$$< G^{R,A}(r,r',\epsilon) >= \int g^{\pm}_{\alpha\alpha}(r,r')\exp(-F^{\pm}(P))DP \tag{10}$$

The function $g^{\pm}(r,r')$ is a solution of the following equation

$$(H_0 \pm iP)g^{\pm}(r,r') = \delta(r-r'). \tag{11}$$

The functional F[P] is equal to

$$F^{\pm}[P] = \int STr\left[-(ln(H_0 \pm iP) + \frac{P^2}{2\gamma}\right] d^2r \tag{12}$$

The supertrace STr in eq. (12) implies the supertrace of the matrix P and the summation over eigenstates of the operator H_0.

The calculation of the integral (10) can be carried out by the saddle point method. If ϵ/ω_c is large (it corresponds to the consideration of high Landau levels) a rather broad region exists where this approximation is applicable. Minimizing F[P] (12) we obtain

$$P(r) = \pm i\gamma\, g(r,r). \tag{13}$$

The eq. (13) has the evident solution $p_0 \cdot 1$ which does not depend on coordinates. For p_0 we obtain the following equation

$$p_0 = \pm\frac{i\gamma}{2\pi a^2}\sum_n \frac{1}{\epsilon - \epsilon_n \pm ip_0} \tag{14}$$

where $(2\pi a^2)^{-1}$ is the degeneracy of a Landau level, a is the magnetic length. The imaginary part of p_0 renormalizes the energy, the real one determines the mean free path τ. In the eq. (14) $\epsilon_n = \omega_c(n+1/2)$ where $\omega_c = (ma^2)^{-1} = \frac{eB}{mc}$.

In the limit $\omega_c\tau >> 1$ one can keep in the sum in eq. (14) only one term corresponding to such n when ϵ_n is most close to ϵ. All calculations presented below correspond to this case.

The real part of p_o is not zero only in the region $|\epsilon - \epsilon_n| < \epsilon_c$ where $\epsilon_c = \frac{1}{a}\left(\frac{2\gamma}{\pi}\right)^{1/2}$. This quantity determines the density of states $\rho(\epsilon)$. Using eq. (14) we obtain

$$\rho(\epsilon) = \frac{1}{2\gamma\pi}\left(\epsilon_c^2 - (\epsilon-\epsilon_n)^2\right)^{1/2}. \tag{15}$$

Eq. (15) coincides with the result obtained in the generalized Born approximation [7]. This result is valid when n is large. If $|\epsilon - \epsilon_n| > \epsilon_c$, Rep_0 from eq. (14) is equal to zero. In this case the solution does not contribute to the density of states. In order to obtain a non zero contribution one should study solutions of eq. (13) depending on coodinates.

Let us represent the matrix P (9) in the form

$$P = \mathrm{w} \begin{pmatrix} p & 0 \\ 0 & ip_1 \end{pmatrix} \bar{\mathrm{w}}. \tag{16}$$

A non trivial solution can be found for the variable p_1. Assuming that w and p in eq. (16) do not depend on coordinates and representing $p_1(r)$ in the form $p_1(r) = p_{10} + q(r)$ where $p_{10} = -\frac{1}{2}(\epsilon - \epsilon_n) + \frac{1}{2}\left((\epsilon - \epsilon_n)^2 - \epsilon^2\right)^{1/2}$ one can reduce eq. (13) to the form (for more details look [11]).

$$\Delta y - y + y^2 = 0 \tag{17}$$

where $q = q_0 y(r/r_0), q_0 = 2\epsilon_c \left(2\frac{|\epsilon - \epsilon_c|}{\epsilon_c}\right)^{1/2}, r_0^2 = \left(n + \frac{1}{2}\right)\left(2\frac{|\epsilon-\epsilon_c|}{\epsilon_c}\right)^{1/2} a_H^2$.

Nontrivial solutions y_0 of eq. (17) have the "instanton" form [9]. They are finite for all r and have the following asymptotics at $r \to \infty$

$$y_0(r) = e^{-r} r^{-1/2}. \tag{18}$$

In order to calculate the integral (10) one should expand the functional F[p] near extrema determined by a solution of eq. (13) and integrate over small deviations.

The integration near the instanton is performed using standard methods. The fact that one should integrate over supermatrices does not make the calculations more difficult. As the result we obtain for the density of states $\rho(\epsilon)$

$$\rho(\epsilon) = \frac{c}{\epsilon_c}\frac{|\epsilon - \epsilon_c|}{\epsilon_c} \exp\left(-4(2n+1)\frac{\epsilon - \epsilon_c}{\epsilon_c} f\right) \tag{19}$$

where $f = \int y_0^3(\xi) d\xi$, c is a number of the order of unity.

Eq. (19) is valid provided the following inequalities are fulfilled

$$\frac{1}{n} < \frac{|\epsilon - \epsilon_c|}{\epsilon_c} < 1. \tag{20}$$

The inequalities (20) can be satisfied provided n is large.

For n = 0 one must use another saddle point directly in the Lagrangian (7). Such a procedure was done in [10,11].

4. Density-density correlation function

The calculation of the density-density correlation function and of the conductivity can be reduced to the calculation of the quantity $K(r, r', \epsilon, \omega)$ (5). Just as in the previous chapter let us decouple the term $\gamma/2(\bar{\Phi}\ \Phi)^2$ in L_2 using the integration over supermatrices P. Now the size of the supermatrices Q is double the size of the supermatrices P. After standard transformations [5] and calculation of the integral over Φ we reduce the correlation function $K(r, r', \epsilon, \omega)$ to the form

$$\begin{aligned} K(r, r', \epsilon, \omega) &= \int g^{12}_{\alpha\beta}(r, r) k_{\beta\beta} g^{21}_{\beta\alpha}(r'r') \exp(-F_2[Q]) DQ, \\ F_2[Q] &= Tr \int \left[\frac{Q^2}{2\gamma} - ln\left(\tilde{H}_0 + iQ + \frac{\omega + i\delta}{2}\Lambda\right)\right]. \\ \tilde{H}_0 &= H_0 - \frac{\omega}{2}, k = \begin{pmatrix} 1 & 0 \\ 0 & -1 \end{pmatrix}. \end{aligned} \tag{21}$$

The Green function g in eq. (21) is the solution of the equation

$$\left(\tilde{H}_0 + \frac{\omega\Lambda}{2} + iQ(r)\right) g(r,r') = \delta(r-r'). \tag{22}$$

At $\omega = 0$ the saddle point expression for Q can be represented in the form

$$W = U \begin{pmatrix} p^{11} & 0 \\ 0 & p^{22} \end{pmatrix} \bar{U}, \text{where} \bar{U}\ U = 1. \tag{23}$$

In eq. (23) the matrices p^{11} and p^{22} satisfy eq. (13) (p^{11} corresponds to the solution of eq. (13) with the sign plus, p^{22} to the one with the sign minus). The parametrization of the matrices U can be found in the review [5].

The subsequent procedure of calculations in the limit of low frequencies is as follows: we substitute eq. (23) for Q into eq. (21), and integrate over p^{11}, p^{22} near the saddle points thus reducing the calculation to the integration over U. This procedure is analogous to the one used in [5]. However, now p^{11} and p^{22} at the saddle point depend on coordinates. In the region determined by (20) just as in the previous chapter the one instanton approximation is applicable. Two possibilities exist: either $p^{11} = p, p^{22} = p_0$ or $p^{11} = p_0, p^{22} = p$, where p contains one instanton, p_0 does not contain instantons. Integrating in (21) over p we fix first a position R_0 of the instanton thus reducing the integral (21) to an integral over R_0 and U.

After integration over the supermatrices U one can obtain finally

$$K(r,r',\epsilon,\omega) = \frac{\rho(\epsilon)}{-i\omega b^2}\phi(r-r') \tag{24}$$

where $\phi(r-r') = \int q(r-R_0)q(r'-R_0)d^2R_0, b = \int q(r)d^2r$.

In eq. (24) q is the instanton solution determined in eq. (17), $\rho(\epsilon)$ is specified by eq. (19).

It follows from eq. (24) that the correlation function $K(r,r',\epsilon,\omega)$ falls off exponentially at large distances. The form of the correlation function K proves the localization of wave functions.

In the considered limit of low frequencies the correlation function (24) does not change its form in a many instanton approximation. The presence of ω in the denominator of the expression for K is a consequence of the existence of the rotational zero mode of a system of instantons. The existence of this zero mode is due to the invariance of F[Q] with respect to rotations in the space of matrices U and is not related to the number of instantons.

5. Longitudinal and Hall conductivities

For the calculation of conductivity let us use eq. (6). Using the commutation relations $v_\alpha = i[H, r_\alpha]$ and the expansion (4) we reduce the expression for $\sigma_{\alpha\beta}$ (6) to the form

$$\begin{aligned} \sigma_{\beta\alpha} = & - \frac{\omega e^2}{\pi V}\int r_\alpha(r_\beta - r'_\beta)[(n(\epsilon)-n(\epsilon-\omega))K(r,r',\epsilon,\omega) - \\ & - n(\epsilon)M(r,r',\epsilon,\omega)]drdr'd\epsilon \end{aligned} \tag{25}$$

where $K(r,r',\epsilon,\omega)$ is defined in eq.(5) and $M(r,r',\epsilon,\omega)$ is equal to

$$M(r,r',\epsilon,\omega) = < G^R_{\epsilon+\omega}(r,r')G^R_\epsilon(r',r) > - < G^A_\epsilon(r,r')G^A_{\epsilon-\omega)}(r',r) > . \tag{26}$$

Let us consider first the longitudinal conductivity σ_{xx}. The main contribution comes from the first term in (25). Using the eq. (24) for the correlation function K we can reduce eq. (25) for the longitudinal conductivity to the form

$$\sigma_{xx}(\omega)=\frac{i\omega e^2\rho(\epsilon_F)}{4\pi b^2}\int\phi(r)r^2dr=i\omega e^2\rho(\epsilon_F)r_0^4c \tag{27}$$

where r_0 is specified in (17), c is a numerical factor of the order of unity.

In order to estimate the contribution of the second term in eq. (25) at $\omega\to 0$ one can use the same method as for the calculation of $K(r,r',\epsilon,\omega)$. It is not very difficult to see that the second term has an additional factor of the order of $\left(\frac{\epsilon-\epsilon_c}{\epsilon_c}\right)^{1/2}$ which is small provided the condition (20) is fulfilled.

At first glance it seems from eq. (25) that at $\omega=0$ the Hall conductivity must also turn to zero. However, such an assertion would be erroneous because near a boundary of the sample the correlation function $K(r,r',\epsilon,\omega)$ is no longer determined by eq. (24). Due to the existence of extended states in these regions the correlation function $K(r,r',\epsilon,\omega)$ can decrease much more slowly (at the distances $|r-r'|$ of the order of ω^{-1}) and need not be symmetric with respect to the replacement $r\to r'$). Correspondingly the integration in eq. (25) with the multiplier $r_x(r_y-r_y')$ can not be reduced to an integration with the multiplier $\frac{1}{2}(r_x-r_x')(r_y-r_y')$. Such a reduction is possible only for the longitudinal conductivity. When calculating the Hall conductivity an additional term with the multiplier $\frac{1}{2}(r_xr_y'-r_yr_x')$ appears. It can give an additional factor of the order of the size of the sample when estimating eq. (25). Therefore the relative contribution of the edge extended states does not tend to zero with increasing the size of the system. Correspondingly the Hall conductivity given by the eq. (25) need not be zero when $\omega\to 0$ and $V\to\infty$.

In order to carry out calculations in an explicit form let us single out the contribution $\sigma_{xy}^{(0)}$ of the configuration without instantons. Then eq. (25) takes the form

$$\sigma_{xy}=\sigma_{xy}^{(0)}+\Delta\,\sigma_{xy} \tag{28}$$

The quantity $\Delta\sigma_{xy}$ can be also expressed in the form of an integral multiplied by ω. However, this integral converges already even in the limit $\omega\to 0$. Therefore $\Delta\sigma_{xy}$ vanished in this limit. The term $\sigma_{xy}^{(0)}$ coincides with the Hall conductivity of a pure system in a gap region.

The influence of impurities results only in a shift of the energy ϵ. It is well known that the Hall conductivity of the pure system in the gap region is quantized [12]. Using the arguments presented before we can conclude that in the region of applicability of the instanton approximation the Hall conductivity of the system with impurities is also quantized and can be written in the form

$$\sigma_{xy}=\frac{e^2}{2\pi}\sum_{n=0}^{\infty}\theta\left(\epsilon_F-(n+\frac{1}{2})\omega_c\right) \tag{29}$$

where $\theta(x)=\begin{cases}1, & x\geq 0\\ 0, & x<0\end{cases}$.

This result holds for an arbitrary number of instantons and for any shape of the instantons (only a fast decrease of the solution at large distances is necessary).

6. Discussion

The main goal of this lecture is to show that in certain regions of energy all physical quantities can be calculated for the model of a two dimensional metal in a strong magnetic field. These regions correspond to so called tails of the density of states, where this quantity is exponentially small. The method of integration over supermatrices is applied. As usual one should search for saddle points and integrate near them. If the Fermi energy is in the region of Landau levels with large n this sadlle point approximation works very well.

In regions where the density of states is large one can derive the σ-model with the Θ-term [13]. On the basis of this model renormalization group arguments were proposed [13, 14]. It is clear that these arguments are not applicable in the considered region of tails of the density of states. However, as it is seen from previous chapters, this region is more simple for an analytical study.

The density of states is not zero in the region under consideration. At the same time the Hall conductivity has steps as a function of energy. Therefore the dependence of the Hall conductivity on the density of carriers has plateaus. The longitudinal conductivity is zero in these regions. Of course these facts agree with experimental results. A study of regions where the density of states is large is much more difficult and this problem is far from to be solved. Apparently new ideas are necessary.

Using the instanton approximation one can calculate not only the static Hall conductivity but also the Hall conductivity at finite frequencies provided the frequency is not very high. This calculation shows [15] that the Hall conductivity can be expanded in ω for small ω. The first correction to the static Hall conductivity is proportional to ω^2. Using the result of the calculation one can estimate how the form of the Hall steps is effected by the frequency. Characteristic frequencies which change the form of the steps considerably correspond to the ones observed experimentally.

References

1 K. von Klitzing, G. Dorda, M. Pepper, Phys. Rev. Lett. **45**, 494 (1980)

2 R.B. Laughlin, Phys. Rev. B**23**, 5632 (1981)

3 H. Aoki, T. Ando, Solid State Commun. **38**, 1979 (1981)

4 W. Brenig, Z. Wysokinski, Phys. B **63**, 149 (1986);
M. Johnson, Phys. Scr. **32**, 435 (1985);
Qian Niu, D.J. Thouless, J. Phys. A **17**, 2453 (1984);
J. Avron, R. Seiler, Phys. Rev. Lett. abf54, 259 (1985);
R. Tao, F.D.M. Haldane, Phys. Rev. B **333**, 3844 (1986);
J. Hajdu, M. Janßen, O. Viehweger, Z. Phys. B **66**, 433 (1987);

5 K.B. Efetov, Adv. in Phys. **32**, 53 (1983)

6 A.J. Mc Kane, Phys. Lett. A **76**, 33 (1980);
G. Parisi, N. Sourlas, J. Phys. Paris **41**, L 403, (1980)

7 T. Ando, Y. Uemura, J. Phys. Soc. Japan **36**, 959 (1974)

8 K.B. Efetov, V.G. Marikhin, Phys. Rev. B **40**, 12126 (1989)

9 J. Langer, Ann. of Phys. **41**, 108 (1967)

10 L.B. Ioffe, A.I. Larkin ZH. EKSP Teor. Fiz. **81**, 1048 (1981); Sov. Phys. JETP **54** 556 (1981)

11 I. Affleck, J. Phys. C**17**, 2323 (1984)

12 L. Smrcka, P. Streda, J. Phys. C**10**, 2153 (1977)

13 H. Levine, S.B. Libby, A.M.M. Pruisken, Phys. Rev. Lett. **51**, 1915 (1983)

14 D.E. Khmel'nitskii, Pis'ma ZhETF **38**, 454 (1983)

15 O. Viehweger, K.B. Efetov (to be published)

Part IV

Tunneling and Localization in Low-Dimensional Structures and Superlattices

Coherence and Localization in Superlattices

E.E. Mendez

IBM Research Division, Thomas J. Watson Research Center,
P.O. Box 218, Yorktown Heights, NY 10598, USA

Abstract. The application of an electric field to a semiconductor superlattice along its direction of periodicity induces a Stark ladder of quantum states and a gradual localization of the electronic carriers, both of which have been observed recently. This article reviews how these field-induced phenomena are being used to determine the coherence of the superlattice wavefunctions and to design new electro-optic devices.

1. Introduction

Since the introduction of the superlattice concept [1] there have been many suggestions to exploit the opportunities offered by a potential whose periodicity can be tailored from a few times to hundreds of times the period of natural inorganic crystals. The original proposal already envisioned the possibility of increasing the kinetic energy of electrons high in the miniband of allowed energies. Such a large gain is bound to give rise to negative-differential-resistance effects if electrons get to the middle of the miniband, and it could conceivably lead to Bloch oscillations were the carriers able to reach the miniband edge.

The existence of electronic states extended throughout the superlattice lies at the core of the miniband formation and therefore it is essential to the realization of any novel proposal. In an ideal superlattice those states would be fully delocalized, as a consequence of the resonant coupling among all the identical quantum-well constituents. In a real superlattice, however, even small deviations from perfect periodicity, be it due to layer thickness or composition fluctuations or to a considerable number of impurities or defects, reduce well coupling and cause a certain localization of the electronic wavefunctions. If those deviations become significant enough they can diminish the coupling to the point where the miniband loses its meaning, becoming simply a set of discrete energy levels whose wavefunctions extend only through a few periods. In the limit of very large potential fluctuations, electrons will be confined to individual quantum wells, energetically decoupled from each other.

Although the formation of a superlattice in this electronic sense was explored, and to a certain degree demonstrated, by early transport [2] and optical [3] measurements, these experiments did not provide information on the actual coherence of the superlattice wavefunctions, as determined by their spatial ex-

Springer Series in Solid-State Sciences, Vol. 97
Localization and Confinement of Electrons in Semiconductors
Editors: F. Kuchar · H. Heinrich · G. Bauer

tension. The application of a magnetic field perpendicular to the superlattice direction has shed some light on this important parameter. From magneto-transport [4], magneto-absorption [5,6] and cyclotron resonance [7] results it has been possible to get an estimate of the number of wells "seen" by the electrons. Since the formation of Landau levels in the "perpendicular" configuration requires a coherence length larger than the classical cyclotron radius, the presence in those experiments of Landau-level related features at fields above 4T implied a coherence of about 260Å, corresponding to a few superlattice periods.

It has been only recently that much more quantitative information on this parameter has been obtained, by taking advantage of the gradual localization effects of an electric field on the superlattice wavefunctions [8]. In this paper we review the basic consequences of the application of a field parallel to the superlattice direction and their use in the determination of quantum coherence.

2. A Superlattice in an Electric Field

Semiclassically, in the absence of any scattering, an electron in a state of a miniband of width Δ, subject to an electric field $\mathcal{E}$, would undergo an oscillatory motion of frequency $\nu = e\mathcal{E}D/h$ and amplitude $\Delta/e\mathcal{E}$, where D is the period of the superlattice. Such a behavior, discussed in detail by James [9] and usually referred to as *Bloch oscillation*, leads to a localization length of the electron inversely proportional to the field strength. In a conventional solid, in which D and Δ are, respectively, of the order of angstroms and electron-volts, for any realistic field ($\mathcal{E}<10^6$ V/cm) the electron would be confined to a region of thousands of angstroms, which in most cases is much larger than the mean free path. On the other hand, in a superlattice with a period of 60Å and a bandwith of 60meV, a field of 10^5V/cm would be sufficient to produce electron localization within about one period.

Quantum-mechanically, an electric field breaks the quasi-continuum of the miniband into a discrete energy spectrum formed by a set of levels with equidistant energies, given by $E_0 + n\mathcal{E}D$ ($n=0,1,2,...$), where E_0 is the eigenenergy of a quantum well of the superlattice, considered isolated. Such a ladder of states, normally called a *Stark ladder* (and sometimes referred to as a Wannier-Stark ladder) was first mentioned by James [9] for any periodic system and studied extensively by Wannier [10]. The existence of the Stark ladder in solids was never convincingly demonstrated experimentally, and even the concept remained controversial for a long time [11] because of the simplifications used in early derivations. More recently, the influence of disorder on the Stark ladder has been considered [12], and the presence of multiple electronic bands has been included in describing Stark localization [13].

In superlattices, it has also been predicted for a long time that a strong field would localize electrons within a unit cell [14,15]. The first detailed calculation on a semiconductor superlattice under an electric field, although limited to a structure of only four periods, showed the gradual localization of the electron and hole wavefunctions and the corresponding reduction of their overlap, which should lead to a drastic change in the optical properties of the superlattice [16]. The optical absorption under field has been calculated using a tight-binding formalism, and a blue shift of the absorption edge as well as oscillations periodic with the inverse of the field have been predicted [17].

The field-induced localization can be easily understood from the following simple physical picture. In the absence of a field, the inherent degeneracy of the quantum states in the N identical wells of the superlattice produces resonant coupling among them. If the barriers between them are narrow enough, wavefunction leakage out of the wells becomes substantial and is enhanced by the coupling. We have then a set of N extended wavefunctions and an energy spectrum that for large N forms a quasi-continuum --a miniband.

When a field is applied along the superlattice direction, each well is shifted in energy by $e\mathcal{E}D$ relative to its neighbor, and the degeneracy is lifted. The coupling of a given state is then limited to those of nearby wells and the enhancement of the penetration of the wavefunctions in the barriers diminishes. As a result, their spatial extensions are reduced and the wavefunctions are gradually centered in individual wells. Since the zero-field coupling is given by Δ, states split by the field by a larger energy become decoupled. The spatial extension of the wavefunction can be estimated to be $\Delta/e\mathcal{E}$, which is the same as the classical localization length. In the high-field limit, that is, when $e\mathcal{E}D >> \Delta$, even the coupling between adjacent wells is destroyed and the wavefunctions are localized in isolated wells.

Since Δ is a measure of the superlattice coupling and this depends on the effective mass of the electronic particle, the high-field limit is in general different for electrons and holes. Figure 1 shows the wavefunctions for the two kinds of carriers in GaAs-GaAlAs superlattices, for electric fields up to 200 kV/cm. For the configuration used in this example (40Å wells, 20Å barriers, and

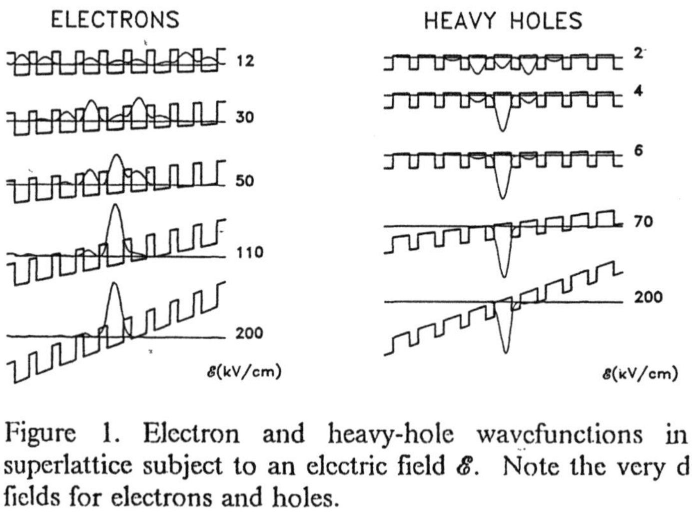

Figure 1. Electron and heavy-hole wavefunctions in a GaAs-GaAlAs superlattice subject to an electric field $\mathcal{E}$. Note the very different localization fields for electrons and holes.

x = 0.35), at 12 kV/cm the electrons are extended throughout almost twenty periods. At 50kV/cm and 110kV/cm the localization lengths are about five and three periods, respectively, and finally at 200kV/cm electrons are fully localized. In contrast, the localization length of heavy holes is just a few periods at 2 kV/cm and they are localized in single wells below 20 kV/cm. The wavefunctions in Fig. 1 were calculated by solving numerically Schrödinger's equation for a 25-well superlattice. The problem of calculating bound states in the presence of an electric field was overcome by placing infinite potential barriers on both ends of the structure.

3. Optical Determination of the Quantum Coherence

The very different field-induced localization lengths of electrons and heavy holes in semiconductors can be used to determine the spatial extension of the electrons at a given field. Let us consider an intermediate field in which electrons are extended for several periods while heavy holes are completely localized, as sketched in Fig. 2. For simplicity we have represented the confined hole states for several wells but only one electron wavefunction, centered in the middle well. The others are similar except that they peak in neighboring wells ($N = \pm 1,2,3,...$). Figure 2 also shows the hole Stark ladder of energy levels and one element of the electron ladder.

The energies for interband transitions between states of the conduction- and valence-band ladders with indices n and p, respectively, are given by [8]

$$E_{np}(\mathscr{E}) = E_0(\mathscr{E}) + (n-p)e\mathscr{E}D \qquad (n,p = 0,\pm 1,\pm 2,...), \tag{1}$$

where E_0 is the intrawell transition energy in the isolated-well limit, which in general depends on the electric field (*Stark effect*). In Fig. 2 we have chosen the origin of indices at the center well, so that the transitions shown there would correspond to $m \equiv n - p$, $n = 0$ and $p = 0,\pm 1,2,3$.

Since the wavefunction overlap between the various holes and an electron is non-zero for as many periods as the electron is extended, the oscillator strength of interband transitions involving those states should be finite and observable in optical experiments such as absorption, reflectance or luminescence. The transitions should exhibit characteristic features: an equal

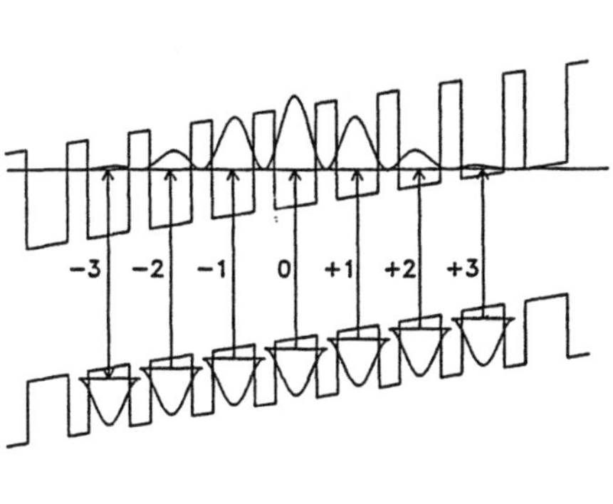

Figure 2. Interband transitions between localized heavy holes and a partially-extended electron whose wavefunction is centered in the middle quantum well. Transition energies are separated by $e\mathscr{E}D$, where D is the superlattice period.

energy separation, proportional to the field, and a field- and index-dependent intensity. With increasing field the intensity of transitions with non-zero index should decrease and eventually disappear, starting with those of the highest index.

The observation of spectral features with these characteristics would be a definitive proof of the existence of the Stark ladder and at a given field the number of observable transitions would measure directly the extension of the electrons. A zero-field extrapolation would then yield the quantum coherence of the superlattice in equilibrium.

4. Observation of the Stark Ladder

The unambiguous demonstration of the formation of the Stark ladder was first provided by photocurrent (PC) measurements in superlattices consisting of 60 periods of 30-Å GaAs wells and 35-Å barriers, in which up to five different transitions were observed [8]. Subsequently, by reducing the width of the barriers to 20Å while keeping the period almost unchanged (60Å), the number of visible transitions increased significantly [18]. For the present purpose, a measurement of the photocurrent is equivalent to measuring the absorption coefficient, in view of the proportionality between the two.

Selected low-temperature (5K) PC spectra for the latter superlattice are reproduced in Fig. 3, for electric fields up to the electron high-field limit. At very low fields the spectra still retain a superlattice-like behavior; the two main peaks correspond to the onset of interband transitions from the top of the heavy- and light-hole minibands to the bottom of the electron miniband. With

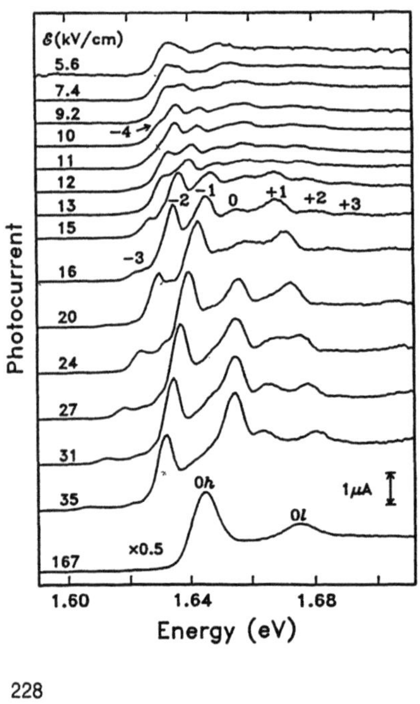

Figure 3. Photocurrent spectra of a 40Å-20Å GaAs-GaAlAs superlattice under a field ℰ. The spectra show the evolution from superlattice to isolated-well behavior with increasing field. At intermediate fields, transitions involving the Stark states are apparent. (After [18].)

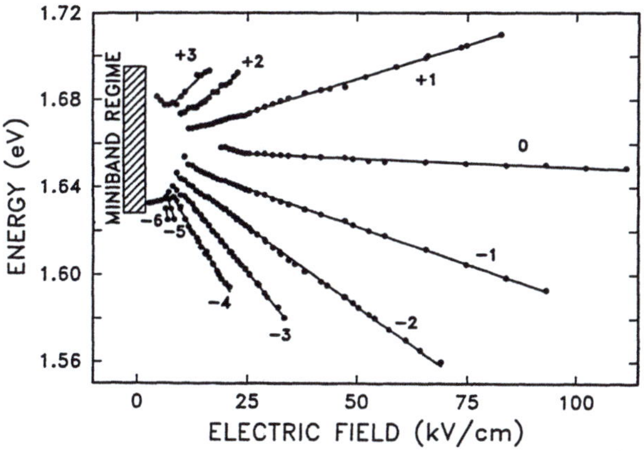

Figure 4. Interband transition energies for a 40Å-20Å superlattice, determined from the peak positions in photocurrent spectra like the ones in Fig. 3. Each branch corresponds to a member of the Stark ladder, labelled by an index *m*. (Adapted from [18].)

increasing fields each peak unfolds into a set of features, some of which shift to higher energies and some to lower energies. Initially strong, those at the wings progressively weaken and ultimately disappear. Finally, at very high fields only two structures remain, one for each group.

The peak energies for the heavy-hole set as a function of electric field are compiled in Fig. 4. (A similar chart was obtained for light holes [18].) At moderate and high fields the energy of each transition, labelled by an index *m*, has a linear dependence on field, with a slope proportional to its index. This characteristic, together with their field-dependent strengths, proves that the transitions indeed involve Stark-ladder levels. The $m = 0$ line represents the transition at which the electron and hole wavefunction overlap is maximum (intra-well transition) and it is the only one that survives at the highest fields. All the others (inter-well transitions) disappear as the electrons become increasingly localized with field.

The observation of up to ten different transitions at low fields gives a *lower* bound of ten periods for the coherence length of the superlattice at $\mathcal{E} = 0$. Although the high positive-index transitions $m = 4,5,6$ were not observed, the symmetry of the Stark ladder implies a minimum coherence of thirteen periods.

The coherence length depends in practice on the superlattice period, decreasing as the latter increases. For an 85Å periodicity the coherence has been determined to be only three periods [18], whereas for 55Å and 36Å, values as high as 17 and 20 periods, respectively, have been reported [19,20]. The unavoidable potential fluctuations of real heterostructures reduce the interwell coupling; however, the effect is much more drastic for long periodicities mainly because of the strong dependence of the coupling on barrier width. For example, the electron miniband width for a 40Å-45Å superlattice is only 10 meV, while for 40Å-15Å it is 95meV. The large coherences that have been observed

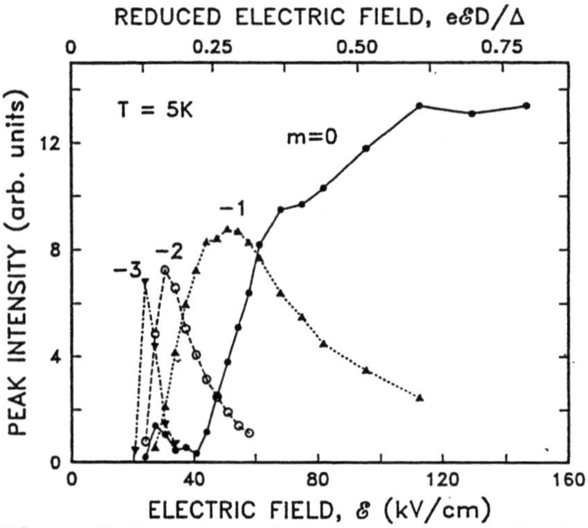

Figure 5. Intensity of the excitonic peaks in the photocurrent spectra of a 40Å-15Å GaAs-GaAlAs superlattice as a function of an external electric field. The inverse of the reduced electric field represents the field-induced localization length, in units of the superlattice period. (After [19].)

are undoubtedly possible by the very good quality of the interfaces formed nowadays with deposition techniques like molecular beam epitaxy. Further advances in this field should result in even larger lengths.

Although the results of Figs. 3 and 4 fit perfectly the behavior expected when a Stark ladder is formed, a careful examination reveals at low fields subtle differences with respect to the simple picture of moderate and high fields. In Fig. 3 the high-index transitions are initially stronger than those with low index. In fact, their amplitudes oscillate before they are finally damped by the field. Both features are common to all the superlattices we have studied. The latter, illustrated in Fig. 5 for a 40Å-15Å superlattice, is a direct consequence of the existence of nodes in the electronic wavefunctions, which change position with field, as seen in Fig. 1.

Another important difference is the deviation at low fields from a linear dependence of the peak position for the $m=0$ structure. (See Fig. 4.) Such nonlinearity results from excitonic effects, which have been ignored so far in this discussion. At $\mathcal{E}=0$ the Coulomb interaction between electrons and holes in a superlattice is quasi-three dimensional since their wavefunctions are extended [21]. However, the field-induced localization (*Stark localization*) brings both types of carriers in proximity and enhances their binding from a three-dimensional to a two-dimensional limit. This increase has been demonstrated experimentally in an 80Å-period superlattice, in which the binding energy of the exciton was enhanced from about 5meV to 12meV [22]. A similar effect is responsible for the nonlinearity of the $m=0$ transition in Fig. 4. As for the interwell excitons, their binding energies gradually decrease with fields since the electrons and holes are increasingly more separated.

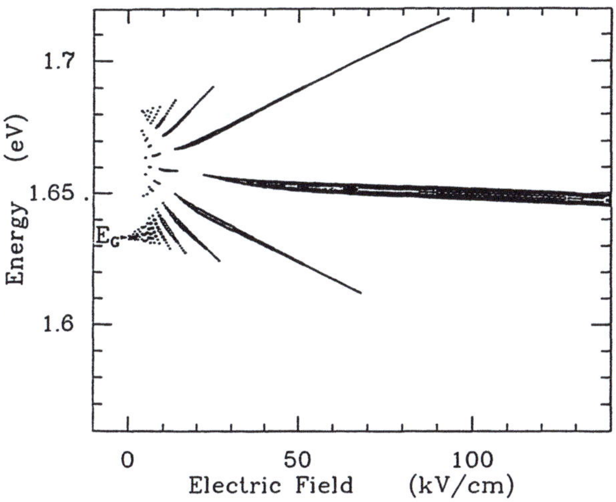

Figure 6. Calculated transition energies for a 40Å-20Å GaAs-GaAlAs superlattice, showing the various branches of the Stark ladder. The calculation, which includes excitonic effects, reproduces very well the experimental results of Fig. 3. The area of the circles is proportional to the oscillator strength of the transition. The oscillatory intensities of the transitions at low fields and the weaker intensities of the branches with positive indices are in agreement with experimental observations. (After [25].)

Several groups have calculated the excitonic binding in superlattices under electric fields [23-25], even for states with large dipole moments [25], making possible a detailed comparison between experimental and calculated absorption spectra. Figure 6 summarizes the computed excitonic transition energies for a superlattice similar to that in the experiment of Fig. 4. The calculations are able to reproduce with high precision not only the details of the measured transition energies but also their relative strengths, indicated in Fig. 6 by the size of the circles representing them. Moreover, these calculations confirm the observation that the absorption strength is larger for states with negative index than for those with positive index. This effect is related to an asymmetry of the electron and hole wavefunctions that decreases their overlap, specially when the Coulomb interaction is considered [25].

The Stark localization of electrons and holes has one more important consequence: the blue shift of a high-field absorption spectrum relative to the zero-field spectrum [17]. In the limit of strong localization the absorption edge is determined by the combined electron and hole confinement energies of isolated wells. On the other hand, in the absence of field the onset of absorption is defined by the energy difference between the top of the valence miniband and the bottom of the conduction miniband. The conduction miniband edge is below the single-well confinement energy by about $\Delta_e/2$, where Δ_e is the miniband width (and analogously for the valence miniband of width Δ_h), and, therefore, the field-induced blue shift is expected to be $\simeq(\Delta_e + \Delta_h)/2$.

The shifts observed experimentally are significant, although smaller than those predicted by that simple argument. Two factors contribute to this reduction: the increase of exciton binding caused by the field, discussed above, and the Stark effect on individual wells [26]. For the spectra of Fig. 3 the blue shift of the PC spectrum at $\mathcal{E} = 167$kV/cm is 12meV, well below the estimated 37meV. As will be mentioned below, although modest, these blue shifts are already finding applications in electro-optic devices. By reducing the superlattice period it would be possible to obtain larger shifts, but at the expense of higher localization fields.

Other optical experiments that have also shown the formation of the Stark ladder include photoluminescence and photoluminescence excitation [8], electroreflectance [27], resonant Raman scattering [28], and absorption [29]. Non-optical techniques, such as deep-level transient spectroscopy [30] and tunneling [31] have also evidenced, to some extent, the Stark states.

5. Temperature Dependence of Coherence

The initial determinations of quantum coherence were done at liquid helium temperatures to minimize thermal processes that could affect it adversely. But it is clear that the main interest of long coherences should reside at 300K, a preferable temperature for most applications. A first study showed that long coherences could be maintained at high temperatures, at least up to 200 K, the highest temperature of that work [18]. Subsequently, coherences of up to five [32] and seven [33] periods were observed at 300K.

Very recently, a detailed investigation between 4K and 300K has evidenced a gradual but moderate decrease of coherence in a 40Å-15Å GaAs-GaAlAs superlattice, from 17 to 9 periods at room temperature [34]. In that experiment, PC spectra similar to those of Fig. 3 were complemented with measurements of the photocurrent-voltage characteristics for various wavelengths of the incident light. The photocurrent is enhanced resonantly whenever the energy of incident photons coincides with that of an interband transition that involves the valence- and conduction-band Stark ladders. Consequently, for fixed energy the current shows an oscillatory behavior with increasing electric field, the period of which is proportional to the inverse of the electric field, in close analogy to the Shubnikov-de Haas oscillations observed in the magnetoresistance of semiconductors.

Figure 7 summarizes the oscillatory characteristics at several temperatures as a function of the voltage applied between p and n GaAs electrodes by which the superlattice was clad. At 5K up to 8 different oscillations can be resolved, corresponding to the $m = -1,-2,..,-8$ interwell transitions, which imply a coherence of 17 $(2m+1)$ periods. With increasing T the oscillations broaden and their intensities are reduced while the background photocurrent increases. As a result, at 292K only four transitions ($m = -1,-2,-3,-4$) are resolved. The entire Stark ladder, obtained from PC spectra and the maxima in the PC oscillations,

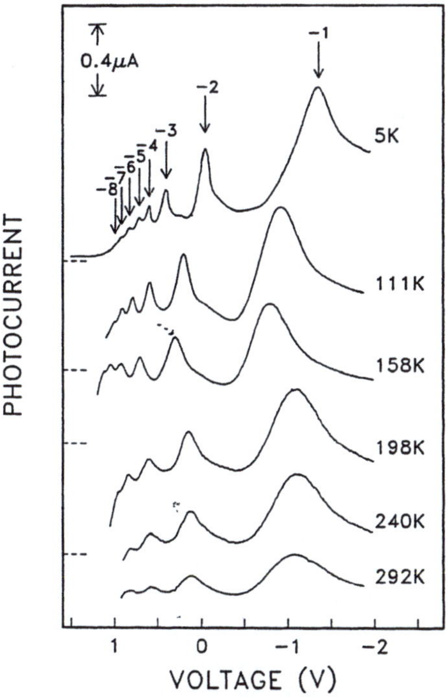

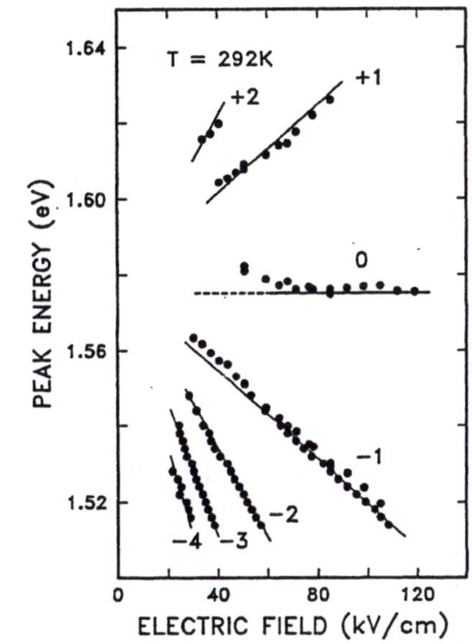

Figure 7. (Left) Oscillatory photocurrent-voltage characteristics, at several temperatures, of a 40Å-15Å superlattice. The incident photon energy was changed for every temperature to compensate for the change in energy gap.

Figure 8. (Right) Transition energies, at room temperature, for the superlattice of Fig. 7. At low fields, energy separations significantly smaller than kT can be resolved.

is mapped out in Fig. 8. The absence of the $m = +3$ and $+4$ transitions is again a consequence of the asymmetry in oscillator strengths.

It should be noted that in Fig. 8 it is possible to resolve transitions separated by $\simeq$15meV, an energy considerably smaller than kT, at fields below 25kV/cm. At these fields, the localization length is limited by Stark localization and not by temperature itself. The coherence at $\mathcal{E} = 0$ may be significantly larger than the nine periods estimated from the observation of Stark-ladder transitions, but at low fields structures below $m = -4$ are smeared out by thermal broadening, the loss of excitonic effects, and a large PC background.

6. Applications of the Stark Ladder

The blue shift in absorption spectra induced by Stark localization and the oscillations in the photocurrent characteristics offer interesting possibilities for opto-electronic applications. The former could be used as a "normally-off" modulator, as proposed by Bleuse et al. [29], the latter for bi-stable or even multi-stable devices [34, 35].

In modulators based on the red shift induced by the Stark effect in single wells, the incident wavelength is selected to coincide with an excitonic transition at $\mathscr{E} = 0$, so that the absorption is maximum. An external field shifts the excitonic peak to lower energies, decreasing the optical absorption. Because the quenching is not complete due to a residual band-to-band absorption, the contrast cannot be very large unless the device operates in the reverse mode, that is, an "off" state in which the field is fixed and the incident wavelength matches the exciton peak, and an "on" state in which the absorption decreases drastically if the field is zero. In contrast, a modulator based in a superlattice would be normally in an "off" state without an applied field, for an incident energy corresponding to the absorption peak. An external field would shift the spectrum towards higher energy, turning the device "on" as it becomes almost transparent to the radiation.

Bar-Joseph et al. have obtained at room temperature changes of up 7700 cm^{-1} in the absorption coefficient of 30Å-30Å GaAs-GaAlAs superlattices at 200kV/cm; 2:1 modulation ratios were predicted at 300 kV/cm [32].

A limitation of a modulator that relies on the blue shift generated by Stark localization is the large operating electric fields it may require. Although with the inherent disadvantage of a normally-on device, it should be possible to make a low-field modulator using the shifts to lower energies of the Stark-ladder transitions ($m = -1, -2, -3, \ldots$). The advantages of such a device are its linearity with voltage and the large energy shift per volt, which should translate in a large contrast at low fields [34].

Several regions of negative differencial resistance like the ones in Fig.7 have also been observed at room temperature in 32Å-9Å GaAs-AlAs superlattices. The positive feedback provided by negative resistance has been exploited to produce optical multi-stability and switching in self electro-optic devices [35].

7. Conclusion

The use of an electric field has been very fruitful in the past to modify the absorption spectra of isolated quantum wells and has created new opportunities in electro-optics. By now it is clear that its effects in superlattices are at least equally interesting for both fundamental and applied physics. The formation of the Stark ladder and the simultaneous electronic localization have allowed direct probing of the spatial extension of electrons in superlattices throughout many periods. The energy shifts induced by the field are beginning to be used in alternative devices for light modulation and switching.

Less progress, however, has been made in our understanding and use of the transport properties of superlattices under fields, an active area of research not discussed here. Some interesting questions still open are the mechanisms of transport at moderate and high fields, when localization is significant, and the possibility of semiclassical effects like negative differential resistance [36] at fields at which the Stark ladder is already formed. But, without any doubt, the most interesting of all is the observability of Bloch oscillations, a concept almost as fascinating for its elusiveness as for its simplicity and elegance.

Acknowledgements

I have enjoyed the collaboration of F. Agulló-Rueda, J. A. Brum, and J. M. Hong in the work on Stark localization and quantum coherence of superlattices, which has been sponsored in part by the Army Research Office. I thank M. Dignam and J. E. Sipe for giving me permission to use Fig. 6.

References

1. L. Esaki and R. Tsu, IBM J. Res. and Develop. **14**, 61 (1970).
2. L. Esaki and L. L. Chang, Phys. Rev. Lett. **33**, 495 (1974).
3. R. Dingle, A. C. Gossard, and W. Wiegmann, Phys. Rev. Lett. **34**, 1327 (1975).
4. L. L. Chang, H. Sakaki, C.-A. Chang, and L. Esaki, Phys. Rev. Lett. **38**, 1489 (1977).
5. R. Dingle, Surf. Sci. **73**, 229 (1978).
6. G. Belle, J. C. Maan, and G. Weimann, Solid State Comm. **56**, 65 (1985).
7. T. Duffield, R. Bhat, Mikoza, F. DeRosa, D. M. Hwang, P. Grabbe, and S. J. Allen, Jr., Phys. Rev. Lett. **56**, 2724 (1986).
8. E. E. Mendez, F. Agulló-Rueda, and J. M. Hong, Phys. Rev. Lett. **60**, 2426 (1988).
9. H. M. James, Phys. Rev. **76**, 1611 (1949).
10. See, e.g., G. H. Wannier in *Elements of Solid State Theory*, p. 190, Cambridge Univ. Press, 1959.
11. See, e.g., J. Zak, in *Solid State Physics*, vol. 27, p. 1, Ed. H. Ehrenreich, F. Seitz, and D. Turnbull, Academic Press, NY 1972.
12. E. Cota, J. V. Jose, and G. Monsivais, Phys. Rev. B**35**, 8929 (1987).
13. D. Emin and C. F. Hart, Phys. Rev. B**35**, 8929 (1987).
14. R. F. Kazarinov and R. A. Suris, Sov. Phys.- Semicond. **6**, 120 (1972).
15. R. Tsu and G. Döhler, Phys. Rev. B**12**, 680 (1975).
16. P. W. A. McIlroy, J. Appl. Phys. **59**, 3532 (1986).
17. J. Bleuse, G. Bastard, and P. Voisin, Phys. Rev. Lett. **60**, 220 (1988).
18. F. Agulló-Rueda, E. E. Mendez, and J. M. Hong, Phys. Rev. B**40**, 1357 (1989).
19. E. E. Mendez, F. Agulló-Rueda, and J. M. Hong, Appl. Phys. Lett. **56**, 2545 (1990).
20. J. W. Little and R. P. Leavitt, Bull. Am. Phys. Soc. **35**, 345 (1990).

21. A. Chomette, B. Lambert, B, Deveaud, F. Clerot, A. Regreny, and G. Bastard, Europhys. Lett. **4**, 461 (1987).
22. F. Agulló-Rueda, J. A. Brum, E. E. Mendez, and J. M. Hong, Phys. Rev. B**41**, 1676 (1990).
23. J. A. Brum and F. Agulló-Rueda, Proc. Int. Conf. Electron. Prop. Two Dimens. Syst., Grenoble, Sept. 1989, to appear in Surf. Sci. (1990).
24. R. H. Yan, F. Laruelle, and L. A. Coldren, Appl. Phys. Lett. **55**, 2002 (1989).
25. M. M. Dignam and J. E. Sipe, Phys. Rev. Lett. **64**, 1797 (1990).
26. E. E. Mendez, G. Bastard, L. L. Chang, L. Esaki, H. Morkoç, and R. Fischer, Phys. Rev. B **26**, 7101 (1982).
27. P. Voisin, J. Bleuse, C. Bouche, S. Gaillard, C. Alibert, and A. Regreny, Phys. Rev. Lett. **61**, 1639 (1988).
28. F. Agulló-Rueda, E. E. Mendez, and J. M. Hong, Phys. Rev. B **38**, 12720 (1988).
29. J. Bleuse, P. Voisin, M. Allovon, and M. Quillec, Appl. Phys. Lett. **53**, 2632 (1988).
30. S. L. Feng, J. C. Bourgoin, and G. G. Qin, Appl. Phys. Lett. **54**, 532 (1989).
31. P. England, M. Helm, J. R. Hayes, J. P. Harbison, E. Colas, and L. T. Florez, Appl. Phys. Lett. **54**, 647 (1989).
32. I. Bar-Joseph, K. W. Goossen, J. M. Kuo, and R. F. Kopf, D. A. B. Miller, and D. S. Chemla, Appl. Phys. Lett. **55**, 340 (1989).
33. K. Fujiwara, Jpn. J. Appl. Phys. **28**, L1718 (1989).
34. E. E. Mendez and F. Agulló-Rueda, IBM Tech. Disclos. Bull., in press.
35. H. Schneider, K. Fujiwara, H. T. Grahn, K. v. Klitzing, and K. Ploog, Appl. Phys. Lett. **56**, 605 (1990).
36. A. Sibille, J. F. Palmier, H. Wang, and F. Mollot, Phys. Rev. Lett. **64**, 52 (1990).

Tunneling Spectroscopy of Low-Dimensional States

J. Smoliner and E. Gornik

Walter Schottky Institut, Technische Universität München,
Am Coulombwall, D-8046 Garching, Fed. Rep. of Germany

We have investigated the tunneling processes between barrier–separated systems of equal and different dimensionality. In two independently contacted 2D–electron gas systems, which are separated by a barrier of only 195 Å, subband resonances are observed as a series of peaks in the derivative of the tunneling current. In transverse magnetic fields, the conservation of the canonical momentum results in a new and giant splitting of the dI/dV peaks. From the resonance peak–splitting the extent of the wavefunctions, even for non–occupied subbands, was determined. In tunneling processes, where 3D electrons tunnel from a Schottky gate contact into quantum wires having different geometrical widths, the 1D subband energies were measured directly. Studying the influence of a magnetic field on the 1D subbands, excellent agreement was found with the predictions of a simple harmonic oscillator model.

Introduction :

Tunneling processes in the presence of additional magnetic fields offer a variety of interesting new effects, especially when low–dimensional states are involved. In double barrier heterostructures, where a two–dimensional system is located inside the barrier, the position of the negative differential resistance is shifted if a magnetic field is applied parallel to the sample [1]. Further, the tunneling current is reduced by an increased effective barrier height [2,3] and the negative differential conductivity region is washed out [4]. Magneto–quantized interface states, corresponding to a transition from 3–dimensional to 2–dimensional states, were investigated both on InP–InGaAs [5,6] and on GaAs–AlGaAs [7–9] single–barrier heterostructures. At extremely high fields, an anticrossing of the energy levels on both sides of a tunneling barrier results in a new series of resonances in the tunneling current [10–12]. Tunneling experiments through one–dimensional and zero–dimensional states are commonly performed on laterally structured samples [13–16]. In lateral double barriers, which were fabricated by a dual gate structure [17,18], tunneling through one–dimensional states is evident as a series of negative differential resistance regions. Due to the small gate widths, these devices might be used for high speed applications. Vertical transport through low–dimensional structures also reflects the existence of quantized states. Quantum dots fabricated on double barrier structures showed a series of negative differential resistance regions due to resonant tunneling processes between the 0D subbands [19,20].

We have studied the tunneling processes between two barrier separated two–dimensional electron gas systems and transitions between a 3D metal contact and 1D–quantum wires. If a transverse magnetic field is applied to the 2D–2D sample, we observe a tremendous splitting of the subband resonance peaks, being more than one order of magnitude larger than the cyclotron energy $\hbar\omega_c$ [21]. To investigate 3D–1D tunneling processes [22], quantum wires having different geometrical widths were fabricated on GaAs–AlGaAs modulation doped heterostructures. Using tunneling spectroscopy, we directly determine the magnetic field dependence of the 1D–subband energies.

Springer Series in Solid-State Sciences, Vol. 97
Localization and Confinement of Electrons in Semiconductors
Editors: F. Kuchar · H. Heinrich · G. Bauer

2D–2D tunneling processes in transverse magnetic fields :

The 2D–2D samples consist of an unintentionally p–doped GaAs layer grown on a semiinsulating substrate ($p<1\cdot10^{15}cm^{-3}$), followed by an undoped spacer (d=50 Å), doped AlGaAs (d=45 Å, $n=4\cdot10^{18}cm^{-3}$), another spacer (d=100 Å) and n–doped GaAs (d=800 Å, $n=1.2\cdot10^{15}cm^{-3}$). An additional GaAs cap layer was highly n–doped (d=150 Å, $n=6.4\cdot10^{18}cm^{-3}$). The resulting band structure is shown in fig.1.
Thus, we have a system where an accumulation layer and an inversion layer are separated by a barrier of only d_b=200 Å. From Shubnikov–de Haas measurements it was deduced that only one subband is occupied both in the inversion layer and the accumulation layer having, electron concentrations of $n_s^{inv}=6.1\cdot10^{11}cm^{-2}$ and $n_s^{acc}=5.8\cdot10^{11}cm^{-2}$, respectively. The contacts to the inversion layer were formed using a AuGe alloy. For a ohmic contact to the upper channel AuGe was used also, but in this case the AuGe was only slightly diffused into the n–doped GaAs. The GaAs layers around the top contact were removed selectively, yielding independent contacts to both electron gas systems. All measurements of the current–voltage (IV) characteristics and its derivative (dI/dV) were made using a 4–terminal conductance bridge [23] with a modulation frequency of 22 Hz and a modulation voltage of 0.3 mV to achieve a high resolution.

Fig.2(a) shows the IV characteristics of sample 1816/21 (T=4.2 K). Several step–like structures are clearly observed, which are washed out at more negative bias voltages. If a transverse magnetic field is applied to the sample, the current resonances broaden drastically, which is shown in fig.2(b). This effect is much better resolved in dI/dV, where the broadening of the current resonances results in a splitting of the B=0 T dI/dV peaks (fig.3). With increasing magnetic field, the peaks in dI/dV are washed out and the tunneling current decreases considerably.

We first discuss in detail the B=0 T results: The total energy of an electron in a two–dimensional system can be written as $E=E_\perp+E_\|$, where $E_\perp$ is the subband energy and $E_\|=\hbar^2k_y^2/(2m*)+\hbar^2k_x^2/(2m*)$. If the momentum of the electrons parallel to the barrier is conserved during the tunneling process, resonances between the two quantized systems will occur when two states are aligned energetically in $E_\perp$. $E_\|$ will have no influence on the tunneling current. This situation is achieved each time a subband in the accumulation layer matches a subband in the inversion layer. In this case resonances appear in the tunneling current, which are better resolved as sharp peaks in the dI/dV curves (fig.3) due to the well–defined $E_\perp$ distribution in the emitter electrode. Thus, we attribute the peak at V_b=+2mV (inversion layer grounded) to a tunneling process from E_0^{inv} into the lowest subband in the accumulation layer, E_0^{acc}. As only one peak in dI/dV is observed in forward bias, we conclude that only one subband exists in the accumulation layer. At negative bias, the first peak in dI/dV is due to a tunneling process from E_0^{acc} into the first subband in the inversion layer, E_1^{inv}. The subsequent peaks in dI/dV are due to tunneling processes into the higher subbands. If the parallel momentum were not conserved, the $E_\perp$ distribution of the

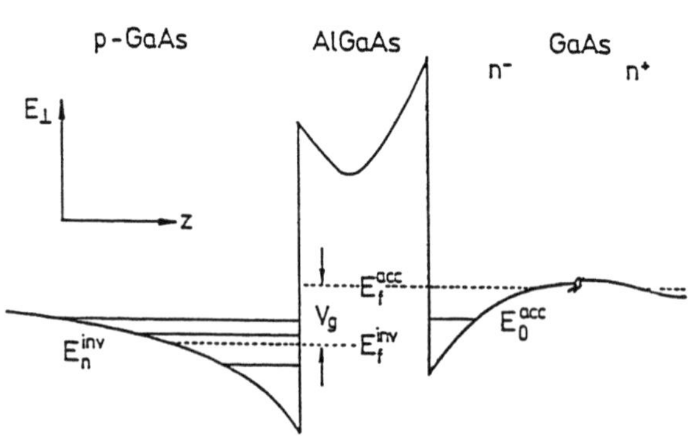

Fig.1: Band structure of a typical sample. E_0^{acc} and E_n^{inv} denote the subbands in the accumulation and inversion layers. E_f^{acc} and E_f^{inv} are the corresponding Fermi levels.

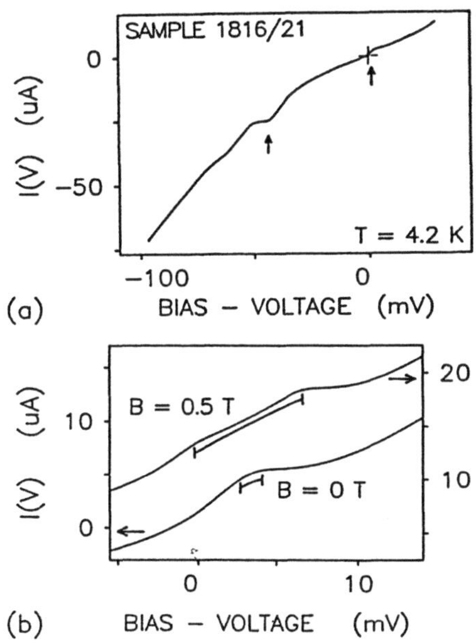

Fig.2(a): IV– characteristics of sample 1816/21 at B=0T. The resonances are marked by arrows. (b): Resonance broadening in transverse magnetic fields. The widths of the resonances are indicated by bars.

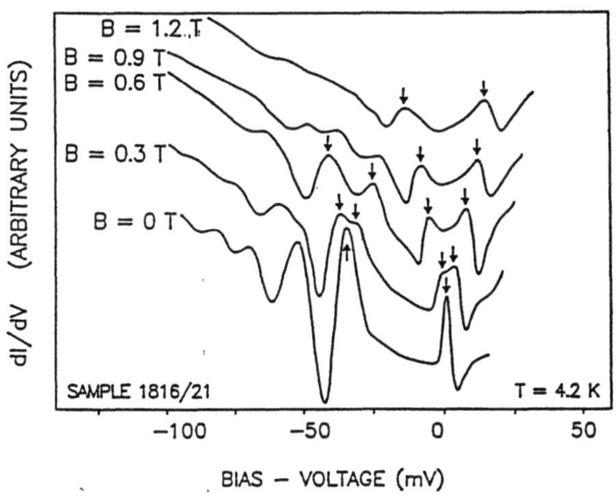

Fig.3: dI/dV curves of sample 1816/21 for various transverse magnetic fields. The arrows indicate the splitting of the B=0 T resonance peaks.

incident electrons would have a width comparable to the Fermi energy, resulting in broad structures in dI/dV. From the experimental data in fig.3, however, we get a value of 8 meV for the linewidth in e·dI/dV of the E_0^{acc}–E_1^{inv} transition. This is much smaller than the Fermi energy E_f^{acc}= 21 meV, which indicates that the k|| conservation is a valid assumption at B=0 T.

In order to gain more information about the momentum conservation rules, a transverse magnetic field is applied to the sample. Classically, the Lorentz force couples the components of momentum in the y– and z–directions, which quantum mechanically is nothing less than the conservation of the canonical momentum. For an electron traveling through the barrier in the z–direction, the wave vector in the y–direction is changed by $\Delta k_y = eBd/\hbar$, where d is the distance traveled. This results in a change of k_z by Δk_z, corresponding to a change in $E_\perp$ by $\Delta E_\perp$. In terms of the total energy, the incident electron is described by $E_\perp + \hbar^2 k_y^2/(2m*) + \hbar^2 k_x^2/(2m*)$. Beyond the barrier, the total energy is conserved and can be written as $(E_\perp + \Delta E_\perp) + (k_y+\Delta k_y)^2\hbar^2/(2m*) + \hbar^2 k_x^2/(2m*)$. Therefore $\Delta E_\perp$ is evaluated as :

$$\Delta E_\perp = (-2\cdot k_y\cdot \Delta k_y - \Delta k_y^2)\cdot \hbar^2/2m^*. \tag{1}$$

Since the wave functions and subband energies in the 2D–systems are only weakly influenced by the occurrence of magneto–electric hybrid states at low magnetic fields, the zero field resonance condition $E_0^{acc}=E_n^{inv}$ can be replaced by :

$$E_n^{inv} = E_0^{acc} + \Delta E_\perp. \tag{2}$$

This means that the resonance condition for a tunneling electron is shifted by $\Delta E_\perp$. Consequently, resonant tunneling now occurs at a bias voltage $V_b+\Delta V_b$, where $\Delta V_b=\Delta E_\perp/e$. From eq.1, it is obvious that $\Delta E_\perp$ is a function of the magnetic field, the length of the trajectory and the value of k_y of the tunneling electrons. Note that the sign of $\Delta E_\perp$ depends on the sign of k_y. If $\Delta E_\perp$ is positive, a tunneling electron will gain k_z and lose k_y, which means that, for this electron, a resonant tunneling process is possible at lower gate voltages than in the case of zero magnetic fields. For negative values of $\Delta E_\perp$ resonant tunneling processes into the n–th subband of the inversion layer can still occur at energies $E_0^{acc} \geq E_n^{inv}$.

Since the electrons in the emitter electrode have k_y values limited by $\pm k_f$, the range of the corresponding $\Delta E_\perp$ values is thus well defined. This results in a broadening of the sharp zero field resonance into a wide but confined resonance range, which is clearly shown in fig.2(b). The double step–like shape of the IV curve is only due to the exponential background signal; both the onset and the end of the resonant tunneling regime are marked by a peak in dI/dV. Thus, the resonance broadening results in a splitting of the dI/dV peaks for increasing transverse magnetic fields. In fig.3 the measured dI/dV curves are plotted for various magnetic fields. The splitting is well resolved for the E_0^{acc}–E_0^{inv} and the E_0^{acc}–E_1^{inv} transitions. Fig.4 shows the positions of the split peaks versus magnetic field. Due to an overlap of the split peaks for higher subbands difficulties appear in assigning peak positions for increasing magnetic fields and the peaks in dI/dV are washed out drastically. This intensity dependence of the individual peaks can be understood classically. When the trajectory of the tunneling particles in high magnetic fields no longer reaches the target electrode, the tunneling current decreases drastically.

In order to verify the conservation of the canonical momentum, the peak positions in dI/dV were calculated using eq.1 for $k_y=\pm k_f$. The distances $d_{0n} = \langle z_0^{acc}\rangle + d_b + \langle z_n^{inv}\rangle$ are used as fitting parameter. The mean value $\langle z\rangle$ denotes the average distance of the electrons from the interface and the indices denote the subband resonances. The d values which give the best fit to the experimental data are d_{00}=350 Å, d_{01}=460 Å, d_{02}=570 Å and d_{03}=650 Å. The corresponding results of $\Delta E_\perp/e=\Delta V_b$ are shown in fig.4 as solid lines. In this fit, the dependence of the wave vectors k_f^{inf} and k_f^{acc} on the applied bias voltage V_b was also taken into account.

At low magnetic fields, Δk_y is much smaller than k_f, and the term $(\Delta k_y)^2$ in eq.1 is small, leading to the linear behavior evident in fig.4. As for the higher subbands the average distance between the electrons and the GaAs–AlGaAs interface is increased, the splitting of the corresponding resonance peaks is larger than the splitting for the transition between the lowest subbands. If the magnetic field increases, the term $(\Delta k_y)^2$ in eq.1 will become important, leading to a nonlinear behavior at higher magnetic fields.

The mean values $\langle z\rangle$ can also be determined theoretically. To check the values of d obtained from the fit, the potential distribution and the eigenfunctions of the system were determined self–consistently [24,25]. All calculations were carried out at B=0 T, since for low magnetic fields the influence of magneto–electric hybrid states can be neglected. Comparing the d_{0n} values obtained from the self–consistent results (d_{00}=340 Å, d_{01}= 450 Å,d_{02}=540 Å, d_{03}=610 Å) with the values from the fit, one gets an excellent agreement. For the lowest two subbands the discrepancy between experiment and theory is only 10 Å; for the higher subbands the theory matches the

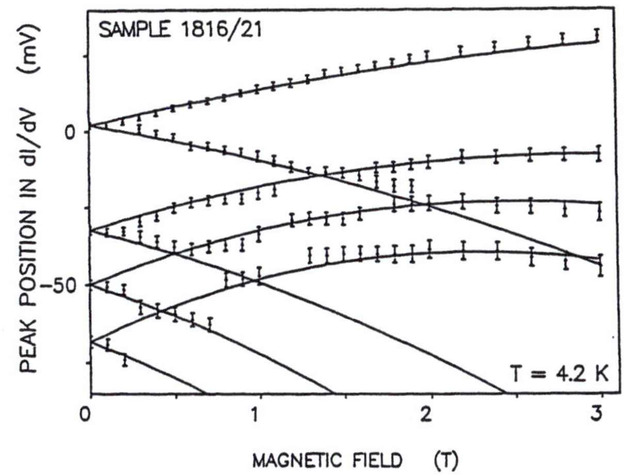

Fig.4: Fan chart of the measured (dots) and theoretically calculated dI/dV peak positions (solid lines).

experiment within 30 Å and 40 Å, respectively. Note that in principle the values of d_{0n} depend on the applied voltage. The numerical results, however, show that this effect is rather small and cannot be resolved within the experimental accuracy. Therefore, the assumption of constant distances d_{0n} is justified when V_b is varied.

Using the fundamental assumption that the canonical momentum is conserved during the tunneling process, we are able to explain the experimental results within a very basic theory. The excellent agreement of the experimental data and the calculated data according to eq.1 indicates that our interpretation is correct. Even the conservation of the component of the wave vector parallel to the applied magnetic field, k_x, is a natural consequence of the presented analysis. The symmetry of the problem implies that, if a magnetic field is applied in the x–direction, k_x is conserved. This leads to the conclusion that at zero magnetic fields, where the problem is completely symmetric with respect to the z–direction, both k_x and k_y are conserved. Thus, the $k_\|$ conservation deduced from the small linewidths of the B=0 T resonance peaks in dI/dV is verified. In addition, our tunneling method not only gives information about the subband energies, but also provides a measurement of the mean values of the electron wave function. From the measured subband energies, the $\langle z \rangle$ values can only be determined by solving Schrödingers equation. The conservation of the canonical momentum, however, for the first time offers a tool to measure the $\langle z \rangle$ values directly, if k_f is known in the emitter electrode.

Tunneling from metal contacts into quantum wires :

To investigate transitions between 3D electrons in a metal contact and 1D electrons, we used conventional modulation doped heterostructures consisting of an unintentionally p–doped GaAs layer grown on a semiinsulating substrate ($p<1\cdot10^{14}cm^{-3}$), followed by an undoped spacer (d=100 Å) having an aluminum concentration of 35 %, and doped AlGaAs (d=100 Å, $n=2\cdot10^{18}m^{-3}$). The additional GaAs cap layer was also highly n–doped (d=200 Å, $n=1.2\cdot10^{18}cm^{-3}$). From Shubnikov–de Haas measurements, the 2D electron concentration n_{2D} was determined to be $3.6\cdot10^{11}cm^{-2}$. Although these sample parameters are favorable for quantum wire fabrication by shallow mesa etching, the mobility (μ=80000 cm^{-2}/Vs) is rather low due to the small spacer thickness and the short distance between the surface and the 2DEG.

Prior to quantum wires fabrication, bar–shaped mesas having an area of 1.5 mm × 0.3 mm were etched into the sample and ohmic contacts were aligned using a AuGe alloy. Then, holographic photoresist–gratings having a period of 1000 nm, 450 nm, and 300 nm were fabricated on the mesas. Finally, these gratings were etched wet chemically into the sample. This procedure enables one to control the wire–etching

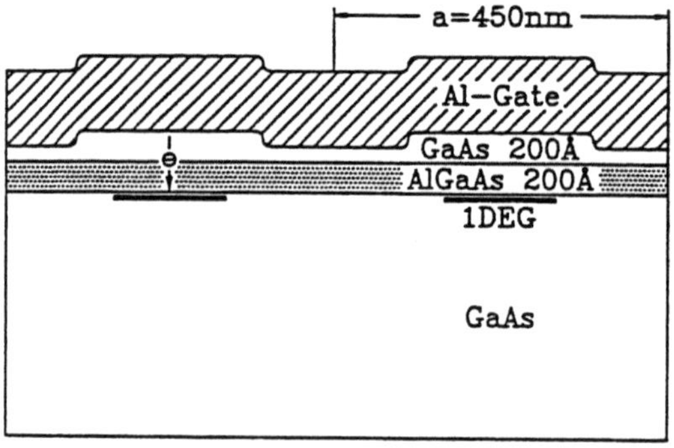

Fig.5: Schematic view of a typical sample.

Fig.6(a): Landau plot for an a=450 nm sample at V_b=0 mV and V_b=–150 mV. (b): Landau plot for samples having different grating periods.

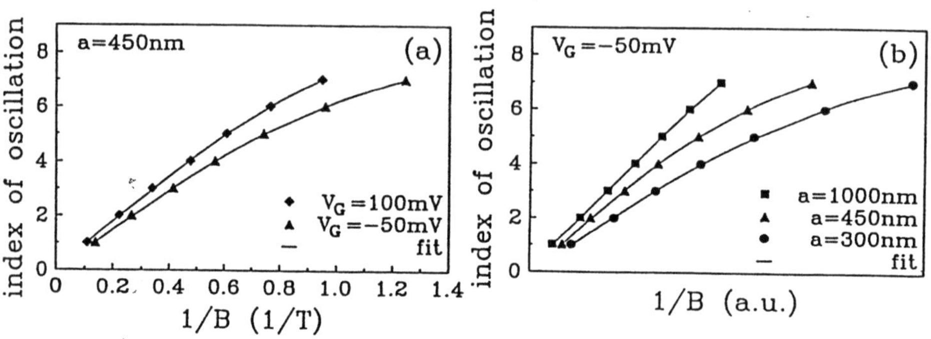

process simply by measuring the sample resistance. Note that, due to the surface depletion, it is not necessary to etch deep into the sample. As a last step, an aluminum gate was evaporated. A schematic view of the sample is shown in fig.5.

To characterize the samples, the differential magneto–resistance (dR/dV_b) was measured as a function of the magnetic field and the applied gate bias voltage V_b. The measurements were carried out at 4.2 K using a drain–source current of 1 μA and a modulation of V_b by 5 mV. This method has the advantage that the influence of the ungated region is eliminated. Fig.6(a) shows a Landau plot of the observed magneto–resistance oscillations for a sample having a grating period of 450 nm at two different bias voltages. Both curves show a clear deviation from a linear behavior at low magnetic fields, demonstrating the existence of one–dimensional subbands. For a gate voltage of V_b=–50 mV the deviation from the linear behavior is considerably larger than in the V_b=+100 mV case. This is consistent with the fact that at more negative bias voltages, the subband energy differences ΔE are larger due to the stronger confining potential. In fig.6(b), a Landau plot is shown for samples having grating periods of 1000 nm, 450 nm and 300 nm at V_b=–50 mV. As expected, the deviation from the linear behavior increases with decreasing grating period.

To extract the 1D electron concentration n_{1D}, ΔE, and the widths w of the conducting channels from the magneto–transport data, a simple harmonic oscillator model $V(x) = 1/2\ m*\ \Omega_0^2\ x^2$ was assumed for the confining potential [26]. The equidistant 1D–subband energy levels in an external magnetic field are given by:

$$E_n = \hbar\Omega\,(n+1/2) + \hbar^2 k_y^2/2m^*(B) \tag{3}$$

with $\Omega^2=\omega_c^2+\Omega_0^2$, $m^*(B)=(\Omega^2/\Omega_0^2)m*$ and ω_c is the cyclotron frequency. Thus, the 1D subbands are separated by $\Delta E=\hbar\Omega$.

The positions of the observed magneto–resistance oscillations due to the depopulation of the 1D subbands in the magnetic field were calculated using a two–parameter (n_{1D}, Ω_0) fit, taking the 1D density of states into account. After

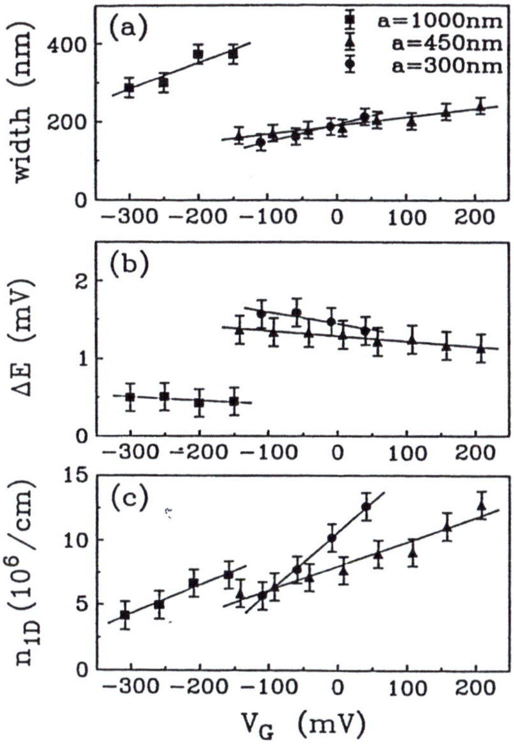

Fig.7(a–c): Dependence of w, ΔE, and n_{1D} on the applied bias voltage for grating periods of 300 nm, 450 nm, and 1000 nm.

Berggren [26,27], the effective width w of the 1D channels is $w=n_{1D}/n_{2D}$. Calculating n_{1D} and n_{2D} from the corresponding density of states one gets [27]:

$$w = 2\pi \, (n_{1D})^{1/3} \left[\frac{2 \;\; \hbar}{3\,\pi \;\; m^* \;\; \Omega_0}\right]^{2/3} . \tag{4}$$

The resulting dependence of the widths of the 1D channels as a function of the applied voltage is shown in fig.7(a). The channel widths decreases slowly at more negative voltages. This can be explained by the fact that the strength of the confining potential is influenced by V_b. Note that, for the 1000 nm sample, 1D–behavior can only be observed if V_b exceeds –150 mV. As expected from the increasing channel widths, the subband energy differences decrease at more positive values of V_b (fig.7(b)). According to Poisson's equation, the electron concentration n_{1D} rises with increasing V_b (fig.7(c)). The different slopes of the curves are due to the variation of the channel width with the applied gate voltage.

To measure the 1D subband energies directly, tunneling experiments were performed, where the electrons tunnel between the 1D channels and the metallic Schottky gate contact, which is indicated schematically in fig.5. All dI/dV measurements were carried out using a 4–terminal conductance bridge [23] at a modulation frequency of 22 Hz to reduce parasitic capacitance effects and a modulation voltage of 0.25 mV to achieve a high resolution. Fig.8 shows the dI/dV curves (T=4.2 K) for an a=450 nm sample at various magnetic fields and the dI/dV curve for an a=300 nm sample at B=0 T. A series of equidistant peaks is evident in all curves, where at B=0T, the period of the observed structures is larger for the 300 nm sample. With increasing magnetic field, the period of the dI/dV structures is also increased. At negative gate voltages, the structures are more pronounced than at positive bias, reflecting the influence of V_b on the confining potential. At very low gate voltages,

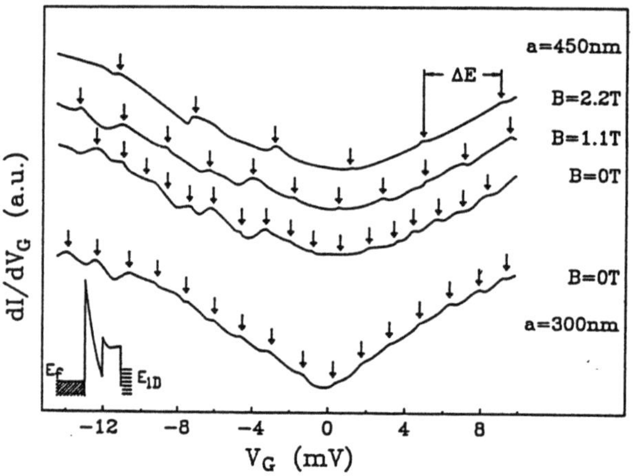

Fig.8: dI/dV characteristics for an a=300 nm sample and an a=450 nm sample at various magnetic fields. The expected resonance spacings are indicated by arrows. All curves are scaled to fit on the same plot.

peaks are difficult to resolve due to the extremely low tunneling current. At too high V_b values, however, the structures in dI/dV are washed out. Tunneling experiments were also carried out on the a=1000 nm samples, but in this case we were not able to detect any structures in dI/dV.

All these effects are well explained by the harmonic oscillator model, assuming that equidistant structures in the tunneling conductance will occur each time the Fermi level in the metal contact crosses a subband in the quantum wires. From fig.7(b) it is obvious that the influence of V_b on the subband spacings is relatively small. Therefore, the 1D subband energy differences ΔE are approximately constant in the small range over which V_b is varied for the tunneling experiments. Thus, the subband spacing ΔE is equal to $\Delta E = e\Delta V_b$. A comparison between the (B=0 T) subband energy differences obtained by tunneling spectroscopy ($\Delta E^{450nm}=1.3 \pm 0.2$ meV, $\Delta E^{300nm}=1.56 \pm 0.2$ meV) and the magneto–transport data ($\Delta E^{450nm}=1.31 \pm 0.18$ meV, $\Delta E^{300nm}=1.47 \pm 0.18$ meV) shows that they agree very well for both samples.

Note, that a V_b variation of 30 mV corresponds to a Fermi energy variation in the range of 1.2 meV. As the typical subband energies are also in the meV range, one can be sure that the observed series of dI/dV oscillations are not due to the filling or depopulation of 1D subbands by the applied gate voltage. This, however, does not exclude that one 1D subband is depopulated at a certain gate voltage, which might explain the nonequidistant peaks at –9 mV in the B=0 T curve of the a=450 nm sample. The small gate voltage variation used in this experiment also makes sure that the potential distribution under the gate contact is homogeneous in the considered voltage range. At higher gate voltages, the tunneling resistance becomes comparable to the quantum wire resistance, resulting in a wash–out of the structures in dI/dV. Due to this effect, it is not possible to determine the number of occupied 1D subbands simply by counting the number of peaks in forward bias. Note, that the observed peaks in dI/dV are weak, but the linewidth of the structures is only in the order of 0.5 meV. In analogy to tunneling processes into 2D subbands, where dI/dV is proportional to the density of states [24], we conclude that the small linewidth is a consequence of the sharp peaks in the density of states in the 1D subbands.

To check the harmonic oscillator model, a magnetic field was applied perpendicular to the sample. As can be seen in fig.8, the spacings between the structures in dI/dV are drastically increased, while the structures tend to wash out at higher magnetic fields. The increased 1D subband energy differences are due to the magnetic contribution to the electrostatic potential (eq.3). As parasitic magneto resistance effects under the gate contact cannot be compensated by the 4–terminal technique, this might be why the resonance peaks are washed out at higher magnetic fields. However, the excellent agreement of the calculated and measured subband energies (see table 1) shows that both the confining electrostatic potential and the

Table 1 : Comparison between the measured and theoretically calculated 1D subband energy differences at various magnetic fields (a=450 nm).

B	ΔE^{450nm} exp.	ΔE^{450nm} calc.
0 T	1.3±0.2 meV	1.3 meV
1 T	2.3±0.3 meV	2.3 meV
2 T	4.0±0.4 meV	4.0 meV

influence of an additional magnetic field is well described by a simple harmonic oscillator model.

Summary :

In summary, we have studied the tunneling processes between states of equal and different dimensionality. In 2D–2D tunneling processes, a giant broadening of the subband resonances is observed in transverse magnetic fields. Through the quantitative explanation of this effect, we verify both the conservation of the canonical momentum and the conservation of $k_{||}$ in the case of zero magnetic fields. In addition, the extent of the wave functions was determined.

On quantum wires, we have investigated the influence of a gate voltage and an additional magnetic field on the 1D subband energies. For the first time, we directly measured the 1D subband energies using tunneling spectroscopy. Excellent agreement was found between the tunneling experiments and the values obtained from magneto–transport measurements. It was shown that both the confining potential and the influence of an additional magnetic field is well described by a simple harmonic oscillator model.

Acknowledgements :

This work was partially supported by "Stiftung Volkswagenwerk" and "Bundesministerium für Forschung und Technologie."

References :

[1] E.E.Mendez, L.Esaki, W.I.Wang, Phys. Rev. B33, 2893 (1986)

[2] L.Eaves, K.W.H.Stevens, F.W.Sheard, Physics and Fabrication of Microstructures and Devices, Springer Proceedings in Physics 13, 343 (1986)

[3] P.Guéret, A.Baratoff, E.Marclay, Europhysics Lett., 3, 367 (1987)

[4] M.L.Leadbeater, L.Eaves, P.E.Simmonds, G.A.Toombs, F.W.Sheard, P.A. Claxton, G.Hill, M.A.Pate, Solid State Electron. 31, 707 (1988)

[5] B.R.Snell, K.S.Chan, F.W.Sheard, L.Eaves, G.A.Toombs, D.K.Maude,G.Hill, J.C.Portal, S.J.Bass, P.Claxton, M.A.Pate, Phys. Rev. Lett. 59, 2806 (1987)

[6] K.S.Chan, L.Eaves, D.K.Maude, F.W.Sheard, B.R.Snell, G.A.Toombs, E.S.Alves, J.C.Portal, S.Bass, Solid State Electron. 31, 711 (1988)

[7] J.A.Lebens, R.H.Silsbee, S.L.Wright, Phys. Rev. B37, 10308 (1988)

[8] E.S.Alves, M.L.Leadbeater, L.Eaves, M.Henini, O.H.Hughes, A.Celeste, J.C.Portal, G.Hill, M.A.Pate, Superlatt. and Microstruc. 5, 527 (1989)

[9] L.Eaves, E.S.Alves, T.J.Foster, M.Henini, O.H.Hughes, M.L.Leadbeater, F.W.Sheard, G.A.Toombs, K.Chan, A.Celeste, J.C.Portal, G.Hill, M.A. Patè, Springer Series in Solid State Sciences Vol.83, 74, (1988)

[10] M.Helm, F.M.Peeters, P.England, J.R.Hayes, E.Colas, Phys. Rev. B39, 3427 (1989)

[11] L.Brey, G.Platero, C.Tejedor, Phys. Rev. B38, 9649 (1988)

[12] P.A.Schultz, C.Tejedor, Phys. Rev. B39, 11187, (1989)

[13] K.Ismail, W.Chu, D.A.Antoniadis, H.I.Smith, J. Vac. Sci. Technol. B6, 1824, (1988)

[14] K.Ismail, W.Chu, D.A.Antoniadis, H.I.Smith, Appl. Phys. Lett. 52, 1071, (1988)
[15] S.Y.Chou, D.R.Allee, R.F.W.Pease, J.S.Harris, Jr., Appl. Phys. Lett. 55, 176, (1989)
[16] K.Ismail, D.A.Antoniadis, H.I.Smith, Appl. Phys. Lett. 55, 589, (1989)
[17] G.Bernstein, D.K.Ferry, Z. Phys. B, 67, 449, (1987)
[18] K.Ismail, W.Chu, A.Yen, D.A.Antoniadis, H.I.Smith, Appl. Phys. Lett. 54, 460, (1989)
[19] J.N.Randall, M.A.Reed, T.M.Moore, R.J.Matyi, J.W.Lee, J. Vac. Sci. Technol. B6, 302, (1988)
[20] J.N.Randall, M.A.Reed, R.J.Matyi, T.M.Moore, J. Vac. Sci. Technol. B6, 1861, (1988)
[21] J.Smoliner, W.Demmerle, G.Berthold, E.Gornik, G.Weimann, W.Schlapp, Phys. Rev. Lett. 61, 2116, (1989)
[22] F.Hirler, J.Smoliner, E.Gornik, G.Weimann, W.Schlapp, submitted to Appl. Phys. Lett.
[23] R. Christanell, J.Smoliner, Rev. Sci. Instrum. 59, 1290 (1988)
[24] J.Smoliner, E.Gornik, G.Weimann, Appl. Phys. Lett, 52, 2136 (1988)
[25] F.Stern, S.D.Sarma, Phys. Rev. B 30, 840 (1984)
[26] K.F.Berggren, T.J.Thornton, D.J.Newson, M.Pepper, Phys. Rev. Lett. 57, 1769, (1986)
[27] K.F.Berggren, G.Roos, H.van Houten, Phys. Rev. B17, 10118, (1988)

Magnetoresistance in One- and Two-Dimensional Lateral Surface Superlattices

D. Weiss

Max-Planck-Institut für Festkörperforschung, Heisenbergstr. 1,
D-7000 Stuttgart 80, Fed. Rep. of Germany

The low temperature magnetoresistance of a high mobility two-dimensional electron gas is dominated by Shubnikov-de Haas oscillations, reflecting the discrete nature of the electron energy spectrum. When a weak one- or two-dimensional periodic potential is superimposed on the two-dimensional electron gas, a novel type of oscillation occurs which reflects the commensurability of the relevant lengths in these systems – the cyclotron orbit diameter at the Fermi energy and the period a of the periodic potential. In addition the electron mean free path l_e also plays a role since the effect is observable only in mesoscopic systems where l_e is significantly longer than the period a of the potential. The essential aspects of these novel commensurability oscillations are sketched. The observed positive magnetoresistance at very low magnetic fields is discussed in terms of magnetic breakdown.

1. Introduction

At low temperatures the magnetoresistance of a degenerate two-dimensional electron gas (2-DEG) exhibits the well known Shubnikov-de Haas (SdH) oscillations reflecting the discrete nature of the degenerate Landau energy spectrum [1]. Using modern lithographic techniques [2] one can superimpose a periodic potential on such a 2-DEG. Now the conduction electrons have to move under the combined influence of a one- or two-dimensional periodic potential and a perpendicular magnetic field B. Such systems have recently attracted some interest since the superimposed periodic potential leads to a novel type of magnetoresistance oscillation periodic in 1/B as long as the period of the modulation is small compared to the mean free path of the electrons [3]. The periodicity of these oscillations is governed by an interesting commensurability problem owing to the presence of two length scales , the period a of the potential and the cyclotron radius R_C at the Fermi energy [3,4]. We have used the persistent photoconductivity effect (PPC) to create one- and two-dimensional periodic potentials in the submicrometer range. A spatially modulated photon flux results in a spatially modulated positive background charge in the Si-doped AlGaAs layer which in turn leads to a modulation in the carrier density. In our measurements a holographic illumination of the heterostructure at liquid helium temperatures is used to produce a periodic potential with a period on the order of the wavelength of the interfering beams, a method first used by Tsubaki et al. [5]. The potential modulation obtained by this technique is on the order of 1 meV where the Fermi energy E_F in our samples is typically 10 meV. Assuming a 1D-periodic potential in x-direction the energy spectrum and the contours of constant energy at the Fermi energy E_F are sketched in Fig.1. Without periodic modulation one obtains the well known parabolic energy dispersion characteristic for a 2-DEG and the contour of constant energy at E_F is a circle with radius k_F. Switching on a weak magnetic field would force the electrons around this Fermi circle corresponding also to a circular motion in real space. The effect of a superimposed periodic potential with period a is that it opens up gaps at the new Brillouin zone boundaries at multiples of $\pm\pi/a$ as is sketched in Fig.1. The existence of gaps at the Bragg planes lead to open orbits in k_x-direction (dashed-dotted line on the right hand side in Fig.1) connected to an electron motion in y-direction in real space. This additional conductivity in y-direction is known to give a positive magnetoresistance ρ_{xx} [6] in x-direction. Increasing the magnetic field increases the probability for an electron

Springer Series in Solid-State Sciences, Vol. 97
Localization and Confinement of Electrons in Semiconductors
Editors: F. Kuchar · H. Heinrich · G. Bauer

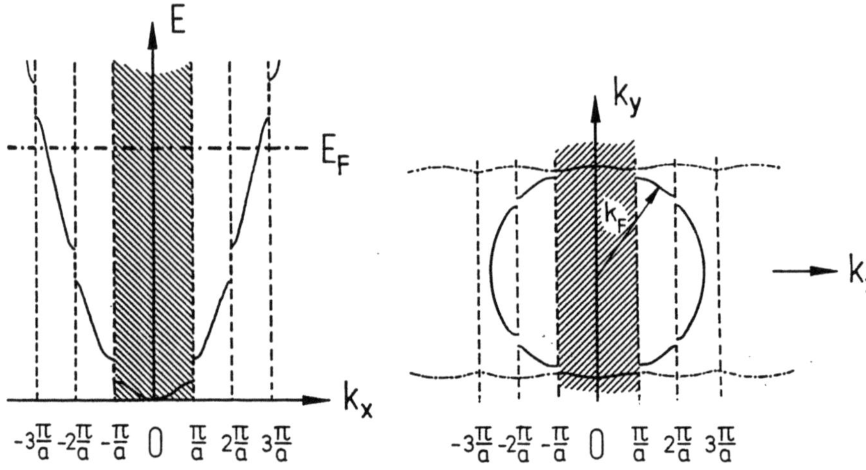

Figure 1: Energy dispersion and contours of constant energy for a weak 1D-periodic potential superimposed on a 2-DEG. The new reciprocal lattice vector is $2\pi/a$ where a is the superlattice period. The first Brillouin zone is hatched. The existence of open orbits at sufficiently low magnetic fields (dashed-dotted line on the right hand side) modifies the low field magnetoresistance.

to tunnel through the classically forbidden region and performing now a circular motion. This effect is known as magnetic breakdown [7]. Under this condition the positive magnetoresistance saturates since the electrons now behave as in a homogeneous 2-DEG. Since we have a weak periodic potential only the gaps between the lowest bands will be significant [8]. For a typical period of $300nm$ and a carrier density $N_S = 2 \cdot 10^{11} cm^{-2}$ we note that we have about 10 bands occupied corresponding to 20 Bragg planes intersecting the original Fermi circle. Therefore we do not expect a significant influence of the periodic potential on the transport properties at zero magnetic field since the Fermi energy is located high in the bandstructure where the dispersion is almost that of a free 2-DEG. The situation changes, however, when a magnetic field is switched on. This will be shown in the next section, where the experiments displaying the novel magnetoresistance oscillations are briefly reviewed.

2. Magnetoresistance Oscillations

The experiments were carried out using conventional AlGaAs-GaAs heterostructures grown by molecular beam epitaxy with carrier densities between $1.5 \cdot 10^{11}$ cm^{-2} and $4.3 \cdot 10^{11}$ cm^{-2} and low temperature mobilities ranging from $0.23 \cdot 10^6$cm^2/Vs to $1 \cdot 10^6$cm^2/Vs. Illumination of the samples increases both the carrier density and the mobility at low temperatures. We have chosen an L-shaped geometry to investigate the magnetotransport properties parallel and perpendicular to the interference fringes. Some of the samples investigated have an evaporated semi-transparent NiCr front gate. A sketch of the experiment exploiting the persistent photoconductivity to periodically modulate the positive background charge in the AlGaAs-layer is shown in Fig.2a and 2b. We used either a HeNe laser ($\lambda = 633$nm) or an Argon-Ion laser ($\lambda = 488$nm). The expanded laser beam entered the sampleholder through a quartz window and a shutter. Two mirrors mounted close to the sample were used to create two interfering plane waves. The advantage of this kind of 'microstructure engineering' is its simplicity and the achieved high mobility of the microstructured sample due to the absence of defects introduced by the usual pattern transfer techniques [2].

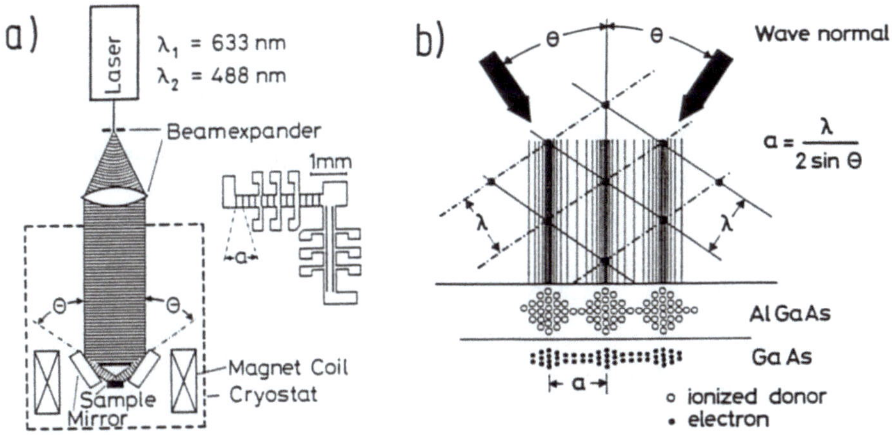

Figure 2: Schematic experimental set up (left hand side) and top view of the L shaped sample geometry where the interference pattern is sketched (a). Sketch of the spatial modulation of the concentration of ionized donors in the AlGaAs layer and of electrons in the 2-DEG produced by holographic illumination using two interfering laser beams with wavelength λ. The interference pattern created is shown schematically (b).

The result of standard magnetoresistance measurements carried out perpendicular ($\rho_\perp = \rho_{xx}$) and parallel ($\rho_\parallel = \rho_{yy}$) to the periodic modulation is shown in Fig.3. In addition to the usual Shubnikov-de Haas oscillations appearing at about 0.5T additional oscillations become visible at even lower magnetic fields. While pronounced oscillations of this new type dominate $\rho_\perp$ at low magnetic fields, weaker oscillations with a phase shift of 180^0 relative to the $\rho_\perp$ data are visible in the $\rho_\parallel$ measurements. No additional structure appears in the Hall resistance. The novel oscillations are, analogous to SdH oscillations, periodic in 1/B as is displayed in the inset of Fig.3. The periodicity is given by the commensurability condition

$$2R_c = (\lambda - \frac{1}{4})a, \qquad \lambda = 1, 2, 3, ..., \tag{1}$$

between the cyclotron diameter at the Fermi level, $2R_c = 2v_F/\omega_c = 2l^2k_F$, and the period a of the modulation. Here $k_F = \sqrt{2\pi N_s}$ is the Fermi wavenumber, $l = \sqrt{\hbar/eB}$ the magnetic length, and $\omega_c = \hbar/m^*l^2$ the cyclotron frequency with the effective mass $m^* = 0.067m_0$ of GaAs. For magnetic field values satisfying Eq.(1) minima are observed in $\rho_\perp$. The periodicity $\Delta(1/B)$ can easily be deduced from Eq.(1)

$$\Delta\frac{1}{B} = e\frac{a}{2\hbar k_F} \tag{2}$$

The validity of Eq.(1) has been confirmed by performing these experiments on different samples, by changing the carrier density with an applied gate voltage, and by using two laser wavelengths in order to vary the period a [3]. To resolve an oscillation an elastic mean free path l_e at least as long as the perimeter of the cyclotron orbit is required. This agrees with our finding that the number of oscillation periods resolved depends on the mobility of the sample: the higher the mobility the more oscillations are observable since the electrons can traverse more periods of the potential ballistically. Consistent experimental results have been obtained by Winkler et al. [9] and Alves et al. [10] using conventionally microstructured samples. The key for the explanation of this novel oscillatory behaviour of ρ_{xx} and ρ_{yy} in such periodically modulated 2-DEG's is a modification of the Landau level spectrum discussed in the next section.

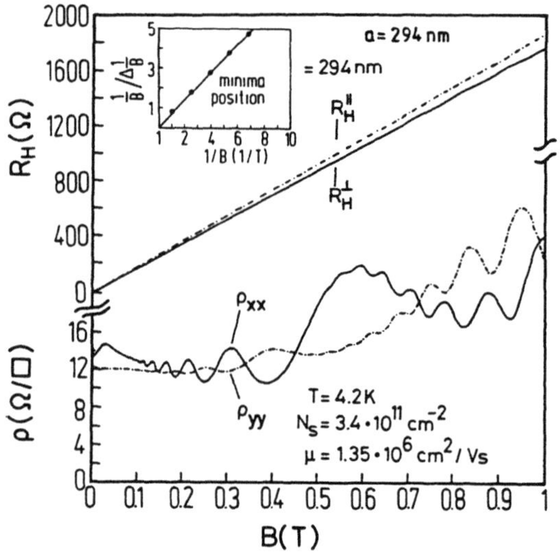

Figure 3: Magnetoresistivity ρ and Hall resistance R_H parallel and perpendicular to the interference fringes. The positions of the minima of $\rho_\perp$ are plotted in the inset demonstrating the 1/B periodicity of the novel oscillations.

3. Origin of the Oscillations

The common origin of both types of oscillations observed is due to a modification of the Landau level (LL) energy spectrum. This has been pointed out by Gerhardts et al. [4] and Winkler et al. [9] who contributed the $\rho_\perp$ oscillations to an additional band conductivity discussed in section 3.2. However, there are other models explaining the $\rho_\perp$ oscillations. Beenakker [11] noticed that these oscillations can be attributed to a guiding center drift resonance. He showed for high temperatures that it is not necessary to start from the LL energy spectrum and to treat the problem quantum-mechanically. In this limit the oscillations in $\rho_\perp$ can be explained semiclassically. Quite recently Streda and MacDonald [12] explained the oscillations by an oscillating probability for magnetic breakdown. However, all the theories above cannot explain the oscillations in $\rho_{\parallel}$. One has to go beyond the constant scattering time approximation as has been pointed out first by Gerhardts [8].

3.1 Landau Levels in a 1D-Periodic Potential

The energy spectrum of electrons subjected to both a magnetic field and a periodic one-dimensional potential has been calculated by several authors [13,14,15] using first order perturbation theory. Starting point is a Hamiltonian of the form

$$H = \frac{1}{2m^*}\left[-\hbar^2\frac{\partial^2}{\partial x^2} + \left(\frac{\hbar}{i}\frac{\partial}{\partial y} + \frac{e}{c}Bx\right)^2\right] + V_0 cos(Kx) \tag{3}$$

containing a periodic potential in x-direction $V(x) = V_0 cos(Kx)$ with period $a = 2\pi/K$. The energy spectrum can be taken in first order perturbation theory in V and is given by

$$E_n(x_0) \approx (n + \frac{1}{2})\hbar\omega_c + \langle nx_0 \mid V(x) \mid nx_0\rangle. \tag{4}$$

The right hand side matrix element (containing the wavefunctions $\mid nx_0\rangle$ of the homogeneous 2-DEG) can be regarded as effective potential acting on an electron averaged over the spatial extent of the wavefunction $\mid nx_0\rangle$ given by $2l\sqrt{2n+1}$ which is equal to the classical cyclotron diameter $2R_c$ for high quantum numbers n. Two extremal situations can be considered. Assuming, for sake of simplicity, a stepfunction-like wavefunction, the matrix element $\langle nx_0 \mid V(x) \mid nx_0\rangle$ at the Fermi energy vanishes if the cyclotron diameter equals an integer of the period a leading to a flat Landau band, independent

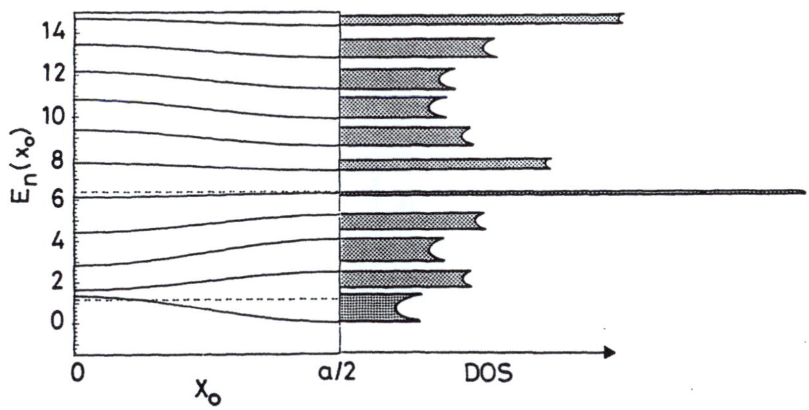

Figure 4: Calculated energy spectrum (in meV, first order perturbation theory) for B=0.8T, $V_0 = 1.5meV$ and a=100nm. The corresponding DOS is sketched. The dashed lines correspond to the flat band situation determined by Eq.(1).

of the center coordinate x_0. On the other hand, a maximum of the matrix element is expected for a cyclotron diameter equal to an odd integer of half the period a leading to Landau bands with strong curvature with respect to x_0. More precisely, the matrix element $\langle nx_0 \mid V(x) \mid nx_0 \rangle$ can be calculated analytically giving

$$E_n(x_0) \approx (n + \frac{1}{2})\hbar\omega_c + U_n cos K x_0 \qquad (5)$$

with $U_n = V_0 \exp(-\frac{1}{2}X)L_n(X)$ where $X = \frac{1}{2}K^2l^2$ and $L_n(X)$ stands for the n-th Laguerre polynomial. $L_n(X)$ is an oscillating function of both its index n and its argument X where the flat band situation is given by $L_n(X) = 0$. This flat band condition can be expressed in terms of the cyclotron radius R_c and is given by Eq.(1) [4]. A typical energy spectrum - calculated in first order perturbation theory is plotted in Fig.4. The corresponding DOS is sketched on the right hand side in Fig.4. Note the double peak structure of the DOS at the band edges which one typically expects for a 1D bandstructure due to the van Hove singularities. These singularities have been recently investigated experimentally [16].

The existence of the shape of the DOS sketched in Fig.4 can be directly proven by magnetocapacitance experiments. The magnetocapacitance C(B) - measured between the semi-transparent top gate and the 2-DEG - depends not only on the thickness of the dielectric layers but also on the thermodynamic DOS [17,18]. When the Fermi energy is located in a LL (maximum DOS) the capacitance displays a maximum; a minimum in C(B) is observed when E_F is between two LL's (minimum DOS). The height of these magnetocapacitance oscillations is directly connected to the thermodynamic DOS and gives therefore information about the LL-width Γ. This has been used previously in homogeneous systems to investigate systematically the LL-width Γ as a function of the electron mobility in such samples [19]. In Fig.5 the magnetocapacitance after an initial holographic illumination (a) is compared with the capacitance measured after an additional illumination which essentially smears out the periodic modulation (b) [20]. In contrast to Fig.5b, where the magnetocapacitance behaves as usually observed in a 2-DEG, the capacitance oscillations in Fig.5a display a pronounced modulation of both the minima and maxima which is easily explained from the energy spectrum plotted in Fig.4. At about 0.69 T (marked by an arrow) the cyclotron diameter at the Fermi energy equals three quarter of the period a and corresponds to the last flat band situation ($\lambda = 1$). Therefore, the magnetocapacitance values near 0.69 T are approximately equal in Fig.5a and Fig.5b. If now the magnetic field is increased, modulation broadened Landau bands are swept through the Fermi level, and cause the nonmonotonic behaviour visible in Fig.5a. At

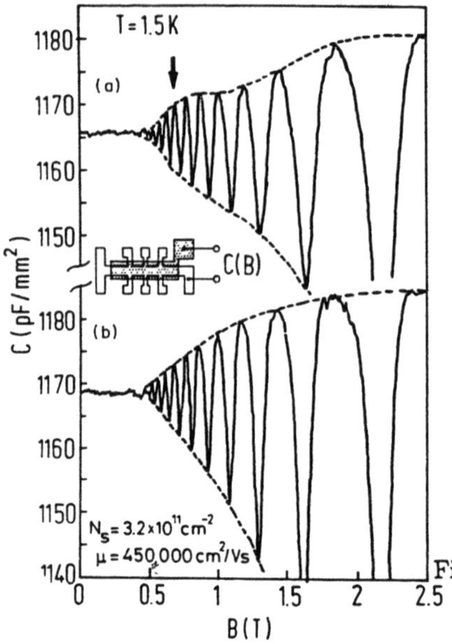

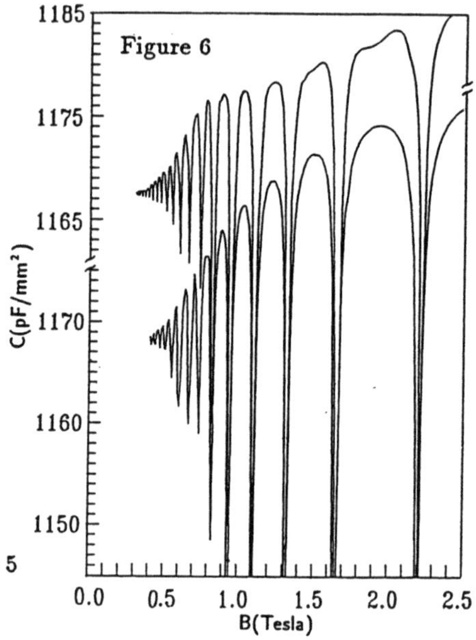

Figure 5: Measured magnetocapacitance (a) of a modulated sample (a=365nm) compared to the capacitance of an essentially unmodulated sample (b). The arrow corresponds to the magnetic field value fulfilling Eq.(1) for $\lambda = 1$.

Figure 6: Calculated magnetocapacitance versus magnetic field for $N_S = 3.2 \cdot 10^{11} cm^{-2}$ and $a = 365nm$. The upper curve is for $V_0 = 0.7meV$, and the lower one for the weak modulation $V_0 = 0.1meV$.

higher magnetic fields the modulation broadening saturates and the usual LL degeneracy again raises the DOS in a LL with increasing field. The discussion of the magnetocapacitance up to now uses qualitative arguments rather than quantitative ones. In order to check the magnetocapacitance data theoretically, microscopic calculations of the DOS based on a generalization of the well known selfconsistent Born approximation [21] have been performed by Zhang [20]. Once the DOS is calculated the data can be easily converted into magnetocapacitance data. Calculating the magnetocapacitance for a modulation amplitude $V_0 = 0.7meV$ (Fig.6) reproduces the experimentally observed modulation of the envelope of the capacitance maxima (Fig.5a). A weaker amplitude of the periodic potential $V_0 = 0.1meV$ leads to the behaviour of the envelope usually observed in homogeneous 2-DEG's. In Fig.6 an intrinsic linewidth broadening $\Gamma = 0.3meV$ has been assumed. The modified energy spectrum which has been proven experimentally in this section is the key for the explanation of the periodic potential induced oscillations given in the next section.

3.2 Oscillations in $\rho_\perp$: Additional Bandconductivity

The theory presented here follows closely the calculations of Gerhardts et al. [4] for a cosine modulation in x-direction. The oscillations in $\rho_\perp$ ($= \rho_{xx}$) can be understood within a simple damping theory which means that electron scattering is described by a constant relaxation time τ. The k_y dispersion of the Landau energy spectrum leads to an additional contribution to the conductivity σ_{yy} which is within the framework of Kubo's formula (see e.g. Ref.[22]) given by

$$\Delta\sigma_{yy} = \frac{2e^2\hbar}{2\pi l^2} \int_0^a \frac{dx_0}{a} \sum_n \left(-\frac{1}{\gamma}\frac{df}{dE}(E_n(x_0)) \mid \langle nx_0 \mid v_y \mid nx_0 \rangle \mid^2 \right) \tag{6}$$

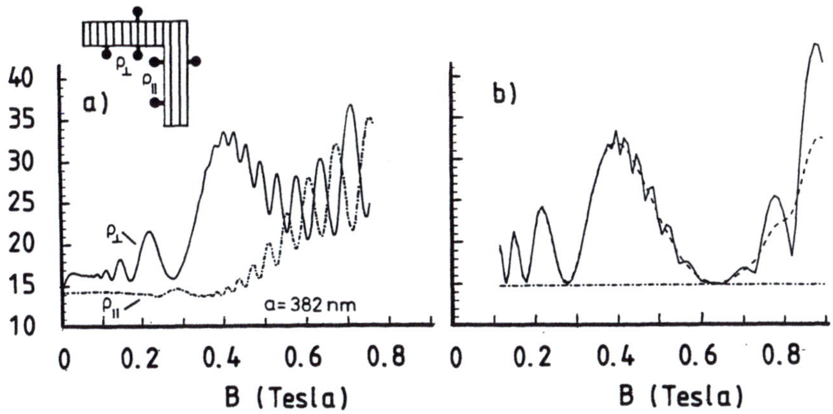

Figure 7: Magnetoresistivities for current perpendicular and parallel to the interference fringes for a sample with $N_S = 3.16 \cdot 10^{11} cm^{-2}$, and $\mu = 1.3 \cdot 10^6 cm^2/Vs$ (a) measured at temperature $T = 2.2K$; (b) calculated for $T = 2.2K$ (solid line) and for $4.2K$ (dashed line) [4].

where $\gamma = \hbar/\tau$, f is the Fermi function, $| nx_0\rangle$ are the eigenstates of Eq.(3) and v_y is the velocity operator in y direction. These eigenstates carry current in the y direction

$$\langle nx_0 \mid v_y \mid nx_0\rangle = -\frac{1}{m^*\omega_c}\frac{dE_n}{dx_0} = \frac{1}{\hbar}\frac{dE_n}{dk_y} \tag{7}$$

but not in x-direction,

$$\langle nx_0 \mid v_x \mid nx_0\rangle = 0, \tag{8}$$

which is the reason for the anisotropic behaviour of σ_{xx} (corresponding to ρ_{yy}) and σ_{yy} (corresponding to ρ_{xx}). In Eq.(7) the modified energy spectrum comes into play. The matrix element $\langle x_0n \mid v_y \mid x_0n\rangle$ vanishes always then when flat Landau bands (see e.g. Fig.4) are located at the Fermi energy, and $\Delta\sigma_{yy} = 0$. Consequently also $\Delta\rho_{xx}$ – the extra contribution to the resistivity ρ_{xx} – vanishes since $\Delta\rho_{xx} \approx \Delta\sigma_{yy}/\sigma_{xy}^2$ if $\omega_c\tau > 1$. On the other hand dE_n/dx_0 displays a maximum value when the Fermi energy is located within the Landau band with the strongest dispersion and therefore $\Delta\sigma_{yy}$ ($\Delta\rho_{xx}$) is at maximum. A calculation based on the evaluation of Eq.6 which is compared to experimental magnetoresistivity data is shown in Fig. 7. Assuming a modulation potential of 0.3 meV in the calculations reproduces nicely the oscillations of ρ_{xx} (solid lines in Fig.7). Note that the weak temperature dependence of the commensurability oscillations in ρ_{xx} is given correctly by the calculation in agreement with the experiment [3]. The temperature dependence of these oscillations is much weaker than that of SdH oscillations, since the relevant energy is the distance between flat bands, which is much larger than the mean distance between adjacent bands. In the high temperature limit ($\hbar\omega_c < kT$) and for high quantum numbers n the expression for the additional conductivity becomes much simpler [9] and can be expressed making use of the fact that $\Delta\rho_{xx} \approx \sigma_{yy}/\sigma_{xy}^2$:

$$\Delta\rho_{xx} \approx \frac{1}{2\pi\hbar}\frac{V_0^2}{\gamma\hbar\omega_c}\frac{4}{ak_F}\frac{B^2}{N_s^2}\cos^2\left(2\pi\frac{R_c}{a} - \frac{\pi}{4}\right). \tag{9}$$

Equation (9) may be used to estimate from the amplitudes $\Delta\rho_{xx}^{max}$ of the commensurability oscillations the amplitude V_0 of the superimposed periodic potential. From the maximum of ρ_{xx} at 0.41T (Fig.7a) one estimates $V_0 = 0.28meV$ in good agreement with the calculation in Fig.7b.

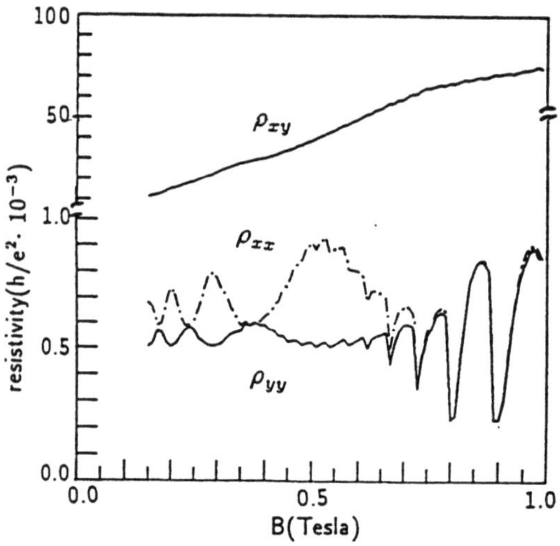

Figure 8: Calculated magnetoresistivity and Hall resistance for T=4.2K, and a=294nm. The oscillations in ρ_{yy} are due to a DOS dependent scattering rate while the oscillations in ρ_{xx} are dominated by the additional bandconductivity (Gerhardts and Zhang [24]).

While the low field oscillations of ρ_{xx} are nicely reproduced by the calculation, the calculated ρ_{yy}-data (dashed-dotted line in Fig.7b) display simply the magnetic field independent Drude result in contrast to the experiment which shows maxima when the Landau bands are flat (DOS maximum at E_F). This is not too surprising since one cannot describe the usual SdH oscillations of a homogeneous 2-DEG within the constant relaxation time approximation; one ends up with the simple Drude result. The same result has been obtained by Beenakker using his semiclassical model [11].

3.3 Oscillations in $\rho_{\|}$: Oscillating Scattering Rate

The results in the previous section have been obtained using the constant scattering time approximation which, however, has no justification. For a homogeneous electron gas in a quantizing magnetic field e.g. it is well known that the scattering time itself depends on the DOS [21]. In order to understand the experimentally observed $\rho_{yy}(=\rho_{\|})$ oscillations one has to go beyond this approximation – in analogy to the description of the SdH oscillations – and consider a density-of-state-dependent scattering rate. In the calculations one has to go through the formalism of the selfconsistent Born approximation [21] using the solutions of Eq.2. A detailed description of this theory has been given by Zhang and Gerhardts [23]. In analogy to the theory of SdH oscillations they find that

$$\sigma_{xx} \propto D_T^2(\mu) = \int dE \frac{df(E-\mu)}{d\mu} D(E)^2 \tag{10}$$

where $D_T^2(\mu)$ is the thermal average of the square of the DOS. It should be emphasized that even for $kT \approx \hbar\omega_c$ where the individual LL's are no longer resolved in ρ_{yy}, $D_T^2(\mu)$ oscillates, reflecting the oscillating DOS sketched in Fig.4. Consequently one can observe low field oscillations in ρ_{yy} even when in this magnetic field range no SdH oscillations can be resolved. From Eq.10 follows that the weak antiphase oscillations in ρ_{yy} are in phase with the density of states oscillations and maxima in ρ_{yy} are always observed when the DOS at the Fermi-energy is at maximum, in contrast to ρ_{xx} which displays minima when the Landau bands are flat since $dE_n/dx_0 = 0$. Calculated curves of ρ_{xx} and ρ_{yy} taken from Gerhardts and Zhang [24] are shown in Fig.8 demonstrating this behaviour in agreement with the experiment (Fig.3). Recent calculations by Vasilopoulos and Peeters [25] show similar results but with much smaller amplitude of the ρ_{yy} oscillations.

4. Positive Low Field Magnetoresistance and Magnetic Breakdown

The low field oscillations in $\rho_\perp$ are accompanied by a positive low field magnetoresistance which saturates below 0.2T. This positive magnetoresistance is absent in $\rho_\parallel$ as is displayed in Fig.3. Applying a negative gate voltage on the semi-transparent top gate one can increase the built in periodic modulation amplitude V_0 [16] which results in turn in an increased positive magnetoresistance step shown in Fig.9. Such a behaviour would be expected for magnetic breakdown, mentioned in the introduction. Open orbits in k_x direction (see Fig.1b or inset Fig.10) give an additional contribution to σ_{yy} and lead to a positive magnetoresistance in ρ_{xx} [6] as long as the electrons can move in open orbits. The probability p for an electron to tunnel from an open orbit (a) to a circular orbit (b) in Fig.10 is given in the literature by $p = exp(-B_0/B)$ with the magnetic breakdown field B_0:

$$B_0 = \frac{\pi m^* E_g^2}{4e\hbar E_F sin2\Theta} \qquad (11)$$

where E_g is the energy gap at the Brillouin zone boundary and Θ the angle defined by $cos\Theta = \pi/ak_F$ [26]. For sufficiently weak periodic potentials we may assume – following Streda and MacDonald [12] – that gaps $E_g \approx V_0$ exist only at the boundary of the *first* Brillouin zone (assuming more a situation sketched in the inset of Fig.10 rather than in Fig.1b). Using Eq.(9) we can extract V_0 from the low field ρ_{xx} oscillations in Fig.9 and compare the calculated B_0 values with those taken from the experiment (the magnetic field where the positive magnetoresistance saturates). The comparison is plotted in Fig.10. The calculated B_0 values are not in perfect agreement with the experimental ones. For higher modulation amplitudes V_0, Eq.(11) overestimates the breakdown field B_0. One should bear in mind, however, that we do not take into account finite temperature effects and the higher gaps sketched in Fig.1b.

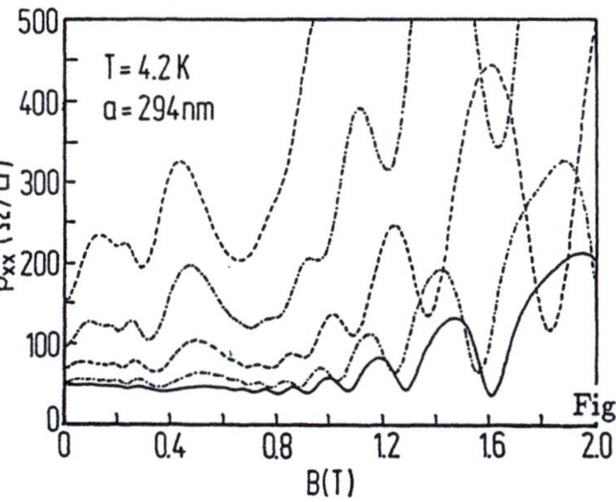

Figure 9

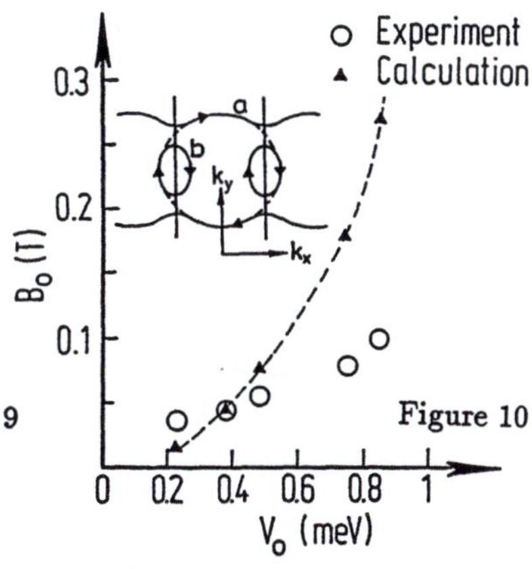

Figure 10

Figure 9: ρ_{xx} for different gate voltages (0mV, -100mV, -200mV, -300mV, -400mV from bottom to top). With increasing negative bias, both the amplitude of the low field oscillations and the magnetic field value where the positive magnetoresistance disappears (taken as B_0) increase

Figure 10: Comparison of calculated (Eq.(11)) and measured (Fig.9) breakdown fields B_0. The inset shows the contours of constant energy at E_F for a weakly modulated 2-DEG in the extended zone scheme.

5. Magnetoresistance in a Two-Dimensional Periodic Potential

In this last section some preliminary results of low field magnetotransport experiments in a two-dimensional periodic potential are presented. In such a potential grid the commensurability problem becomes more severe as compared to the 2-D case and results in a complicated energy spectrum [27]. In a weak 2D-periodic potential the LL spectrum depends on the flux $\Phi = Ba^2$ penetrating one unit cell. If $\Phi/\Phi_0 = q/p$ where Φ_0 is the flux quantum and p/q is a rational number, the 2D-periodic potential splits one LL into q subbands separated by gaps (see e.g. [28]). The LL width ,on the other hand, is modulated by Laguerre polynominals very similar to the 1D case (Eq.(4)), so that the flat band condition for 1D-and 2D-periodic potentials is the same [29]. The two-dimensional periodic potential ($V_0 \ll E_F$) with $a = 365nm$ is created by successively illuminating holographically a high mobility heterostructure ($\mu = 1.2 \cdot 10^6 cm^2/Vs$). Holographic illumination of type (a) in Fig.11 produces additional oscillations in the magnetoresistance due to an additional bandconductivity (dashed-dotted line in Fig.11). An additional holographic illumination where the sample has been rotated by 90° results then in a grid potential sketched in Fig.11 (c). The magnetoresistance obtained under such conditions (solid line in Fig.11) displays a weak oscillating behaviour also corresponding to the commensurability condition Eq.(1), with maxima where ρ_{xx} –measured for situation (a)– shows minima. If one starts with an illumination of type (b) followed by (a) one ends up with the same result. The result obtained for the magnetoresistance in a two-dimensional periodic potential is therefore very close to the result one gets when the current flows parallel to the potential grating, discussed as additional oscillations in ρ_{yy} above (see section 3.3). Therefore one can conclude that the 2D-periodic potential destroys the bandconductivity oscillations triggered by $\langle nx_0|v_y|nx_0\rangle \propto dE_n/dk_y$ for the 1D case. If the collision broadening for the 2D case is small compared to the gaps between the LL subbands the corresponding matrix elements are significantly reduced so suppressing the bandconductivity contribution [29]. Since this seems to be the case in our experiments we can only observe the scattering rate oscillations displaying maxima when the flat band condition Eq.(1) is fulfilled.

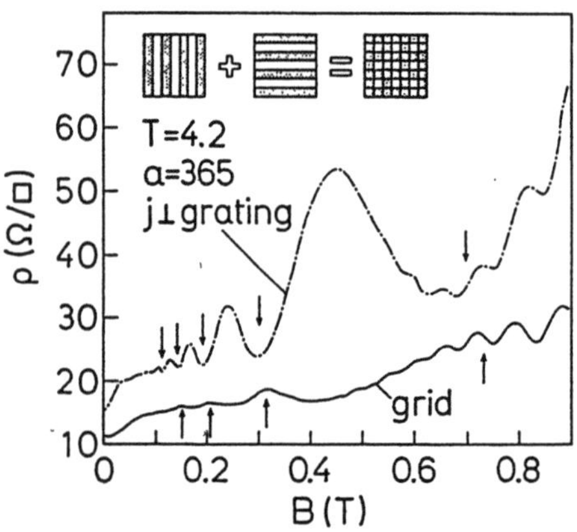

Figure 11: Magnetoresistance in a grating (j ⊥ grating) and grid. The creation of the holographically defined pattern is shown schematically. The arrows correspond to Eq.(1).

I would like to thank K. von Klitzing , R. R. Gerhardts, C. Zhang, U. Wulf, G. Müller, and D. Heitmann for valuable discussions and I am grateful to K. Ploog and G. Weimann for providing me with high quality samples.

6. References

[1] L. Shubnikov, W. J. de Haas, *Leiden Commun.* **207a**, **207c**, **207d**, **210a** (1930)

[2] For recent work in this field see *Physics and Technology of Submicron Structures*, ed. by H. Heinrich, G. Bauer, and F. Kuchar, Springer Series Solid-State Sciences **83**, (1988)

[3] D. Weiss, K. v. Klitzing, K. Ploog, G. Weimann *Europhys. Lett.* **8**, 179 (1989); also in *The Application of High Magnetic Fields in Semiconductor Physics*, ed. G. Landwehr, Springer Series in Solid-State Sciences **87** , 357 (1989)

[4] R. R. Gerhardts, D. Weiss, K. v. Klitzing; *Phys. Rev. Lett.* **62**, 1173 (1989)

[5] K. Tsubaki, H. Sakaki, J. Yoshino, Y. Sekiguchi, *Appl. Phys. Lett.* **45**, 663 (1984)

[6] C. Kittel; Quantum Theory of Solids, Wiley, New York, (1967)

[7] M. H. Cohen, L. M. Falicov; *Phys. Rev. Lett.* **7**, 231, (1961)

[8] R.R.Gerhardts in "Science and Engineering of 1- and 0-Dimensional Semicondcutors", edited by S.P. Beaumont and C.M. Sotomayor Torres (Plenum, London), Proc. NATO ARW, 29 March - 1 April 1989, Cadiz, Spain, to be published

[9] R. W. Winkler, J. P. Kotthaus, K. Ploog; *Phys. Rev. Lett.* **62**, 1177 (1989)

[10] E. S. Alves, P. H. Beton, M. Henini, L. Eaves P. C. Main, O. H. Hughes, G. A. Toombs, S. P. Beaumont, C. D. W. Wilkinson; *J. Phys.: Condens. Matter* **1**, 8257, (1989)

[11] C.W.J. Beenakker; *Phys. Rev. Lett.* **62** , 2020, (1989)

[12] P. Streda, A. H. MacDonald; preprint

[13] G. R. Aizin, V. A. Volkov; Sov. Phys. JETP **60**, 844 (1984) [Zh. Eksp. Teor. Fiz **87**, 1469 (1984)]

[14] H. J. Kelly; *J. Phys. C* **18**, 6341 (1985)

[15] A. V. Chaplik; *Solid State Commun.* **53**, 539, (1985)

[16] D. Weiss, K. von Klitzing, K. Ploog, G. Weimann; *Surface Science* **229**, 88 (1990)

[17] T. P. Smith, B. B. Goldberg, P. J. Stiles, M. Heiblum, *Phys. Rev.* **B32**, 2696 (1985)

[18] V. Mosser, D. Weiss, K. v. Klitzing, K. Ploog, G. Weimann, *Solid State Commun.* **58**, 5 (1986)

[19] D. Weiss, K. v. Klitzing in *High Magnetic Fields in Semiconductor Physics*, ed. G. Landwehr, Springer Series in Solid-State Sciences **71**, 57 (1987)

[20] D. Weiss, C. Zhang, R. R. Gerhardts, K. von Klitzing, G. Weimann; *Phys. Rev.* **B39**, 13030 (1989)

[21] T. Ando, Y. Uemura, *J. Phys. Soc. Jpn.* **36**, 959 (1974); R. R. Gerhardts, *Z. Physik* **B21**, 285 (1975)

[22] R. Kubo, S. J. Miyake, N. Hashitsume; *Solid State Physics* **17**, 239 (1965)

[23] C. Zhang, R. R. Gerhardts; submitted to Phys. Rev. B

[24] R. R. Gerhardts, C. Zhang; *Surface Science* **229**, 92 (1990)

[25] P. Vasilopoulos, F. M. Peeters *Phys. Rev. Lett.* **63**, 2120 (1989)

[26] E. I. Blount;*Phys. Rev.* **126**, 1636 (1962); A. B. Pippard; *Proc. Roy. Soc. (London) A*, **270**, 1 (1962); J. R. Reitz; *J. Phys. Chem. Solids* **25**, 53 (1964)

[27] D. R. Hofstadter; *Phys. Rev.* **B14**, 2239 (1976)

[28] D. J. Thouless; in *The Quantum Hall Effect* ed. by R. E. Prange and S. M. Girvin, Springer (1987)

[29] R. R. Gerhardts, D. Weiss, U. Wulf; unpublished

Extended and Localized States in Fibonacci Quasiperiodic Superlattices: A Probe of the Electronic Coherence

F. Laruelle [1], D. Paquet [2,†], M.C. Joncour [2], B. Jusserand [2], J. Barrau [3], F. Mollot [1], and B. Etienne [1]

[1]L2M, CNRS, 196 Avenue Henri Ravera, F-92220 Bagneux, France
[2]Laboratoire de Bagneux, CNET, 196 Avenue Henri Ravera, F-92220 Bagneux, France
[3]Laboratoire de Physique des Solides, INSA, Avenue de Rangueil, F-31077 Toulouse Cedex, France

Abstract: We investigate physical properties of Fibonacci superlattices (SLs) using a series of successive periodic approximations. The electronic band structure is studied in the envelope function approximation using the recurrence properties of the Fibonacci law. The wavefunction localization can be predicted with the help of an invariant. We establish experimentally the nature of optical transitions as being due to bulk states, which is confirmed via photocurrent spectroscopy in an electric field. For periodic SLs the same experiment leads to a wavefunction spectroscopy where a relatively extended electron state is probed by a Wannier-Stark ladder of localized heavy hole states.

I) INTRODUCTION

The discovery of quasiperiodic order in rapidly quenched Al-Mn alloys and of their three dimensional icosahedral symmetry [1] stimulated a lot of work on the consequences of such a non-translational invariant order for the structural and electronic properties of these solids.

The one-dimensional quasiperiodic case, with the archetypal Fibonacci sequence, was studied a long time ago by theoreticians because of the particular structure of the energy spectrum [2,3] and of the electronic localization [4-6]. Quasiperiodicity, with critical states, is intermediate between periodicity with extended Bloch states and random disorder with exponentially localized states. R. Merlin et al. [7] realized the first quasiperiodic semiconductor using GaAs/GaAlAs superlattices. Using Raman scattering and high-resolution X-ray diffraction they studied their structural properties [8] and paid only little attention to their electronic properties [9].

We use SLs of increasing complexity, which are successive periodic approximations of the quasiperiodic SL, to study the evolution of structural and electronic properties from periodic order to quasiperiodicity. We can also extract from our experimental work an estimation of the interface abruptness and of the electronic coherence and localization lengths. These limitations imply that fractal aspects, which are involved in the theory of quasiperiodic systems [10], remain out of reach in the study of GaAs/GaAlAs SLs. Furthermore we study electric field effects on the electronic structure.

Springer Series in Solid-State Sciences, Vol. 97
Localization and Confinement of Electrons in Semiconductors
Editors: F. Kuchar · H. Heinrich · G. Bauer

II) HOW DO YOU BUILD A FIBONACCI SUPERLATTICE IN REAL SPACE?

Let us consider two different elements A and B and the Fibonacci inflation law [11] : A $\rightarrow$ AB and B $\rightarrow$ A. The sequence S_j is obtained after j applications of this law. Starting with S_{-1} = B we get S_0= A, S_1=AB, S_2= ABA, S_3=ABAAB, S_∞ = ABAABABA... We immediately see that S_{j+1} is the concatenation of S_j and S_{j-1} so that F_j the number of elements in S_j follows the Fibonacci series: $F_{j+1} = F_j + F_{j-1}$ with $F_{-1} = F_0 = 1$. The ratio F_{j+1}/F_j converges toward the golden mean $\tau = (1 + \sqrt{5}) / 2$. A quasiperiodic SL is obtained by stacking along the growth axis elements A and B, made of semiconductor layers, according to S_∞ whereas a periodic approximation of order j is obtained by growing a periodic SL whose unit cell is based on S_j.

Structural properties of periodic approximations and of quasiperiodic SLs can be understood quantitatively using the projection method [12-15]. We have demonstrated that the structure factor has an expression very similar to that for periodic SLs [13]. The rich structures of the diffraction spectra contain a lot of information, including the interface abruptness.[13,14].

III) ELECTRONIC PROPERTIES

Qualitatively simple tight-binding considerations and renormalization arguments can be used to describe band structure and also short and long range aspects of electronic wavefunctions [10,16]. However, GaAs/GaAlAs periodic SLs are well described within the envelope function approximation. Using the transfer matrix method we build a bridge between these two separate pieces of work to get for realistic SLs [17] the quantities required for the fractal analysis and localization theories developed on model Hamiltonians.

1. Transfer matrix analysis

We choose the A and B elements to be made of the same GaAs quantum well (QW) and of a GaAlAs barrier either thicker in B than in A, case α, or higher in B than in A, case β. We preserve a resonant coupling between adjacent QWs and modulate the strength of the tunnel coupling. We solve the Schrödinger equation in the envelope function approximation for electrons moving along the growth axis. The envelope function of an electron in a QW is written as a linear combination of two left and right propagating plane waves. The electron transfer into an adjacent QW is described by a matrix connecting the two coefficients of the linear combination in the originating QW to those in the neighbouring QW. This transfer matrix results from continuity conditions of the envelope function and of its current at the interfaces. For an electron of energy ε moving in a QW of thickness l_w, transferred through a potential barrier of height V_b and thickness L_b, into a QW of identical thickness the transfer matrix can be written as [17]

$$M = \begin{bmatrix} X + i\,Y & -i\,R \\ i\,R & X - i\,Y \end{bmatrix}$$

where X, Y, R are functions of the electron energy and structural parameters [18].

For periodic boundary conditions allowed energy bands are given by $| x_j | \leq 1$ where x_j is the half-trace of M_j. However, the calculation of M_j is not required since we can get x_j from the recurrence relation between S_j [16]:

$$M_j = M_{j-2} M_{j-1} \text{ and } x_j = 2 x_{j-1} x_{j-2} - x_{j-3} .$$

With the knowledge of $x_{-1} = X_B$, $x_0 = X_A$, $x_1 = X_A X_B + R_A R_B - Y_A Y_B$ we can deduce x_j for any order j and therefore the band structure evolution with j. The traces, x_js, are bound by the following relation for any order j:

$$I = -1 + x_j^2 + x_{j-1}^2 + x_{j-2}^2 - 2 x_j x_{j-1} x_{j-2} .$$

I only depends on the energy and on the structural parameters. It can be expressed with the matrix elements of M_A and M_B:

$$I = (Y_A R_B - Y_B R_A)^2$$

and is connected to the commutation properties of M_A and M_B through

$$\left[M_A , M_B \right] = 2 (Y_A R_B - Y_B R_A) \begin{bmatrix} 0 & 1 \\ 1 & 0 \end{bmatrix} .$$

I is the relevant parameter to describe the fractal evolution of the band structure toward the devil's staircase [19]. The main difference with previous tight-binding models is the energy dependence of **I**, which is relevant for characterizing the wavefunction localization [20]. Band structure and wavefunction localization can be understood by considering the invariant energy dependence.

2. Band structure

We display in fig. 1 the band structure evolution with order j in case α. The structural paramaters are A: 28.3 Å GaAs, 31.1 Å $Ga_{0.77}Al_{0.23}As$, B: 28.3 Å GaAs, 67.9 Å $Ga_{0.77}Al_{0.23}As$. At order j = 1 two bands labelled 2 and 4 correspond to the bonding and anti-bonding states of the two coupled QWs of the SL unit cell. At order j = 2 three mini-bands are associated with the bonding (1), non-bonding (3) and anti-bonding states (5). Mini-bands of even and odd index do not overlap in energy. At order j = 3 we get a superposition of the two previous structures. At order 4 odd index minibands split into two while those of even index are not affected: in S_4 the three coupled QWs pattern is doubly degenerate and the corresponding mini-bands split into bonding and anti-bonding states. The hierarchy of two and three coupled QWs in S_j is the reason for the band structure evolution toward the devil's staircase as j increases [16]. The invariant **I** slowly varies over the energy range of the entire spectrum and is always large (fig. 1). There is a direct link between the band structure (reciprocal space) and the wavefunction localization (real space).

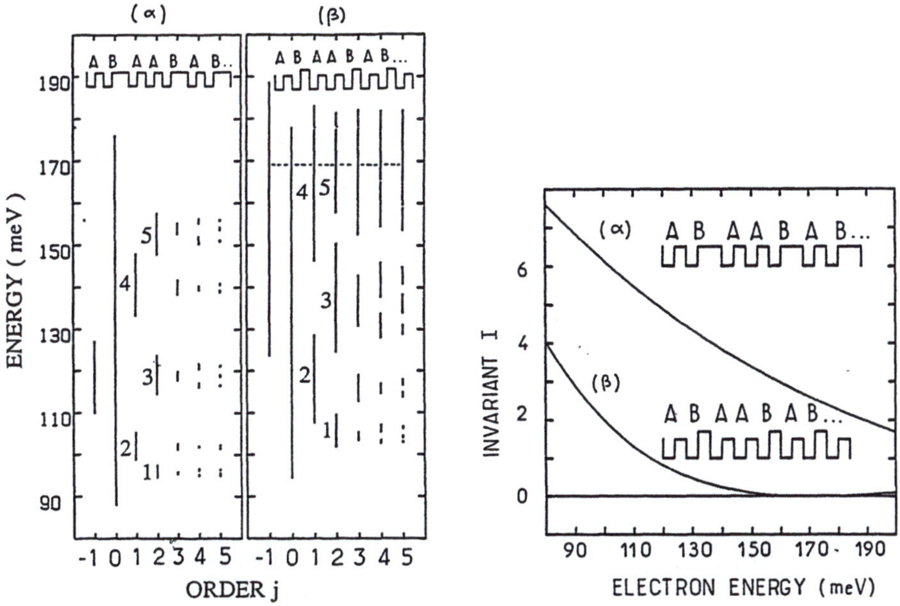

Figure 1: Left: Electron band structure as a function of order j in cases α and β. Energy ranges from 1 to 5 correspond to mini-bands at order 1 and 2. The insert depicts the SL conduction band profile. The dashed line indicates the energy for which **I** *is equal to zero.*

Right: Electron energy dependence of the invariant in cases α and β.

3. Wavefunction localization

Despite resonance between all QWs in case α the tunnel coupling modulation leads to a rapid decrease of the mini-band's width. Wavefunctions peak on different sequences depending on their energies: states in odd index energy ranges are peaked on BAAB patterns whereas states lying in those of even index are peaked on the BAB sequence (see in fig. 2).

Is this phenomenon universal in quasiperiodic SLs? It is known that some particular quasiperiodic potentials are associated with extended states [4-6]. They are quite unrealistic in the SL growth point of view. We have studied a more realistic case in which the tunnel transfer is modulated via the barrier height, that is, by changing the Al content of the GaAlAs alloy between A and B (case β). We choose the following parameters: A: 28.3 Å GaAs, 31.1 Å $Ga_{0.75}Al_{0.25}As$, B: 28.3 Å GaAs, 31.1 Å $Ga_{0.65}Al_{0.35}As$. The band structure evolution is displayed in fig. 2. The energy spectrum is very similar to case α except in the high energy part where mini-bands 4 and 5 overlap and where minigaps are small. Wavefunctions are no longer peaked on short SL sequences as is the case in the low energy range (see fig. 2). We find in fig. 1 that **I** is very small in the common energy range of minibands 4 and 5 and that **I** equals zero, for a particular value of the energy (≈ 169 meV). When **I** equals zero matrices M_A and M_B commute: the wavefunction is not affected by the relative position of A and B elements in S_j: the wavefunction is extended. The arrangement of A and B elements can be periodic, quasiperiodic or random: this

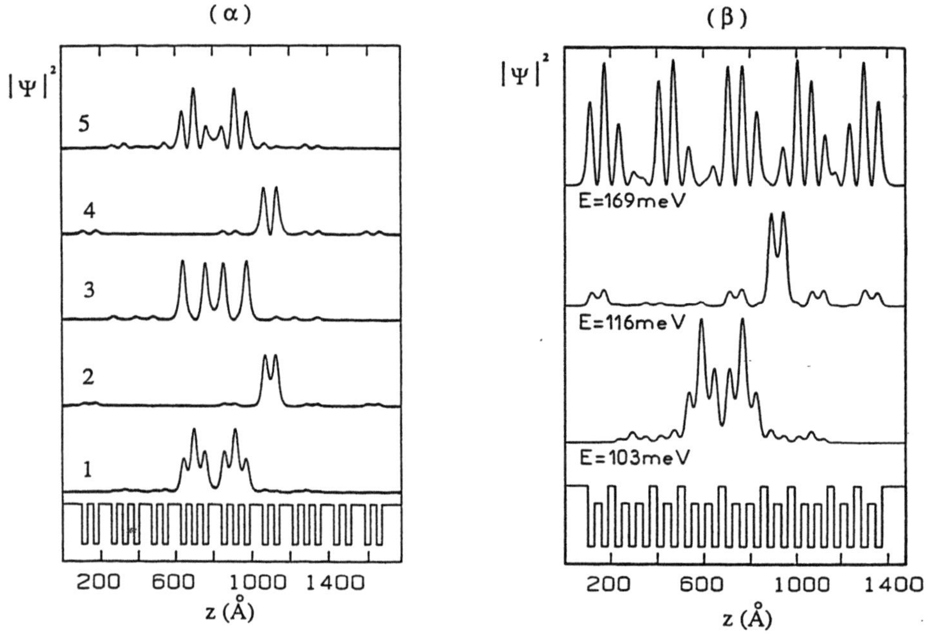

Figure 2: Left: Square modulus of the electron wavefunction in case α (j = 6) for the five different energy ranges indicated on the left hand side.
Right: Square modulus of the electron wavefunction in case β (j = 6) for three different energies. **I** *equals zero for the top wavefuction.*

state is not sensitive to the SL order [21] as long as only two distinct elements are combined. This has also been discussed by Xie and Das Sarma [22] and by Ananthakrishna and Kumar [23].

4. Optical transitions: selection rules

Rough selection rules can be established in case α since the wavefunction is peaked on short SL sequences. There is no overlap in real space between states lying in odd and even energy ranges. States of identical parity in energy range overlap on the same kind of patterns but with a different symmetry (see fig. 2) which warrants a zero overlap. We expect a $\Delta n=0$ selection rule where n is the energy range defined in the band structure of electrons or holes as well.

In case β the situation is somewhat more involved. Due to very different effective masses between electrons and heavy holes, the heavy hole band structure can be quite different from the electron one (in other words the variation of **I** for electron and holes may be different in the spectrum energy range). The steepest dispersion relation will probably be smoothed in the absorption profile by the smoothest one.

5. Experiments

The available experimental data on this matter before this work was an absorption spectrum of a quasiperiodic GaAs/GaAlAs SL [9]. The structures observed in the

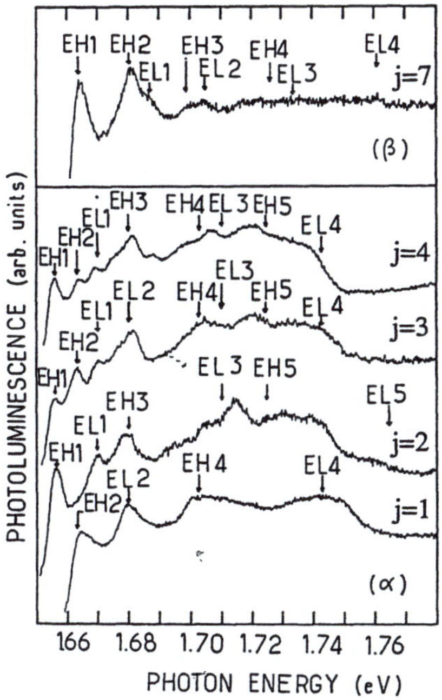

Figure 3: Low temperature (T = 2 K) photoluminescence excitation spectra in cases α and β. The different transitions are labelled according to the text.

spectrum were not clearly identified. Conversely, our series of SLs with increasing j establishes that structures seen in the absorption spectrum are related to bulk states and not to surface states. In fig. 3 we display the photoluminescence excitation spectrum of samples corresponding to case α: j = 1, 2, 3, 4 and to case β: j = 7. The labels EHn refer to transitions between an electron (E) in the energy range n and a heavy (H) or light (L) hole in the same energy range. At order j = 1 we observe four transitions assigned to bonding and anti-bonding states for heavy and light holes. At the next order we observe six transitions as expected from the band structure evolution. At order j = 3 we observe the superposition of the spectra of order j = 1 and j = 2 as is the case for the band structure. We do not see any appreciable difference beyond this order. This is probably due to the fact that the coherence length is not large enough to ensure the formation of new gaps. In better samples (grown at a substrate temperature of 640 °C instead of 670 °C) we observe an additional splitting of EH1 at order 4, which is the signature of a larger coherence length, 500 Å in that case. In case β similar structures appear in the low energy part as expected from the electronic band structure. However, the high energy part of the spectrum is structureless as it is for periodic SLs. This is consistent with the fact that minigaps are small as expected from our calculations.

IV) ELECTRIC FIELD EFFECTS

An electric field applied to a periodic SL localizes electronic wavefunctions [24]. The consequence of this effect for the absorption spectrum has recently been treated theoretically by Bleuse et al. [25], then experimentally by Mendez et al. [26] and by Voisin et al. [27]. We use these ideas to study the optical spectrum of quasiperiodic

SLs in an electric field. In periodic SLs we investigate the early stage of the electric field induced localization.

1. Quasiperiodic superlattices in an electric field

The quasiperiodic SL parameters were slightly different from the previous case α: A: 30 Å GaAs, 30 Å $Ga_{0.75}Al_{0.25}As$, B: 30 Å GaAs, 50 Å $Ga_{0.75}Al_{0.25}As$, order j = 9. As we apply an electric field *F* we observe a strong change of the photocurrent spectra displayed in fig.4 The energy dependence of the different lines is summarized in the adjacent plot. In this plot of energy versus *F* the slope gives the separation of the carriers in real space contributing to a given transition.

Transitions A and B shift at a rate corresponding to - 110 Å: electron and hole (heavy hole for line A, light one for line B) are located on opposite ends of a three coupled QW pattern. Another set of two lines C and D shifts at a rate corresponding to - 60 Å. As lines A and B, lines C and D are respectively attributed to heavy and light holes. The electron is in the first neighbouring QW. The smaller electron-hole separation is also evidenced by a slower decrease of the oscillator strength and of the excitonic behaviour. The doublet on peak C is characteristic of the three coupled QW pattern. With lines A and B this is the distinctive spectroscopic feature compared to previously studied double coupled QWs [28,29]. Lines E and F (heavy and light holes) do not shift at all as *F* is increased and their excitonic behaviour increases with *F* in sharp contrast with A-D transitions. This is interpreted as a

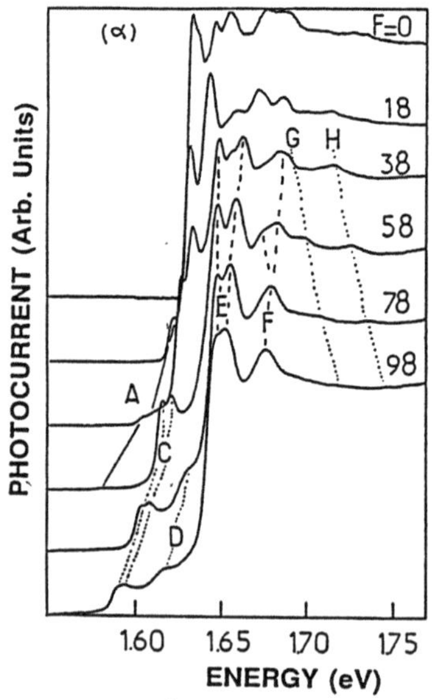

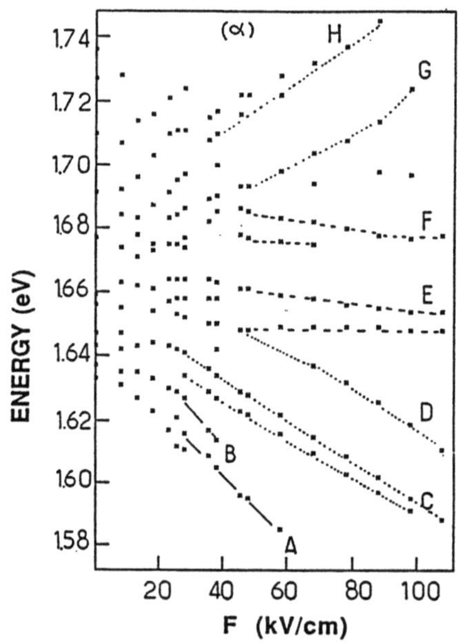

Figure 4: Left: Low temperature (T = 4.2 K) photocurrent spectra of the case α sample described in the text. Note the correspondence of the continuous, dotted and dashed line with the right plot. Letters refer to transitions discussed in the text. The electric field is indicated for the different spectra in kV/cm.

Right: Plot of the various transitions as a function of the electric field.

transition between electron and holes located in the same QW either in BAB patterns or in BAAB patterns. Lines G and H are the replica of lines C and D but shift at a rate corresponding to + 60 Å. This electro-optical study confirms the zero field analysis of optical transitions. A simple tight-binding model of two or three coupled QWs in an electric field leads to a good fit with no adjustable parameters of the electric field dependence of the various transitions [30].

2. Periodic SL in a weak electric field: wavefunction spectroscopy

We studied a similar structure consisting of a periodic SL with a unit cell made of the A element of the previous case. For this sample and the investigated electric field range (10-17 kV/cm) we find theoretically that heavy holes are localized in one QW whereas electrons still spread over 15-10 periods of the SL as can be seen in fig.5. The electric field breaks the SL translational invariance and lifts the optical selection rules [25]. We restrict our analysis to one electron and the heavy hole Stark ladder. As the heavy hole is localized in one QW, the electron-hole overlap is directly proportional to the electronic probability density in the QW in which the hole is localized. Since the different hole levels are on a Stark ladder the energy scale can be transformed in a real space scale: we observe the changes of the electronic probability density as F is varied [30].

Wavefunctions displayed in fig. 5 have been calculated using the transfer matrix method in which the linearly varying potential has been discretized into small steps.

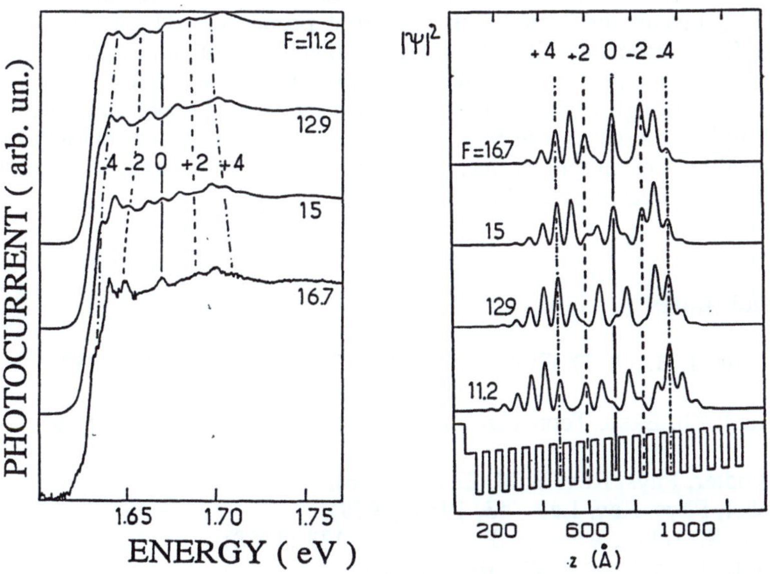

Figure 5 Left: Low temperature (T = 4.2 K) photocurrent spectra of the periodic sample described in the text. Note the correspondence of the continuous, dashed and dash-dotted line with the right figure The electric field is indicated for the different spectra in kV/cm.

Right: Square modulus of the electron wavefunction for different electric fields. SL periods corresponding to transitions labelled on the left are indicated with the same type of lines.

The disappearance of different transitions is due to the movement of wavefunction nodes swept by the electric field. As transitions reappear they can rise either symmetrically (± 1 transitions) or asymmetrically (± 2 , ± 3 transitions) (index refers to the number of SL periods separating electron and hole for a given transition). This cannot be explained through analytical expressions derived by Bleuse et al. [25] since the overlap integral between nearest neighbours is neglected. Either a numerical solution of the Schrödinger equation or a complete tight-binding model is required to explain these results [30]. As we observe transitions up to ± 6 the coherence length is 800 Å in this sample. This is consistent with our previous estimation in the Fibonacci SL since in the latter case the smaller coupling through thick B barriers tends to reduce the coherence [32].

V) CONCLUSION

Since the work by Belle, Maan and Weimann [33] the electronic coherence in SLs is a wide open question. A recent examination of their magneto-optic experimental results leads to a coherence length of 400 Å [34]. The introduction of non-periodicity either in the structure (Fibonacci SL) or with an external electric field (Wannier-Stark ladders) or a magnetic field (Landau levels) is required in order to get information about the coherence length in these coupled QW systems. It seems that the limitation lies in the uncontrollable residual SL fluctuations: alloy scattering, interface roughness scattering, thickness fluctuations [32, 34]. The advantage of quasiperiodic SLs lies in their very rich structural properties, which allows a correlation with the electronic quality whereas in periodic SLs structural information is more concentrated in reciprocal space. We have shown that the electronic structure can be characterized through very simple calculations and fits the experiments well. High quality samples behave as true SLs: the coherence length is large enough to allow the observation at room temperature of the negative effective mass within the SL miniband [35] as predicted by Esaki and Tsu in 1970 [36].

References and notes

[1] D. Shechtman, I. Blech, D. Gratias, and J.W. Cahn, Phys. Rev. Lett., **53**, 1951 (1984).

[2] M. Ya Azbel, Zh. Eksp. Teor. Fiz., **46**, 929 (1964) [Sov. Phys. JETP **19**, 634 (1964)].

[3] D.R. Hofstadter, Phys. Rev. B **14**, 2239, (1976).

[4] M. Ya Azbel, Phys. Rev. Lett., **43**, 1954 (1979).

[5] G. André and S. Aubry, Ann. Isr. Phys. Soc. **3**, 133 (1980).

[6] J.B. Sokoloff, Phys. Rev. B, 22, 5823 (1980).

[7] R. Merlin, K. Bajema, R. Clarke, F. Y. Juang, and P.K. Bhattacharya, Phys. Rev. Lett., **55**, 1768 (1985).

[8] J. Todd, R. Merlin, R. Clarke, K.M. Mohanty, and J.D. Axe, Phys. Rev. Lett., **57**, 1157 (1986).

[9] R. Merlin, K. Bajema, R. Clarke, J. Todd, Proc. 18 th International Conference on the Physics of Semiconductors, O. Engström editor, Stockholm , Sweden 1986, p. 675.

[10] M. Kohmoto, C. Tang, and L.P. Kadanoff, Phys. Rev. Lett., **50**, 1871 (1983).
[11] L. Fibonacci, in Liber Abaci (Pisa, 1202).
[12] M. Duneau and A. Katz, Phys. Rev. Lett., **54**, 2688 (1985).
[13] D. Paquet, M.C. Joncour, F.Mollot , and B. Etienne, Phys. Rev. B, **39**, 10973 (1989).
[14] D. Paquet, M.C. Joncour,B. Jusserand, F. Laruelle, F.Mollot , and B. Etienne, Spectroscopy of semiconductor microstructures, NATO ASI Series, plenum B 206, 223 (1989).
[15] B. Jusserand, D. Paquet, F.Mollot , M.C. Joncour, and B. Etienne, Phys. Rev. B, **39**, 3683 (1989).
[16] Q. Niu and F. Nori, Phys. Rev. Lett., **57**, 2057 (1986).
[17] F. Laruelle and B. Etienne, Phys. Rev. B, **37**, 4816 (1988).
[18] $X = \cos(k_w l_w)\, \mathrm{ch}(\kappa_b L_b) + \frac{1}{2}(\xi - \frac{1}{\xi}) \sin(k_w l_w)\, \mathrm{sh}(\kappa_b L_b)$

$$Y = \sin(k_w l_w)\, \mathrm{ch}(\kappa_b L_b) - \frac{1}{2}(\xi - \frac{1}{\xi}) \cos(k_w l_w)\, \mathrm{sh}(\kappa_b L_b)$$

$$R = (\xi + \frac{1}{\xi})\, \mathrm{sh}(\kappa_b L_b),\quad \xi = \frac{\kappa_b m_w}{k_w m_b},\quad \varepsilon = \frac{h^2 k_w^2}{2m_w},\quad V_b - \varepsilon = \frac{h^2 \kappa_b^2}{2m_b}$$

[19] M. Kohmoto and Y. Oono, Phys. Lett., **102 A** , 145 (1984).
[20] V. Kumar and G. Ananthakrishna, Phys. Rev. Lett., **59**, 1476 (1987).
[21] F. Laruelle, D. Paquet and B. Etienne, Superlatt. and Microstruct., **5**, 495 (1989).
[22] X. C. Xie and S. Das Sarma, Phys. Rev. Lett., **60** , 1585 (1988).
[23] G. Ananthakrishna and V. Kumar, Phys. Rev. Lett., 60, 1586 (1988).
[24] R.F. Kazarinov and R.A. Suris, Sov. Phys. Semicond., **6**, 121 (1972).
[25] J. Bleuse, G. Bastard and P. Voisin, Phys. Rev. Lett., **60**, 220 (1988).
[26] E.E Mendez, F. Agullo-Rueda, and J.M. Hong, Phys. Rev. Lett., **60**, 2426 (1988).
[27] P. Voisin, J. Bleuse, C. Bouche, S. Gaillard, C. Alibert, and A. Regreny, Phys. Rev. Lett., **61**, 1639 (1988).
[28] Y.J Chen, E.S. Koteles, B.S. Elman, and C.A. Arniento, Phys. Rev. B, **36**, 4562 (1987).
[29] S.R. Andrews, C.M. Murray, R.A. Davies, and T.M. Kerr, Phys. Rev. B, **37**, 8198 (1988).
[30] F. Laruelle, B. Etienne, J. Barrau, K. Khirouni, J.C. Brabant, T. Amand, and M. Brousseau, Surf. Sci. (to be published).
[31] F. Agullo-Rueda, E.E Medez, and J.M. Hong, Phys. Rev. B, **40**, 1357 (1989).
[32] J. Barrau, K. Khirouni, Do Xuan Than, T. Amand, M. Brousseau, F. Laruelle, and B. Etienne, Solid State Comm. (to be published).
[33] G. Belle, J.C. Maan, G. Weimann, Solid State Comm., **56**, 65 (1985).
[34] D. Paquet, Phys. Rev. B, **39**, 7776 (1989).
[35] A. Sibille, J.F. Palmier, H. Wang, and F. Mollot, Phys. Rev. Lett., **64**, 52 (1990).
[36] L. Esaki and R. Tsu, IBM J. Res. Develop, **14**, 61 (1970).

Periodic Magnetic Field Induced Localization in Superlattices

J. Ma, R.A. Puechner, A.M. Kriman, and D.K. Ferry

Center for Solid State Electronics Research, Arizona State University,
Tempe, AZ 85287, USA

Abstract. Over the past decade, several experiments have demonstrated the quantum interference of electrons in condensed matter systems. These observations have principally come from simple, multiply-connected structures such as Aharonov-Bohm rings. Similarly, interferences leading to a magnetoconductance periodic in the magnetic field coupled through a unit cell of a two-dimensional superlattice has been hypothesized for many years. In these latter structures, the presence of two characteristic lengths – the lattice periodicity and the magnetic length – gives rise to a complicated energy spectrum. The presence of magnetic-field induced oscillations in the transport parameters due to one- and two-dimensional superlattices will be reviewed. Measurements will be described of low-mobility quasi-two-dimensional systems, subject to a two-dimensional superlattice, in which the existence of weak-localization induced negative differential magnetoresistance is replicated at multiples of a quantum flux coupled through each unit cell. The presence of a significant source-drain potential shifts the resonances in magnetic field.

1. Introduction

The two-dimensional motion of electrons that are subjected to both a two-dimensional periodic potential and a perpendicular magnetic field is a problem that has been studied for a great many years. In general, the solution of this problem is complicated by the fact that there are two characteristic lengths in the problem – one is the period *a* of the superlattice potential, while the second is the magnetic length $L_m=(h/eB)^{1/2}$. The complete problem can be solved generally in an infinite domain only when these two lengths are related by the ratio of two integers. However, there are two distinct limits in which perturbation theory can be used to obtain solutions. In one limit, the Landau regime, the periodic potential is regarded as a weak perturbation on the usual magnetic Landau level structure. When the flux coupled through each unit cell of the periodic potential is given by $\phi=ea^2B/h=p/q$, each Landau level splits into p subbands of equal degeneracy. It has been known for some time that a series of oscillations, periodic in 1/B, should arise in magneto-transport measurements [1-3]. These oscillations have an appearance similar to, but an origin different than, the Shubnikov-de Haas oscillations. In fact, these predicted oscillations have recently been observed [4-6] with periodic potentials macroscopically produced by a superlattice, and are now known as Weiss oscillations. When the strength of the superlattice periodic potential is increased, the Landau levels are broadened significantly, the subbands merge, and the Weiss oscillations, *as well as the normal Shubnikov-de Haas oscillations,* are heavily damped [7]. This has also been observed experimentally [8].

In the opposite limit to that above, the magnetic field is treated as a perturbation on the periodic potential. In this regime, known as the Onsager regime, the magnetic transport properties are expected to be periodic in the magnetic flux coupled through each unit cell, i.e., periodic in magnetic field (as opposed to 1/B in the previous case) [9,10]. While the required magnetic field is unreasonably high in normal semiconductor lattices, it is an observable effect in lateral surface superlattice (LSSL) periodic potentials [12,13]. Clearly, the observation of these effects, which are linear in the magnetic field, requires the phase coherence length of the electrons to be larger than the superlattice period. On the other hand, the limit being taken here is the tight-binding limit of the superlattice potential, and the transport can be expected to have strong similarities to more localized types of transport.

Springer Series in Solid-State Sciences, Vol. 97
Localization and Confinement of Electrons in Semiconductors
Editors: F. Kuchar · H. Heinrich · G. Bauer

The structure of the bands can be obtained by solving Schrödinger's equation with the magnetic field as a perturbation. In the absence of the latter field, the energy structure of the superlattice mini-bands are given by

$$E = 2E_o\left[\cos(k_x a) + \cos(k_y a)\right] \quad . \tag{1}$$

With a magnetic field, described in the Landau gauge **A**=(0,Bx,0), the Peierl's substitution leads to the equation

$$\left[\cos(k_x a) + \cos\{(k_y - \frac{eBx}{\hbar})a\}\right]\Psi(x,y) = \frac{E}{2E_o}\Psi(x,y) \quad . \tag{2}$$

The introduction of the wave function *ansatz* $\Psi(x,y)=g(x)\exp(ik_y y)$, and the substitutions x=ma, y=na [10], leads then to the iterative equation

$$g(m+1) + g(m-1) + 2\cos(2\pi m\alpha - \nu)g(m) = \varepsilon g(m) \,, \tag{3}$$

where $\alpha=eBa^2/h$, $\eta=k_y a$, and $\varepsilon=E/E_o$. This equation is the Harper equation [10], and a study of the solutions of this equation have been discussed by Hofstadfer in some detail [11]. The energy structure is periodic in α, which means that it is periodic in magnetic field, as this quantity is the ratio of the flux coupled through a unit cell to the quantum unit of flux h/e.

The source of the periodicity in magnetic field in the tight-binding limit can be understood by considering the presence of magnetic translation operators connected with the periodicity of the lattice. In a periodic lattice, it is known that $\psi(x+a,y)=\exp(ik_x a)\psi(x,y)$, where $\psi(x,y)$ is the Bloch function corresponding to the superlattice in the absence of the field. If the magnetic field is normal to the layer, and the vector potential is taken (as above) in the Landau form A=(0,Bx,0), the motion of successive translations about a rectangle of unit cells (returning to the original point) leads to

$$T(-naj)T(-mai)T(naj)T(mai) = T(-naj-mai+naj+mai)\exp(i\int_0^{na}dy\int_0^{ma}dx\frac{eB}{\hbar})$$

$$= \exp(2\pi inm\alpha) \;, \tag{4}$$

where **i** and **j** are unit vectors in the directions of the LSSL. In fact, the group theoretical arguments for the magnetic translation group have been worked out in some detail [14].

In high-mobility structures, upon which the superlattice potential is imposed, the oscillations periodic in B are observed as very weak oscillations [12], as even in these structures, the inelastic mean free path for carriers in the superlattice appears to only be of the order of 0.7 μm, which is just a few superlattice spacings. In lower mobility material, such as heavily-doped material normally used for MESFETs, the localization behavior of the superlattice is more strongly pronounced. In many cases, weak localization is observed in the absence of the magnetic field, which induces a periodic replication of this weak localization when an integer number of flux quanta h/e are coupled through each unit cell. In these latter structures, the weak localization is not always observed at zero magnetic field. In the absence of this effect, the separation of the magnetic field induced replicas of the weak localization still corresponds to a single flux quantum coupled through each cell.

In this paper, the presence of the magnetic field induced weak localization effects that arise in the transport of carriers through a low mobility quasi-two-dimensional layer, subject to a two-dimensional superlattice potential, will be treated in some detail. In the next section, the material and sample preparation is described. The measurements will be treated in Sec. 3, while the role of the source-drain potential in these structures will be presented in Sec. 4.

2. Sample Preparation

We have incorporated the LSSL gates into structures fabricated on material normally used for MESFETs. Here, the active layer is typically a 60 nm thick, epitaxial layer grown by vapor-phase epitaxy on a lightly-doped substrate. The epitaxial layer is doped to 1.5×10^{18} cm^{-3}. It has been demonstrated previously that such layers will show quasi-two-dimensional behavior at low temperatures when the channel is biased near pinchoff [15]. Structures were prepared by etching a mesa in the sample, and writing the superlattice grid by electron-beam lithography and subsequent lift-off metallization [16]. The individual lines of the grid gate (Fig. 1) were 40 nm wide on a nominal center-to-center spacing of 160 nm. The active area measures 10 μm × 20 μm. We estimate that the inelastic mean free path in these structures is of the order of 0.2–0.6 μm at 5 K, which is considerably less than either dimension in the plane of the sample. In this structure, the electrons will be confined in three-dimensional quantum boxes lying below the open regions of the grid, and it is this total confinement which produces the full superlattice effects.

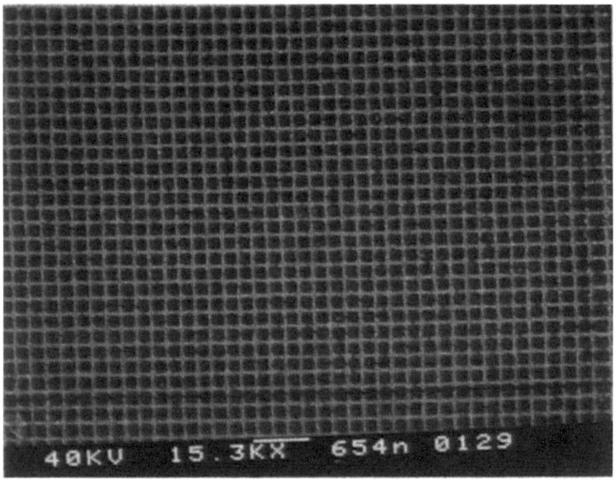

Fig. 1 The grid gate, pictured here, is placed on the mesa.

3. Experimental Results

The first set of measurements to be described concerns devices with relatively low mobility, in which weak localization is observed in the absence of the magnetic field. In Fig. 2, we show the source conductance so that the presence of the negative magnetoresistance at low magnetic fields can be seen. This is clear evidence of weak localization. Other effects can also be seen in this latter figure. The conductance shows sharp drops at regularly spaced values of the magnetic field, and sharp changes in the Hall voltage which are correlated with these. However, the sign of the change in the Hall voltage seems random. The source current drops occur at integral multiples of a flux quantum coupled through each unit cell of the surface superlattice, which we interpret to be periodic replicas of the negative magnetoresistance at zero magnetic field, and hence periodic replicas of the weak localization of the electrons. The current drops at the first and third flux quanta are very weak, being a reduction of only about 1-2% in the current, and are not easily distinguished in the data of Fig. 2, but can be measured by separating the data from the background, which is done for the evaluation described below. It appears from Fig. 2 that the basic periodicity in the conductance is about 1500 Gauss (±100 Gauss), which corresponds to about 165 nm periodicity, quite close to the measured periodicity of the grid.

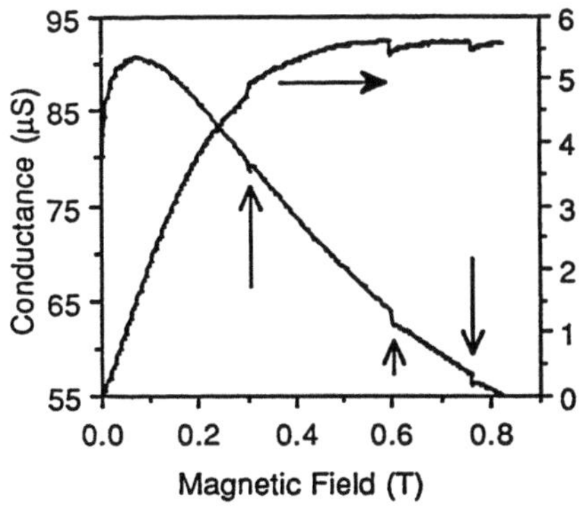

Fig. 2 Conductance and Hall voltage for a MESFET at 4.2 K.

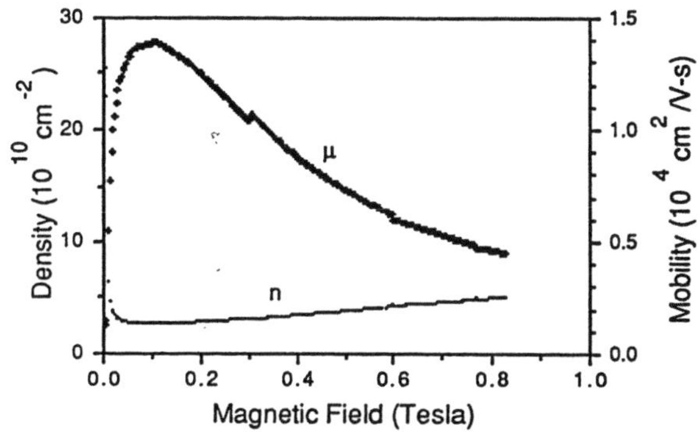

Fig. 3 The density and mobility obtained from Fig. 2 are shown.

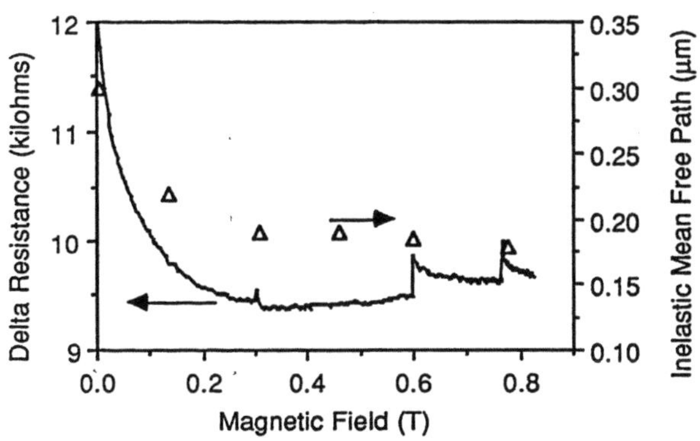

Fig. 4 The values obtained for the inelastic mean free path.

In Fig. 3, the decomposition of the Hall measurements are shown as a function of the magnetic field for this sample. The density is relatively constant for all values of the field above 0.1 T, and the major variation in the Hall voltage and conductance arises in the mobility itself. It is this variation in the mobility which produces the non-linear (in magnetic field) behavior of the Hall voltage. The density is approximately 5×10^{10} cm^{-2} over the entire magnetic field range.

We can estimate the inelastic mean free path near these jumps from the negative magnetoresistance in the source current following the jump. The value of the inelastic mean free path inferred from a data fit to several drops in the conductance at the various integer multiples of a flux quantum (coupled to each well) is nearly the same for a given sample and gate voltage. In Fig. 4, the value of the inelastic mean free path computed for each of the drops (six), found in the data of Fig. 2, are plotted on the same graph with the magnetoresistance data. The value found in this sample is about 0.2 µm and the uncertainty of the fit makes the data consistent with a constant value. However, it also appears that the value of the mean free path is tending to lock to the periodicity distance of the superlattice itself, reinforcing the idea that localization within the superlattice is important for the transport properties.

4. Role of the Source-Drain Potential

In Fig. 5, measurements made on a MESFET layer with a slightly thinner epitaxial layer, and a significantly, higher mobility are shown. Here, negative magnetoresistance occurs only at higher values of the source-drain voltage (2 mV source-drain potential corresponds to approximately 11 µV drop across each cell of the superlattice, while the superlattice potential is estimated to be about 100 mV peak-to-peak amplitude). The onset of the magnetic field induced localization generally does not occur at integer values of the flux coupled through each unit cell (except for the curves with 2 mV and 5 mV potential drop). Rather, the localization dips occur approximately periodically through the magnetic field range. Their periodicity is found to be approximately 0.14 T for all of the curves (Fig. 6), which corresponds to a

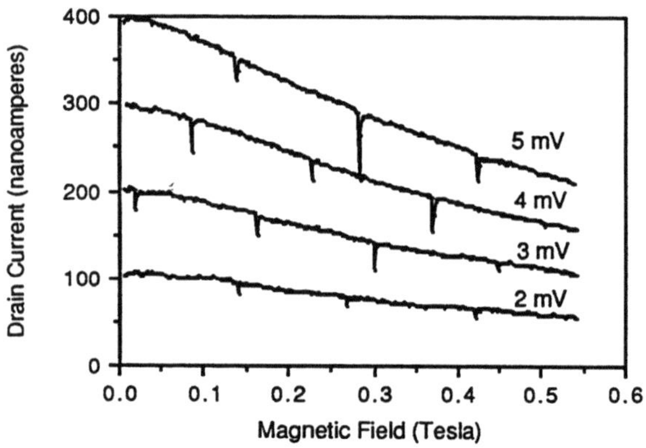

Fig. 5 Current flow in a structure with inelastic mean free path of about 0.55±0.1 µm. The parameter is the source-drain potential (1 mV corresponds to 5.7 µV per unit cell potential drop).

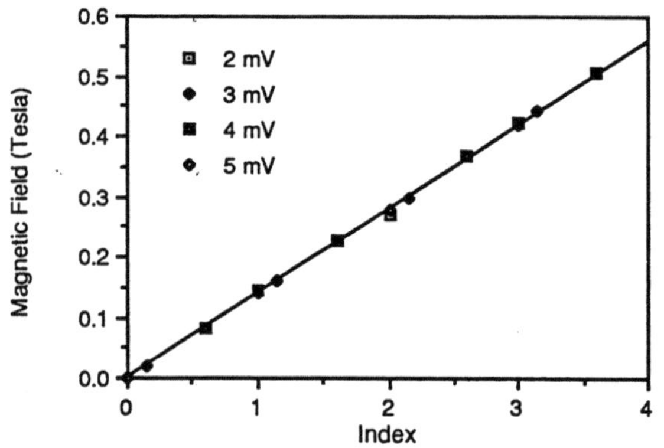

Fig. 6 The value of the magnetic field at which a conductance dip is observed is plotted as a function of its "index" – the rank order of the dips plus an offset in magnetic field. The offset is a constant for each value of the source-drain bias.

periodicity of about 170 nm in the superlattice. The inelastic mean free path, estimated from the negative magnetoresistance that is found at each local minimum in conductivity, is about 0.55±0.1 μm for this sample. The conductance, at zero magnetic field, varies from about 40 μS, at 2 mV, to 80 μS at 5 mV. The (spin degenerate) conductance expected from the Landauer formula for a single channel is 77.5 μS [17], so that it is clear that the structure is not conducting strongly in a free electron fashion. Rather, the conductance is quite likely to be localized. The behavior shown in Fig. 4 is for a temperature 5.7 K, and the effect is not observed for a temperature of 15 K. It is also interesting that the increase of the drain potential from 2 mV to 5 mV has shifted the spectrum almost exactly one cell, bringing the spectra back into commensurability.

One model for the shifts introduced by the source-drain potential can be easily understood. The source-drain bias, not only shifts the energy levels down the channel, in reference to their values at the source end, but also distorts them slightly within each unit cell. It is the latter effects that can give rise to the shifts noted above. For this, we consider the electric field in the vector potential gauge, and oriented along the y-axis, so that $A=(0,Bx+eE\tau,0)$, where the time dependence has been averaged over the momentum relaxation process. For this vector potential, equation (3) is modified to

$$g(m+1) + g(m-1) + 2\cos(2\pi m\alpha - \omega_B\tau - \nu)g(m) = \varepsilon g(m) , \qquad (5)$$

where $\omega_B=eEa/\hbar$ is the Bloch frequency. Here, the presence of the electric field, and the resulting drift velocity shifts the y-momentum contained in ν, and shifts the zero of the cosine function yielding the energy levels. This shift is reflected in the conductance itself. This can be checked by plotting the shift introduced for Fig. 6 as a function of the source-drain potential. This is shown in Fig. 7.

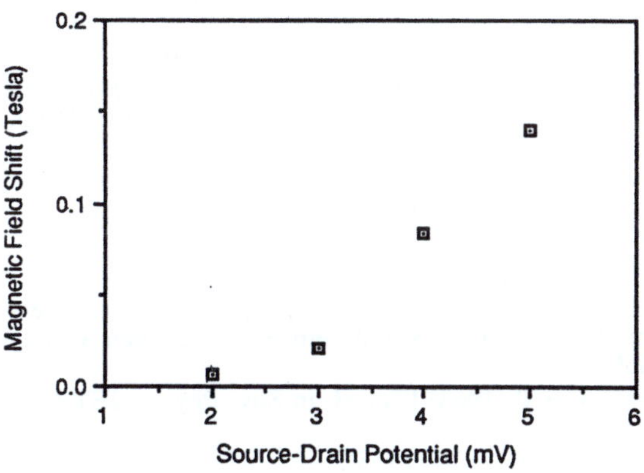

Fig. 7 The value of the magnetic field shift required for the data of Fig. 5 in order to align the curves in Fig. 6.

The shift in Fig. 7 becomes almost linear in source-drain potential for values of this latter quantity above 3 mV. It was remarked above that the conductance of the channel did not turn on linearly until this value as well, so that it is likely that the transport is strongly localized at the lower values of the source-drain potential. The linear portion of the curve in Fig. 7 can be used with the behavior of equation (5) to estimate the scattering time τ, and this gives a value of approximately 2.8×10^{-10} sec. If this value is taken together with the inelastic mean free path, a diffusion constant of 11 cm^2/sec is inferred, which corresponds to a mobility of about 2.5×10^4 cm^2/V-s. The observed mobility is about 25% larger than this, but the numbers give a consistent order-of-magnitude argument to suggest that the shifts observed in the weak localization effects (in magnetic field) are related to the increasing drift momentum caused by the source-drain potential.

5. Summary

We have measured the magnetoconductance in lateral surface superlattices, in which a two-dimensional electrostatic superlattice potential is imposed upon a quasi-two-dimensional electron gas. These measurements show the existence of weak localization at zero magnetic field. At integer multiples of the flux quantum coupled through each unit cell of the lateral superlattice, drops in the conductance are observed. These drops have an incremental magnetic field behavior that fits the current theories of weak localization, and are interpreted to be periodic replicas of the (zero magnetic field) weak localization at integer multiples of the fundamental flux quantum coupled through each unit cell of the superlattice. The sharp drop in conductance that accompanies this localization phenomenon is quite large, and can be used to estimate the inelastic mean free path. This latter is found to be in the range of 0.2–0.6 μm in these structures. In addition, the conductance through the samples exhibits conductance fluctuations. It is worth noting that the sample size here is many times the inelastic mean free path.

The authors are indebted to P. Newman and R. C. Curby for providing the material used in this study. In addition, the authors have enjoyed fruitful discussions with E. Paris, W.–P. Liu, F. H. Claro, R. A. Webb, G. J. Iafrate, J. P. Kotthaus, P. C. Main, and L. Eaves. This work was supported in part by the U. S. Office of Naval Research.

6. References

1. A. B. Pippard, Phil. Trans. Roy. Soc. (London) **A68**, 317 (1964).
2. M. Ya. Azbel, J. Exp. Theor. Phys. **46**, 929 (1964) [Transl. in Sov. Phys. - JETP **19**, 634 (1964)].
3. A. Rauh, G. H. Wannier, and G. Obermair, Phys. Stat. Sol. (b) **63**, 215 (1974).
4. D. Weiss, K. v. Klitzing, K. Ploog, and G. Weimann, in *High Magnetic Fields in Semiconductor Physics II*, Ed. by G. Landwehr (Springer-Verlag, Heidelberg, 1988) p. 357.
5. R. R. Gerhardts, D. Weiss, and K. v. Klitzing, Phys. Rev. Letters **62**, 1173 (1989).
6. R. W. Winkler, J. P. Kotthaus, and K. Ploog, Phys. Rev. Letters **62**, 1177 (1989).
7. H. J. Schellnhuber and G. M. Obermair, Phys. Rev. Letters **45**, 276 (1980).
8. P. Beeton, E. S. Alves, M. Hennini, L. Eaves, P. C. Main, O. H. Hughes, G. A. Toombs, S. P. Beaumont, and C. D. W. Wilkinson, Proc. Symposium on New Phenomena in Mesoscopic Systems, Kona, Hawaii, 6–9 December 1989, Japan Society for the Promotion of Science (unpublished).
9. M. Ya. Azbel, J. Exp. Theor. Phys. **44**, 980 (1963) [Transl. in Sov. Phys. – JETP **17**, 665 (1963)].
10. P. G. Harper, Proc. Phys. Soc. (London) A **68**, 874 (1955).
11. D. R. Hofstadter, Phys. Rev. B **14**, 2239 (1976).
12. D. K. Ferry, G. Bernstein, R. Puechner, J. Ma, A. M. Kriman, R. Mezenner, W.-P. Liu, G. N. Maracas, and R. Chamberlin, in *High Magnetic Fields in Semiconductor Physics II*, Ed. by G. Landwehr (Springer-Verlag, Heidelberg, 1988) p. 344.
13. J. Ma, R. A. Puechner, W.-P. Liu, R. Mezenner, A. M. Kriman, G. N. Maracas, D. K. Ferry, P. Chu, and H. H. Wieder, *Proc. Indium Phosphide and Related Materials for Advanced Electronic and Optical Devices*, Ed. by R. Singh and L. J. Messick (SPIE, Bellingham, WA) **1144**, 619 (1989).
14. J. Zak, Phys. Rev. **134**, A1602 (1964); **134**, A1607 (1964).
15. M. Pepper, J. Phys. C **10**, L173 (1977).
16. G. B. Bernstein and D. K. Ferry, Superlatt. Microstruct. **2**, 373 (1986).
17. R. Landauer, IBM J. Res. Develop. **1**, 223 (1957).

Part V

Optical Properties of Confined Systems

Isoelectronic Ultrathin Layers: A Probe for Heterostructure States

J.-Y. Marzin and J.-M. Gerard

Laboratoire de Bagneux, Centre National d'Etudes des Télécommunications, 196 Avenue Henri Ravera, F-92220 Bagneux, France

Abstract. We discuss localized doping of semiconductor heterostructures with ultrathin layers of isoelectronic species. We show in a first part that it allows an unambiguous determination of the type of a superlattice. In a second part, more quantitative information on the wave functions is extracted from the optical study of GaAs/GaAlAs quantum wells using such probes.

1 Introduction

Probing localized properties of solids on an atomic scale constitutes a difficult experimental challenge. The insertion of isoelectronic impurities inside the material and the study of the perturbation they yield has proven to be an efficient approach to extract information on local arrangements of atoms inside disordered semiconductor alloys, as $GaAs_XP_{1-X}$:N [1], Ga_XIn_{1-X}P:N [2], InGaAsP:Bi [3]. These impurities are isovalent to one of the material components, and are incorporated dominantly in substitution sites. Because of this isovalence, the potential they create is a short range potential. In the case of three-dimensional materials, it may or not bind an electron, a hole or an exciton, depending on the sign and strength of the perturbative potential. In such bound complexes, at least one of the two particles is strongly bound to the isoelectronic impurity and feels mainly the local arrangement of the first nearest neighbour atoms around the impurity. By comparing the luminescence and excitation of the photoluminescence of the N related states in GaP and in GaAsP alloys, Mariette et al. [1] could evidence specific effects of the disorder in the alloy.

However, in the case of three-dimensional materials, isoelectronic probing suffers from strong limitations. The probe potential has to be strong enough to result in states in the band gap; whereas this is achieved rather easily in indirect materials, InP:Bi is the only example in direct semiconductors. In practice, the substituting element has to be far away from the matrix elements, in the periodic classification, so that its size is also very different, and this results in a strong and complicated distortion of the host lattice. Finally, the equilibrium position of the neighbouring atoms of an isolated impurity depends on the presence of the particle it binds, due to its strong localization: the optical spectra of isoelectronic impurities in three-dimensional material is complicated by the resulting strong coupling with the phonons[4].

Most of these drawbacks vanish if one considers a complete atomic plane of isoelectronic substituent. The situation will be far simpler than for the isolated (3D) case: bound states will be more easily obtained and the matrix will remain unstrained.

Springer Series in Solid-State Sciences, Vol. 97
Localization and Confinement of Electrons in Semiconductors
Editors: F. Kuchar · H. Heinrich · G. Bauer

Moreover, the luminescence spectra of the structures containing such planes excited with moderate powers will not be dominated by coupling to phonons, due to a small difference in the average charge per impurity in the ground and excited states.

As will be discussed in this paper, those planes of isoelectronic impurities are particularly interesting if one wants to test (or modify) the local electronic densities in the levels of semiconductor heterostructures, whose properties are dominated by the reduced dimensions along the stucture's growth axis. However, this technique is not restricted to semiconductors, and Hsieh et al. used one gold plane to determine the extension of a surface state on silver [5].

Let us now examine our probe in a short theoretical part.

2 A simple model of isoelectronic ultrathin layers

Considering a semiconductor heterostructure, as a quantum well (QW) or a superlattice (SL) of axis Z, we will focus on levels which are close to the band gap, and which are well described in a one band effective mass framework.

The eigenfunctions (for example for electrons) for state $|i,k_x,k_y\rangle$ of energy $E_i(k_x,k_y)$ will be of the form

$$\Psi_{k_x,k_y,i}(x,y,z) = F_i(z)e^{ik_x x + ik_y y}u_0(x,y,z) \quad , \tag{1}$$

where u_0 stands for the Bloch function at the considered extremum and $F_i(z)$ is the envelope function, eigenfunction of the Hamiltonian H_0:

$$H_0 = -\frac{\hbar^2}{2m^*}\frac{\partial^2}{\partial z^2} + V(z) \quad , \tag{2}$$

where m^* is the effective mass for the electrons and where the potential $V(z)$ is the position-dependent energy of the conduction band extremum of the materials used to build the structure.

The specific properties of the QWs or SLs are due to the departure of F_i functions from simple plane waves and these functions F_i contain the information on confinement, miniband formation and so on: these are the main quantities that can be tailored when the heterostructure is designed.

The associated quantities to which we want access to are the probability densities $|F_i(z)|^2$, which can be measured through the operator $\delta(z-a)$, whose expectation value on $F_i(z)$ yields $|F_i(a)|^2$. The F_i's are supposed to be slowly varying functions so that the insertion at a location $z=a$ of a very thin layer of a semiconductor of the same family as those used to build the heterostructure would be described in this framework by transforming H_0 into

$$H = H_0 + W\delta(z-a) \quad , \tag{3}$$

where the weight of the δ function W is the conduction band discontinuity between the thin layer material and the material it is inserted in, multiplied by the thickness of the probe layer.

In first order perturbation, the change in E_i (at $k_x=k_y=0$) due to this perturbation is just $W.|F_i(a)|^2$. The study of the effect of an insertion at a position a in the structure on a given level will thus give us information on the value of the associated probability density at a. Clearly, this can be used in the case of superlattices to determine in which material the electrons or holes are mostly confined, i.e. to determine the type of the superlattice [6], and two examples are given in sect. 3.

Moreover, if the eigenenergies of H_0 form a discrete set (E_i), and if these energies and the eigenenergies of H (E'_i) are all known, then, beyond this approximation, the $|F_i(a)|^2$ are exactly given by

$$\sum_{\substack{j=1 \\ E_j \neq E'_i}}^{\infty} \frac{|F_j(a)|^2}{E'_i - E_j} = \frac{1}{W} \quad , \tag{4}$$

and $F_i(a)=0$ if one of the E'_js is equal to E_i.

In practice, only a finite set of p perturbed and unperturbed levels are known and equation (4) can be replaced by

$$\sum_{\substack{j=1 \\ E_j \neq E'_i}}^{j=p} \frac{|F_j(a)|^2}{E'_i - E_j} = \frac{1}{W} - \sum_{\substack{m=p+1 \\ E_m \neq E'_i}}^{\infty} \frac{|F_m(a)|^2}{E'_i - E_m} = \frac{1}{W_i} \quad . \tag{5}$$

As a first approximation, one can assume the W_i's to be independent of a. If the location a of the thin probe layer is scanned over an appropriate region of the heterostructure (one period for a superlattice), the variables W_j can be further eliminated by writing the normalization conditions for the F_i functions. This procedure yields approximate values for the probability densities, whose precision depends on the number of considered levels and spacing between the sampling points a.

To apply this, one has first to extract the $k_x=k_y=0$ electron levels. This can be done with reasonable assumptions from the optical transition energies. This kind of procedure has indeed been used, as described in sect. 4, to measure the envelope probability densities in the first three electron levels in a GaAs/GaAlAs quantum well, used as a test system [7].

One can argue about the validity of the effective mass treatment of such a localized perturbation, yielding all but a slowly varying potential. Following the results obtained in short period superlattices [8], the departure from the simple effective mass model, due to the rapid variations of the potential is the coupling of solutions corresponding to different extrema in the Brillouin zone. The values obtained in our simple model will remain relevant as long as the levels that are taken into account are far enough in energy from the other conduction band extrema, and provided that the energy changes are small compared to the band gap or energy distance between the considered extremum and other conduction band extrema. One should note that a simple Slater-Koster model, with a one band, first nearest neighbours tight binding description of the materials, would yield the same equation (1), with $F_j(a)$ replaced by the coefficient of the state at site a in the development of Ψ'.

Finally, the sign of W does not appear in the treatment we have performed and the potential created by the probe material can be either attractive or repulsive. Let us now examine some applications of this approach.

3 Superlattice type determination

Fig. 1 shows the effect on the first electron and hole levels of the insertion of a thin perturbative layer at the center of the larger band gap material layers of a multiple quantum well or superlattice structure, for different band extrema lineups. Clearly, this effect will remain small for type I superlattices (Fig. 1a), where the electron and hole are mainly localized in the unperturbed smaller gap material. On the other hand, for type II superlattices, the probability density for one of the two levels (hole in Fig. 1c or b, electron in d) is maximum at the location of the perturbation, and thus the energy of this level will be strongly affected. A particularly interesting case for superlattice type determination is that where the perturbing material is the smaller gap material (homoperturbation). In the type I situation, the potential corresponding to a homoperturbation is weakly attractive for both electron and hole levels: one expects a small decrease of the transition energy between these two levels due to the perturbation.

On the other hand, in the type II situation (Fig. 1 c or d), the homoperturbation will result in a strong repulsion of the level confined in the perturbed layer and in a weak attraction of the level confined in the well material: one then expects an increase of the transition energy (at least in the limit of a wide perturbed layer).

This technique has been tested on the GaAs/AlAs system, which is known [9,10] to be a type II system for electron states built from X conduction band minimum and Γ hole states (see Fig. 2 a). Fig. 3 shows the low temperature optical spectra obtained

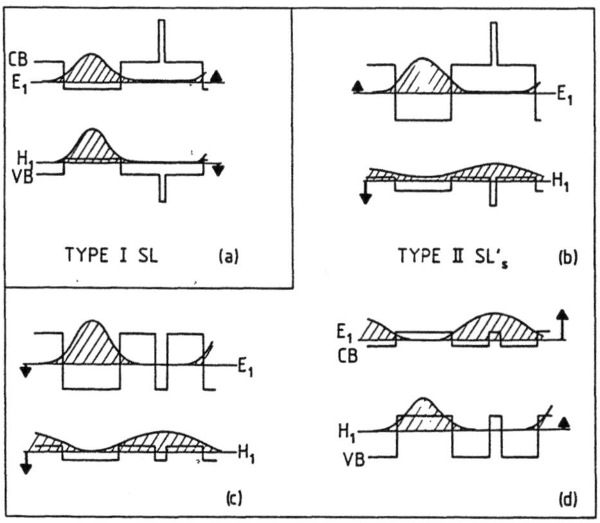

Fig. 1. Schematic view of the effect of a localized perturbation in a superlattice structure, for various band configurations. The arrows indicate the amplitude and sign of the levels' shifts due to the perturbation.

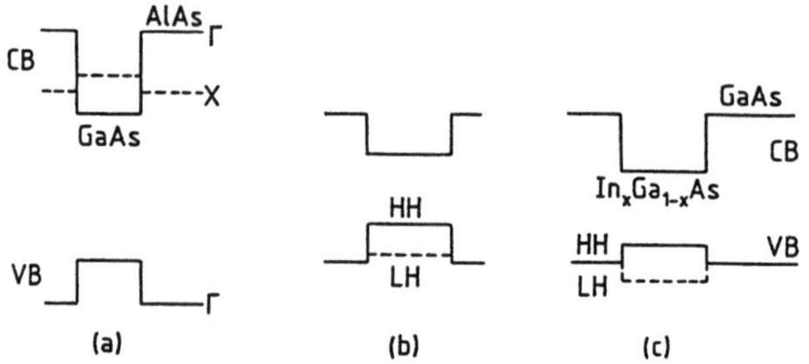

Fig. 2: a) Energy band configuration in a GaAs/AlAs SL for the conduction band (at Γ and X) and valence band extrema. b) and c): Two possible configurations for the Γ conduction and valence band extrema in a GaInAs/GaAs superlattice.

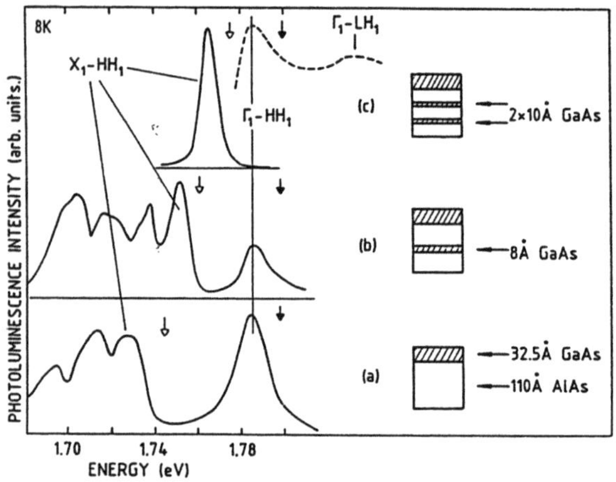

Fig. 3: 10K PL (solid line) and PL excitation (dashed line) spectra for a GaAs/AlAs superlattice, unperturbed in a) and perturbed in b) and c). The composition of one period is shown for each SL. Open (full) arrows mark the calculated energy of the X_1-HH_1 (Γ_1-HH_1) transitions. (Exciton binding energies are not included.)

on a reference (a) and on homoperturbed GaAs/AlAs samples. While the type I Γ_1-HH_1 transition is insensitive to the perturbation, the type II X_1-HH_1 transition is strongly affected by the insertion of one (b) or two (c) thin GaAs layers in the AlAs layers. Moreover, this latter transition indeed shifts to higher energy, as expected from the above analysis.

The band discontinuities are more controversial in the $In_xGa_{1-x}As$/GaAs system. In structures grown on GaAs, the $In_xGa_{1-x}As$ material is strained. The confining potentials for heavy and light holes are different, due to the strain-induced lifting of the $In_xGa_{1-x}As$ bulk material zone center valence band degeneracy. Concerning the electron and heavy hole, this system is of type I, as evidenced by the large oscillator strength of the first excitonic transition (E_1-HH_1). Two valence band configurations are still possible (Fig. 2b and c), leading to a type I or II situation for the electron and the light holes. Menendez et al. [11] reported that for x=0.05, the band configuration is that displayed in Fig. 2

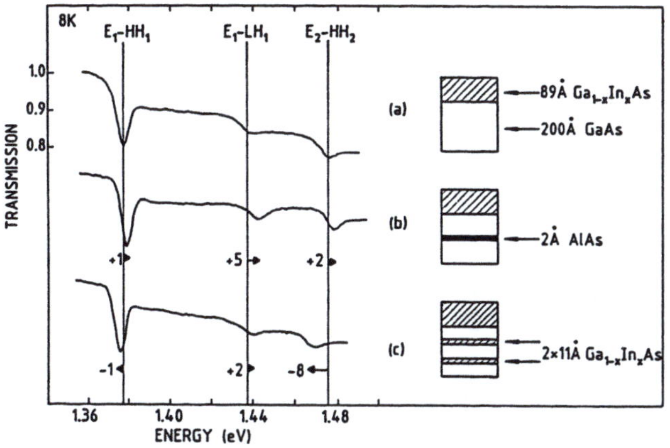

Fig. 4: 10K transmission spectra of a $Ga_{0.85}In_{0.15}As$/GaAs superlattice, unperturbed in a) and perturbed in b) and c). The composition of one period is also shown for each SL.

b, whereas the alternate type II configuration (Fig. 2 c) for light holes was found for x=0.15 [12]. However, for this latter indium content, the height of the potential barrier for the light holes is very small (about 15 meV). We applied the perturbation techniques described above to samples with x=0.15. Figure 4 shows the low temperature transmission spectra for a reference $In_{0.15}Ga_{0.85}As$/GaAs multiple quantum well structure (a), and two perturbed structures. In (b) a repulsive AlAs layer was introduced in the GaAs layers whereas a homoperturbation (here with two layers) was used in (c). The spectrum of Fig.2 b) shows that the E_1-LH_1 transition is much more sensitive to the insertion of AlAs in the GaAs layers than the E_1-HH_1 transition: the LH_1 level has to leak much more in the GaAs layers than E_1 or HH_1 (or even E_2). The transmission spectrum of Fig 4 b) obtained on the sample with a homoperturbation confirms directly that the electrons and light holes are confined in $In_xGa_{1-x}As$ and GaAs respectively . In this case, the shift of the E_1-LH_1 transition on the one hand and of E_1-HH_1 and E_2-HH_2 on the other hand are of opposite sign: this is clear evidence of a type II situation, as discussed above.

These two examples show that this perturbation technique can be applied successfully to the determination of the band configuration in superlattices. Other techniques have yielded direct results on the location of a given confined state in a superlattice. Among these, the study of the in-plane mobility as a function of the location of charged impurities was used successfully in Si/SiGe superlattices to determine the confining layer for the first electron level [13]. This elegant method is however restricted to the lowest electron (or hole) level, whereas the perturbation technique can yield also information on the excited levels, as shown in the case of $In_xGa_{1-x}As$/GaAs.

4 Extraction of the envelope probability densities

As discussed in section 2, the probability densities (PDs) for a discrete set of levels can be probed locally, and extracted from the shifts of the levels due to a localized

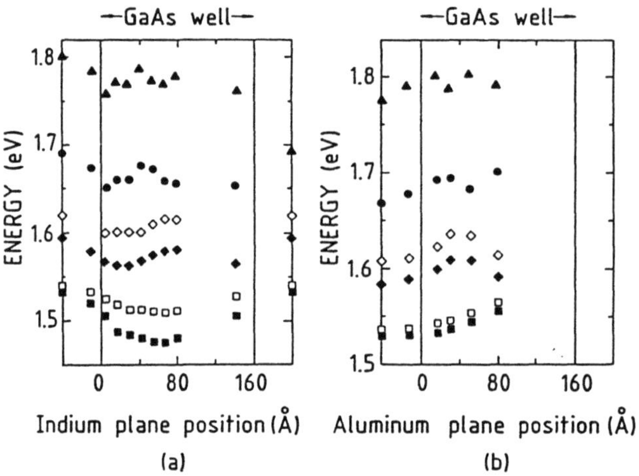

Fig. 5: Experimental transition energies at 10K, as functions of the position of the isoelectronic probe (a):In plane; b):Al plane). The dots plotted at -40Å and 200Å are those of the reference sample.■, E_1-HH_1;□, E_1-LH_1;◆, E_2-HH_2;◇, E_2-LH_2;●, E_3-HH_3;▲, E_4-HH_4. For clarity, the E_1-HH_3 transitions are not indicated.

perturbation. When the position *a* of the probe is varied, the spatial dependance of the PDs is mapped. This technique has been successfully applied to the case of a GaAs/$Ga_{0.7}Al_{0.3}As$ QW [7] (GaAs: 200Å; GaAlAs: 85Å); 1 monolayer thick attractive (InAs) and repulsive (AlAs) probes have been used in two independent series of experiments. For each kind of probe, a series of 6 period multi-QW samples has been grown by MBE (one for each probe position) and studied by PL excitation and wavelength-modulated reflectivity at low temperature (10K). The experimental transition energies of the QWs are plotted in figure 5 as functions of the location of the probe. The energy shifts of E_n-HH_n or E_n-LH_n transitions already reflect the shapes of the PD in the nth electron or hole level (since these are roughly similar for a type I system), and are as expected of opposite signs for attractive and repulsive probes. Finally, they are large compared to the experimental uncertainties (1 meV) but also of the order of the spacing between the transitions so that the coupling of the QW levels induced by the perturbation has to be taken into account. In this experiment, we have obtained the PDs for the first three bound electron levels. For that purpose, one has first to extract for each location *a* of the probe the energy differences E'_n-E_m (n,m $\leq$3) which appear in equ. (5) of section 2 from the E_n-HH_n (n$\leq$3), E_n-LH_n (n$\leq$2), E_1-HH_3 and E_2-LH_1 optical transition energies in the corresponding sample, and in a reference (unperturbed) QW. Details of this procedure have been extensively described elsewhere [7,14]. We have used Equ. (5), with p=3, and normalized each $|F_i|^2$ on one period of the multi-QW structure to eliminate the W_i's. This constitutes a major advantage of this technique; no detailed knowledge of the perturbing potential is required to extract the PDs. The reproducibility of the planar probe from sample to sample (at a submonolayer scale) is far more important in this experiment than the precise strength of the perturbing potential.

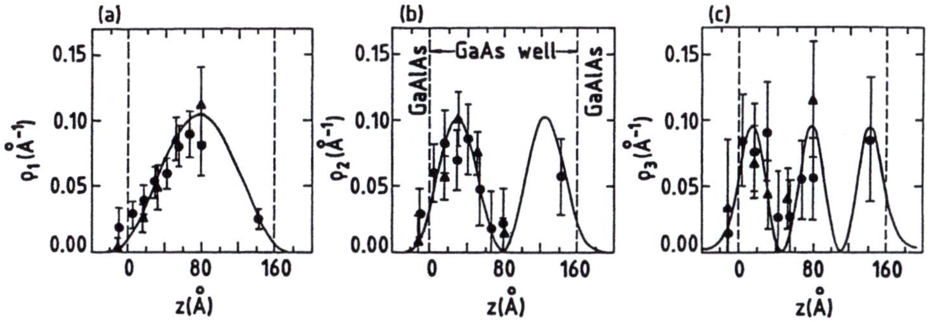

Fig 6: Experimental PDs $\rho_i = |F_i|^2$ for the first three bound electron levels, compared to the theoretical effective-mass result. Dots and triangles indicate the data obtained using the In and Al probes, respectively.

The "experimental" PDs which result from this procedure are in good agreement with the theoretical effective mass estimate, for both kinds of probes, as shown in figure 6. The error bars on the PDs are plotted for 3 standard deviations; they include the effect of some simplifying assumptions in the treatment of the data and the errors in the determination of the transition energies. The lack of precision of the higher transition energies is mainly responsible for the increasing error bars on $|F_1|^2$, $|F_2|^2$ and $|F_3|^2$.

The fundamental limitations of this technique are related to the cut-off in the spatial frequencies of the experimental PDs, which is introduced by the sampling of *a* (10-15 Å between successive values), and the fact that a finite set of levels is considered in our treatment of the data. The latter point introduces an error in the experimental PDs of the same typical variation length as the first neglected level F_4 ($d/(4\pi)$= 12 Å), and whose amplitude is smaller for the lower energy levels. The effect of the finite width (3Å) of the probe is negligible, when compared to these limiting factors.

5 Conclusion

We have shown that the insertion of built-in isoelectronic planar probes is a powerful technique to study locally the electronic properties of heterostructures. It is beyond the scope of this paper to discuss its numerous possible other applications. In addition to the experiments described in Sections 3 and 4, it can for example be used to compare the behaviour of the two interfaces of a QW, by studying a series of samples with probes located near one or other of the interfaces. Information on the probe itself can also be obtained when it is inserted in a well-controlled quantum well, and we were in fact able to characterize indium segregation from the results presented in Section 4.

Moreover, the insertion of isoelectronic ultrathin layers can clearly be used to tailor the electronic properties (energy levels, oscillator strengths) of the hetrostructures. Fig. 5 shows that the energy of the levels of a quantum well can be tuned somewhat independently by inserting one or several attractive and (or) repulsive perturbations.

Due to this flexibility, ultrathin isoelectronic layers constitute an attractive tool for heterostructure studies and design.

References

1 H. Mariette, J. Chevallier and P. Leroux Hugon, Phys.Rev B **21**, 5706 (1980).
2 H. Mariette, Solid State Comm. **38**, 1193 (1981).
3 W. Ruhle, W. Schmid, R. Meck, N. Stath, J.U. Fischbach, I. Strottner, K.W. Benz and M. Pilkhun, Phys. Rev B **18**, 7022 (1978) and references therein.
4 P.J. Dean, Journal of Luminescence **1,2**, 398 (1970).
5 T.C Hsieh, T. Miller and T.C. Chiang, Phys. Rev. Lett. **55**, 2483 (1985)
6 J-M Gerard and J-Y Marzin, Phys Rev B **40**, 6450 (1989)
7 J-Y Marzin and J-M Gerard, Phys. Rev. Lett. **62**, 2172 (1989)
8 See e.g. M. Jaros and K.B. Wong, J. Phys. C **17**,L765 (1984)
9 E. Finkman, M.D. Sturge and M.C. Tamargo, Appl. Phys. Lett. **49**, 1299 (1986)
10 G. Danan, B. Etienne, F. Mollot, R. Planel, A.M. Jean-Louis, F. Alexandre, B. Jusserand, G. Le Roux, J-Y Marzin, H. Savary and B. Sermage, Phys. Rev B **35**, 6207 (1987)
11 J. Menendez, A. Pinczuk, D.J. Werder, S.K. Sputz, R.C. Miller, D.L. Sivco and A.Y. Cho, Phys. Rev. B **36**, 8165 (1987)
12 J-Y Marzin, M.N. Charasse and B. Sermage, Phys. Rev. B **31**, 8298 (1985)
13 G. Abstreiter, H Brugger, T Wolf, H. Jorke and J.H. Herzog, Surf. Science 174, 640 (1986)
14 J.Y. Marzin and J.M. Gerard, in "Spectroscopy on Semiconductor Microstructures" (Venice, May 1989), G. Fasol, A. Fasolino and P. Lugli editors, Proc. Adv. Res. Work. B **206**, p305, Plenum (1990).

Band-Gap Renormalization in Quantum Wells Studied with High Excitation Luminescence in High Magnetic Fields

J.C. Maan [1], *M. Potemski* [1], *K. Ploog* [2], *and G. Weimann* [3]

[1] Grenoble High Magnetic Field Laboratory,
Max-Planck-Institut für Festkörperforschung, 166X, F-38042 Grenoble, France
[2] Max-Planck-Institut für Festkörperforschung, Heisenbergstr. 1,
D-7000 Stuttgart 80, Fed. Rep. of Germany
[3] Walter Schottky Institut, Technische Universität München,
Am Coulombwall, D-8046 Garching, Fed. Rep. of Germany

Abstract. Luminescence spectra under high excitation in quantum wells in high magnetic fields show pronounced Landau levels which are shifted to lower energy with respect to the separately measured absorption maxima. This shift does not depend on the excited carrier density and suggests a rigid, density independent, renormalization of the effective gap, contrary to theoretical and experimental understanding. A possible explanation based on the assumption of short lived pair correlated states embedded in a renormalized continuum is proposed.

1. Introduction

Under high power excitation, GaAs quantum wells between GaAlAs cladding layers can be filled with an appreciable number of electrons and holes. For excitation pulse lengths longer than the relaxation time within the bands and longer than the recombination time of electrons and holes, this excess carrier density may be considered as being in a dynamic quasi-static equilibrium. Due to many-body effects the energy level structure and thereby the optical properties will change (effective mass and energy-gap renormalization). These effects have been extensively studied in quantum wells, both theoretically [1-4] and experimentally [5-11], because of the fundamental interest of how such effects are affected by the bi-dimensionality of the carriers.

Theoretically various models predict a density dependent renormalization of the band-gap, which for purely two-dimensional systems is given by [1]

$$\Delta G=5.33\; r_C^{-\left(\frac{2}{3}\right)}, \qquad (1)$$

where r_C is the interparticle spacing expressed in units of the 3D Bohr radius $a_B{}^{*}=4\pi\epsilon\epsilon_0\hbar^2/\mu e^2$ and ΔG is in units of the 3D Rydberg energy

Springer Series in Solid-State Sciences, Vol. 97
Localization and Confinement of Electrons in Semiconductors
Editors: F. Kuchar · H. Heinrich · G. Bauer © Springer-Verlag Berlin, Heidelberg 1990

$Ry^*=e^2/2(4\pi\epsilon\epsilon_0)a_B^*$, μ being the reduced effective mass and ϵ the dielectric constant of the material. For GaAs these parameters correspond to 16 nm and 3.7 meV respectively. Different theoretical models give a similar dependence of ΔG on n_C [1-3] for purely 2D, while ΔG is shown to decrease at a given density when the finite thickness of the wells is taken into account [4]. It is known that at low densities the spectrum is dominated by the pairing of electrons and holes in an excitonic state, below the band-edge. Schmitt-Rink et al. [12] have calculated the expected luminescence lineshape under high excitation as a function of the density, taking into account both the gap renormalization and the behaviour of the exciton. Their results show that the exciton remains approximately at the same energy while the band-edge luminescence decreases with density. From a certain density the exciton merges with the decreasing band-edge, loses oscillator strength and remains visible as a slight enhancement of the luminescence at the Fermi energy ("Fermi-edge singularity" or "Mahan exciton"[12,13]).

Experimentally, however, a broadening of the luminescence line, both on the low and on the high energy side with increasing pump intensity is observed, while the position of its maximum remains approximately at the same energy [5-11], which apparently contradicts the theory. The broadening on the high energy side is easily explained as due to the band filling and finite carrier temperatures. However, the luminescence tail on the low energy side, which occurs at energies within the fundamental gap, can be interpreted as an experimental indication for bandgap renormalization. The only way to obtain quantitative information from results of luminescence is to analyze the lineshape [14-16]. This can be accomplished by filling hole and electron band with an equal number of carriers, taking into account Fermi statistics and summing all possible recombination transitions. With band filling a tail on the high energy side is obtained which has a decay given by the effective electron and hole temperature and a position related to the carrier density. The fit parameters are then the carrier densities, the carrier temperatures and the renormalized gap. Although the high energy tailing can be fitted easily, to explain the low energy tail further assumptions are needed. This is because the joint density of states for a two-dimensional system, taking into account momentum conservation, shows a sharp onset at the band-edge and remains constant at higher energies, which is not observed experimentally. To explain this phenomenon, it is assumed by some authors [17,18] that the many-body effects destroy momentum conservation. This way the number of possible transitions increases quadratically with increasing energy and explains the low energy tail. Another mechanism is the so called Landsberg broadening [19,15]. In this mechanism it is assumed that an electron at the conduction band-edge recombining with a hole at the valence band-edge leaves a hole in the Fermi sea of the remaining electrons (and similarly for the electron on the top of the valence band); these empty states within full bands will be filled up very rapidly by other particles and are therefore supposed to have a short lifetime, which implies that the states are broadened. This broadening is very important for states far away from the Fermi energy (and thus close to the band-edge) and vanishes for states close to the Fermi energy.

In the absence of renormalization of the gap, both mechanisms would yield a luminescence lineshape with a maximum which moves to higher energy as the density is increased. Since experimentally the maximum remains roughly at the same energy the line is shifted to lower energy by reducing the gap and the renormalization is obtained in this way. In both cases it is assumed that the exciton does not exist anymore as a consequence of the presence of the other carriers which cause band-filling [1] and screening effects.

The agreement between theory and experiment analyzed in this fashion is rather convincing and renormalization of the band gap in 2D appears to be a closed subject [6]. In the following we will show how optical measurements in high magnetic fields can hardly be reconciled with this picture and give rise to new questions.

2. Results in a magnetic field

It is interesting to perform the same experiment, i.e. luminescence under high excitation in quantum wells, in a magnetic field for different reasons. In the first place because the featureless density of states at zero field will be split in discrete Landau levels, separated by the cyclotron energy $\hbar\omega_C$ (ω_C being the cyclotron frequency) and having the same degeneracy for electrons and holes (eB/h). These facts make the interpretation of the data much easier because the highly structured joint density of states shows peaks at each Landau level transition. Therefore if the gap is reduced each peak will move to lower energy and the renormalization can be read directly from the spectra. Furthermore, the density can be obtained from the saturation intensity of each interband Landau level transition, occurring when both states are full. The second reason is that it has been theoretically argued [20,21] that there are interesting effects due to carrier-carrier interaction for particles in discrete Landau levels of electron and holes, and that these effects depend strongly on the filling factor ν=n_C (eB/h) of the Landau levels. In fact this situation is analogous to that of the fractional quantum Hall effect, with the difference that there there is only repulsion between carriers with the same charge, whereas here there is both repulsion and attraction.

Fig. 1 shows the luminescence lineshape under high excitation as a function of the magnetic field [22]. At zero field the lineshape is similar to that obtained earlier [5-11]. In a magnetic field a clear structure

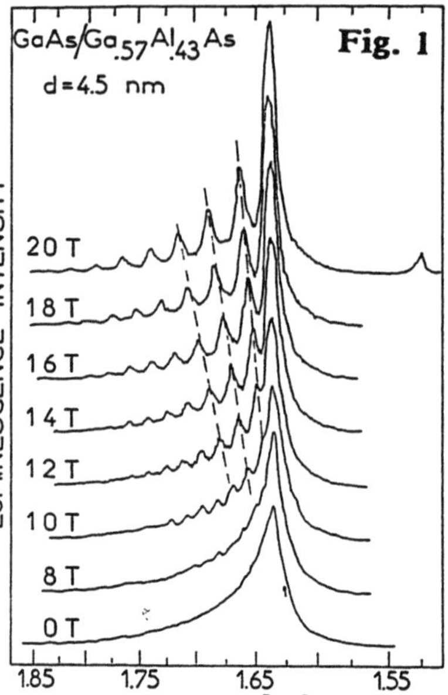

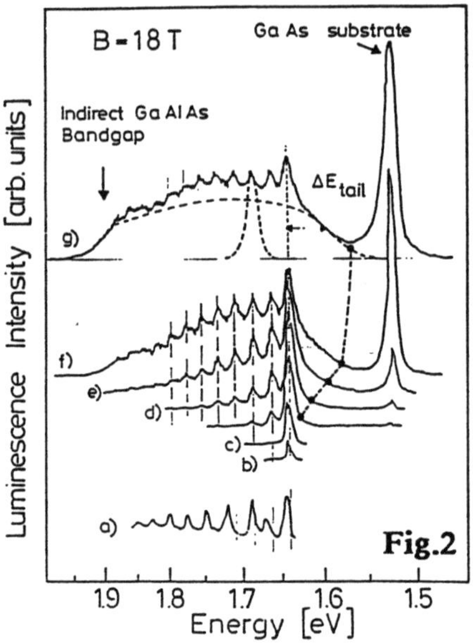

Fig. 1
Luminescence spectra under high excitation of a 4.5 nm GaAs quantum well as a function of the magnetic field.

Fig.2
Photoluminescence excitation (pseudo-absorption) a) and luminescence spectra at 18 T for different excitation intensities b) 3.2×10^3 W/cm^2, c) 10^4 W/cm^2, d) 3.2×10^4 W/cm^2, e) 10^5 W/cm^2, f) 3.2×10^5 W/cm^2 and g) 10^6 W/cm^2. The dashed lines illustrate the possible deconvolution of the spectrum as described in the text. Note that b) to g) are all drawn to scale and that the maximum intensity saturates.

develops at the high energy side which evolves in a field as expected from Landau quantization. In Fig. 2 results are shown as function of excitation intensity (≡carrier density) at fixed magnetic field; it can be seen that with increasing intensity higher Landau levels are successively filled and become visible in emission. Furthermore a clear development of a tail towards lower energies and extending appreciably into the energy gap can also be observed at higher intensities. This tail does not show any structure and is rather broad compared to the relatively sharp structure on the high energy side. For comparison the conventional quasi-absorption spectra (photoluminescence excitation) are also shown in the lower trace. From this comparison it can be seen that the peaks visible in the emission from filled states are quite different from those measured in absorption into empty states, which

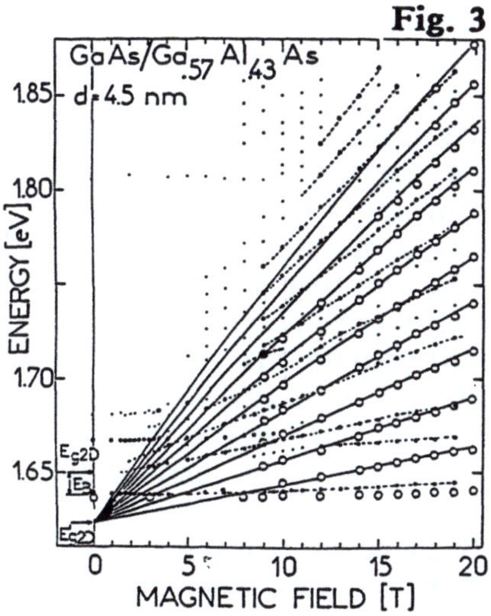

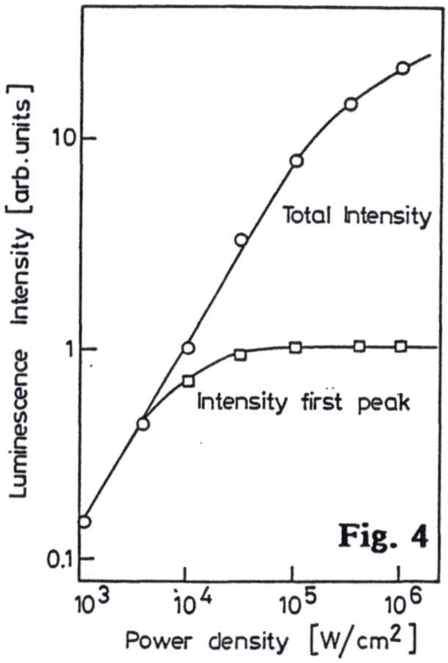

Fig. 3
Magnetic field dependence of the maxima in emission under high excitation (open circles joined by solid lines which represent a fit using a non-parabolic bands [25]) and in pseudo absorption spectra (dark circles, joined by dashed lines as a guide to the eye).

Fig. 4
Dependence of the integrated luminescence intensity and that of the first peak as a function of the exciting power density, illustrating the saturation of the first peak.

demonstrates a change in the optical properties due to the presence of a high carrier density. This change is visible even more clearly in Fig. 3 where the field dependence of the absorption maxima is compared to that of the emission maxima.

The absorption spectra show the usual [23,24] behaviour. For instance at around 4T one can distinguish with increasing energy, respectively, the heavy hole exciton ground state (1s); its first excited state (2s) increasing in energy with field and approaching the behaviour of a free interband Landau level transition; the second excited state (2s); the light hole ground state exciton and all higher excited states. The unrenormalized band gap can be estimated rather precisely from these data as being slightly above the extrapolation to zero field of the higher excited states and is indicated in the figure. The field dependence of the high intensity luminescence peaks can be quite well described with calculated interband Landau level transitions taking into account the GaAs CB non-parabolicity and this calculated dependence is shown as the drawn lines [25]. All transitions, except for the ground state are well described in this manner. From this agreement a clearly lower band-edge is obtained and it is tempting to attribute the change in the band-edge between low and high excitation as the renormalization of the gap. Different samples with different well thicknesses show the same behaviour and the change in the gap is found to be proportional to the heavy hole exciton binding energy; the proportionality constant being two for samples with 3nm up to 14 nm well thickness.

Fig. 4 shows the dependence of the total luminescence intensity and that of the first peak as a function of the exciting power density. It can be seen that the total intensity is linearly dependent on the power, showing that the carrier density is proportional to the power. At the highest intensities, the well is becoming full with carriers and they start to escape into the barrier. In this particular sample the barrier material was indirect gap GaAlAs and therefore these carriers disappear from the luminescence. The high energy cut-off in Fig. 2 corresponds to the energy at which this starts to occur. At the same time deviations from linearity in Fig. 4 are also observed. The intensity of the lowest peak saturates at a certain power and remains constant showing that these states are full. Once the Landau level degeneracy is known, the number of carriers in this state can be obtained. This saturation behaviour of the height of the peaks is important because it proves that the luminescence does indeed come from the same spot on the sample and at the same time. Both spatial and temporal inhomogeneities would manifest themselves as a stronger, never saturating, luminescence at the low energies. For spatial inhomogeneity this is because carriers having diffused away from the region of high density in the center to a lower density region at the edge of the spot may relax to lower energies and then recombine. In the case of temporal inhomogeneity this is so because after the pulse the high energetic carriers have already recombined and the low energy ones still give luminescence.

It is important to exclude these spurious effects since the behaviour described before is in flagrant contradiction with conventional wisdom [1-11], because the gap renormalization obtained from the data in the described manner is completely independent of the density. This can be seen directly from Fig. 2 where none of the luminescence peaks move with intensity. If the gap is decreasing with density one would expect that all Landau level peaks would decrease in energy with increasing density, instead they stay remarkably constant. This contradiction is shown more clearly in Fig. 5 where results of experimental data and several theoretical models are compared with the shift in the band-edge determined from the magneto-optical data. The densities in the magneto-optical data are determined either from a lineshape analysis [25,26] using the appropriate density of states in a magnetic field (Lorentzian broadened Landau levels including selection rules [25,26]) or from the saturation behaviour and the Landau level degeneracy, the two techniques giving similar results within 25%.

3. Discussion

A notable feature of Fig. 2 is the development of the low energy tail, extending into the gap, also in the presence of Landau quantization. This tail, as illustrated in Fig. 2 trace g), is much broader than the distance between the Landau level peaks, it is surprisingly featureless as compared to the well-pronounced structures at the higher energy side and does show a clear dependence on the exciting intensity. Previous experimental work [5,6], which determined the renormalized gap from the lineshape analysis at zero magnetic field, has shown that the fitted position of the gap was somewhere in the low end of this low energy tail. In order to compare the magnetic field data with these results, the shift of the position where the luminescence intensity is 10% of its maximum (see Fig. 2), with respect to that of the unrenormalized gap determined from the pseudo-absorption spectra is plotted in Fig. 5. This quantity does indeed behave rather similarly to the renormalization of the gap determined previously experimentally or theoretically. The choice of 10% is of course somewhat arbitrary, but any other definition would give a similar density dependence although somewhat different absolute values.

When this tail is interpreted as done originally by Landsberg[19], namely as being due to the lifetime broadening of the empty state left in the Fermi sea when an electron and a hole recombine, it is tempting to estimate this lifetime broadening from the width of the tail. For the curve g) in Fig. 2 this corresponds to a lifetime of about τ=1.2 10^{-14} s. If all the states are broadened due to scattering between the carriers due to this scattering time, then the $\omega_C\tau$ would be less than unity, and it would be impossible to see the Landau quantization. On the other hand clear Landau levels are observed in emission, and from the fit of the emission spectra using the density of states with Lorentzian broadened Landau levels, their lifetime can be determined from the broadening. In Fig. 6 the fitted level width and the corresponding lifetime are shown. The τ_{LL} is found to decrease with increasing densities and for the highest densities a value of about 10^{-13}s is obtained. If all states are broadened with these time constants then there should not be such a broad tail at the low energy side as is experimentally observed.

This apparent contradiction leads therefore to the suggestion that the data can perhaps be understood by assuming that there are two coexisting phenomena in the electron hole plasma. On one hand, some of the electrons and holes are continuously and rapidly scattered by each other and have a very short momentum lifetime and a very short mean free path. These behave like a normal plasma and show a low

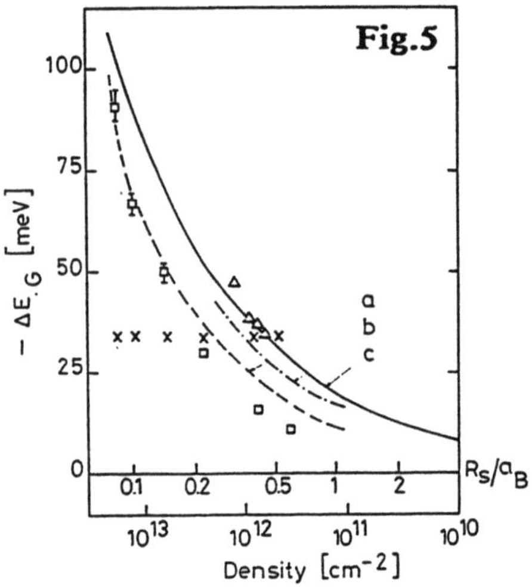

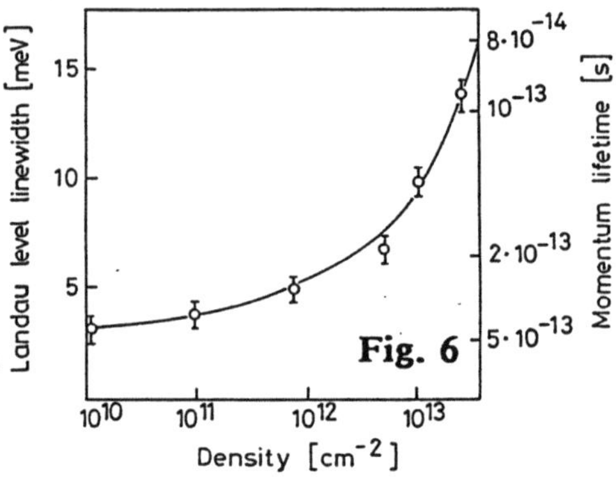

Fig.5
Bandgap renormalization as a function of density or interparticle spacing. a) theoretical ref. [1] and [2]; b) theoretical ref. [4]; c) empirical ref. [11]; from lineshape analysis [5], data points (Δ); experimental, from the position of 10% of the peak luminescence intensity (■) (see Fig. 2); and from the shift of the magneto-optical luminescence spectra with respect to the pseudo-absorption spectra (x) (see Fig.3).

Fig. 6
Landau level linewidth in the luminescence spectra and corresponding momentum lifetime as a function of the carrier density.

energy tail which can be interpreted as a renormalized gap. On the other hand, some of the carriers must have a relatively long momentum lifetime and perform a cyclotron orbit several times without being scattered, and in some sense do not interact with the others. These particles give rise to sharp peaks at fixed, density-independent energies. Since the cyclotron orbit is of the order of 10 nm, the motion of these particles sees only a small part of the two-dimensional area over a comparatively long time. This behaviour is illustrated pictorially in Fig. 7, and the contribution to the spectra of these two classes of particles is sketched in Fig. 2. The density dependence of the Landau level momentum lifetime τ_{LL} as plotted in Fig. 6 can then be interpreted as a demonstration that the lifetime of the coherent motion in an increasingly dense plasma becomes shorter and shorter and eventually approaches the momentum relaxation time τ of the uncorrelated plasma. In this sense the experiment in a magnetic field is only sensitive to those states which live sufficiently long in a local, well-defined, orbit and the number of those states diminishes with density. In the same way the anomalous behaviour of the exciton ground state under

Fig.7
Pictorial representation of the dynamics of the plasma in a magnetic field showing the two coexisting behaviors as described in the text.

high excitation (see Fig. 3) which shows a quadratic field dependence, which is characteristic for a bound electron-hole state, can also be understood, because this state also corresponds to a local (the Bohr radius of the exciton is also about 10 nm) correlation. Eventually, as the density increases these 'local, short lived, resonances within a continuum' will have a shorter and shorter lifetime and merge with the continuum.

4. Conclusions

Luminescence under high excitation in quantum wells in high magnetic fields has shown a clear Landau-level-like structure which is linearly dependent on the magnetic field but which does not depend on density. If the interband Landau level transitions occur at energies of the renormalized gap plus the energy of the quantized bands in a field then the experiment shows a density independent renormalization of the gap, which contradicts present theoretical and experimental understanding. A more speculative explanation is given which is based on the assumption of short lived local resonances at fixed energy observed as peaks in the spectra which coexist within a strongly interacting plasma giving rise to the renormalized gap.

References

1. S.Schmitt-Rink and C.Ell, J. Lumin.30,585,(1985).
2. D.A.Kleinman, Phys. Rev. **B33**,2540,(1986)
3. D.A.Kleinman and R.C.Miller, Phys. Rev. **B32**,2266,(1985)
4. G.E.W.Bauer, Proc. 19th Int. Conf. Phys. Sem. Warsaw 1988, p.143, ed. W.Zawadzki, IPPAS, Warszawa, (1989)
5. G.Tränkle, H.Leier, A.Forchel, H.Haug, C.Ell and G.Weimann, Phys. Rev. Lett. **58**,419,(1987)
6. G.Tränkle, E.Lach, A.Forchel, F.Stolz, C.Ell, H.Haug, G.Weimann, G.Griffiths, H.Kroemer and S.Subbanna, Phys. Rev. **B36**, 6712,(1987)

7. G.Tränkle, E.Lach, F.Scholz, P.Wiedemann, K.W.Benz, and M.Razeghi, J. Superl. and Microst. **3**,21,(1987)
8. R.C.Miller, D.A.Kleinman, O.Munteanu and W.T.Tsang, Appl. Phys. Lett. **39**,1,(1981)
9. Z.Y.Xu, V.G.Kreismanis and C.L.Tang, Appl. Phys. Lett. **43**,415,(1983)
10. R.L.S.Devine and W.T.Moore, Solid State Commun. **55**,177,(1988)
11. S.Tarucha, H.Kobayashi, Y.Horikoshi and H.Okamoto, Jpn. J. Appl. Phys. **23**,874,(1985)
12. S.Schmitt-Rink, C.Ell and H.Haug, Phys. Rev. **B33**,1183,(1986)
13. G.D.Mahan, Phys. Rev. **153**,882,(1967)
14. G.Tränkle, H.Leier, A.Forchel and G.Weimann, Surface Sci. **174**,211,(1986)
15. R.W.Martin and H.L.Störmer, Solid State Commun. **22**, 523,(1977)
16. H. Lobentanzer, H.J. Polland, W.W,Rühle, W.Stolz and K.Ploog, Phys. Rev. **B36**, 1136,(1987)
17. S.Tanaka, H.Kobayashi, H.Saito and S.Shionoya, J. of Phys. Soc. Jpn **49**, 1051, (1980)
18. R.W.J.Hollering, T.T.J.M.Berenschot, H.J.A.Bluyssen and F.Roozenboom, Phys. Rev. **B38**, 13323,(1988)
19. P.T.Landsberg, Phys. Stat. Solidi **16**,623,(1966)
20. D.Paquet, T.M.Rice and K.Ueda, Phys. Rev **B32**, 5208, (1985)
21. G.E.W.Bauer, Phys. Rev. Lett. **64**, 60, (1990)
22. M.Potemski, J.C.Maan, K.Ploog and G. Weimann, Proc. 19th Int. Conf. Phys. Semicon., Warsaw 1988, 119, ed. W.Zawadzki, IPPAS, Warszawa, (1989)
23. F.Ancilotto A.Fasolino and J.C.Maan, Phys. Rev. **B38**, 1788, (1988)
24. L.Vina, M.Potemski, J.C.Maan, G.E.W.Bauer, E.E.Mendez and W.I.Wang, J. Superl. and Microstr. **5**, 371, (1989)
25. J.C.Maan, M.Potemski, K.Ploog and G.Weimann, Proc. NATO workshop on "Spectroscopy of Semiconductor Microstructures", Venice 1989, ed. by G.Fasol, A.Fasolino and P.Lugli, ASI series B, Plenum Press, New York to be published.
26. M.Potemski, J.C.Maan, K.Ploog and G.Weimann, Proc. EP2DS8, Grenoble 1989, Surface Sci. **229**, 380, (1990).

Excitons in Doped Quantum Wells

G.E.W. Bauer

Philips Research Laboratories, 5600 JA Eindhoven, The Netherlands

Abstract. A mean-field theory of excitons in charged electron-hole plasmas is presented, with emphasis on the effects of confinement, finite temperature and external magnetic fields. Absorption and photoemission spectra are found to be strongly influenced by bound excitonic states.

1. Introduction

Investigations of the optical properties of modulation-doped quantum wells with [1 – 15] and without [16 – 27] applied magnetic fields are largely motivated by the hope of finding novel many-body effects in quasi-two-dimensional systems. Here attention is focussed on the optical properties of negatively charged plasmas due to n-type modulation doping under weak excitation conditions. In two recent communications [14,15] a mean-field (or Hartree-Fock) theory of electron-hole plasmas in quantum wells has been introduced. In the present article new material is presented and the theoretical treatment and the results are discussed in more depth than was possible in Refs. 14,15. The electron-hole pairing effects, which can be important in neutral electron-hole plasmas as created by optical excitation, vanish when the density of one carrier overwhelms that of the other and are not considered here. A discussion of neutral plasmas is deferred to a forthcoming workshop [28].

Expressions for the absorption and photoemission spectra in mean-field theory are given in Section 2, taking into account the effects of confinement, magnetic fields, finite temperatures and excitonic bound states. Numerical calculations of the obtained formulae are carried out and the results are presented in context with published experimental results in Section 3. Mean-field and other theories are critically discussed in Section 4.

2. Optical spectra of modulation-doped quantum wells

The optical properties, *i.e.* the linear response to a weak photon probe, can be expressed in terms of the two-particle electron-hole Green function [29,30]. Assuming thermal equilibrium ($\beta = k_B T$) at an applied magnetic field normal to the interfaces and disregarding higher subband transitions the gain/absorption spectra are proportional to ($\ell = c\hbar/eH$ is the magnetic length and m and n Landau level indices of free electron-hole pair states):

$$I^{abs}(\hbar\omega) = \frac{1}{2\pi\ell^2}\frac{1}{\pi}\,\mathrm{Im}\sum_{mn} G_{eh}(m,n;\,\hbar\omega + i\delta) \tag{1}$$

Springer Series in Solid-State Sciences, Vol. 97
Localization and Confinement of Electrons in Semiconductors
Editors: F. Kuchar · H. Heinrich · G. Bauer

and the photoluminescence is obtained as [30]

$$I^{\mathrm{lum}}(\hbar\omega) = I^{\mathrm{abs}}(\hbar\omega)/(e^{\beta(\hbar\omega-\mu_e-\mu_h)}-1), \tag{2}$$

where μ_e and μ_h are the (quasi) chemical potentials of electrons and holes.

The electron-hole Green function in the mean-field approximation is conveniently calculated by making use of a biorthogonal expansion described by Stolz [31]. The final equations for the optical spectra at arbitrary particle densities (but in the absence of electron-hole pairing [15]) are summarized below in a form which can be directly coded for numerical computations. The sum over states in the electron-hole Green function becomes

$$\sum_{mn} G_{eh}(m,n;\hbar\omega) = \sum_i \frac{\mathrm{sgn}\{\lambda_i-\mu_e-\mu_h\}}{\hbar\omega-\lambda_i} \left|\sum_n \phi_i(n)\sqrt{|F(n)|}\right|^2 \tag{3}$$

where ϕ_i (assumed normalized) and λ_i are the solutions of the eigenvalue problem

$$\sum_n [E(m)\delta_{mn} - \mathrm{sgn}\{F(m)\}\sqrt{|F(m)|}\,V_{eh}(m,n)\sqrt{|F(n)|}\,]\phi_i(n) = \lambda_i\phi_i(m). \tag{4}$$

The factors

$$F(n) = 1 - f_e(n) - f_h(n), \quad f_a(n) = \frac{1}{e^{\beta(\varepsilon_a(n)-\mu_a)}+1} \tag{5}$$

take account of the phase space occupation of states with energies

$$E(n) = \varepsilon_e(n) + \varepsilon_h(n), \tag{6}$$

where

$$\varepsilon_a(n) = \varepsilon_a^{\mathrm{sub}} + \frac{eB\hbar}{m_a}(n+\frac{1}{2}) - \sum_m f_a(m)V_{aa}(n,m). \tag{7}$$

The third term of Eq. (7) is the Hartree-Fock exchange self-energy of particle type a. The Coulomb matrix elements read

$$V_{ab}(n,m) = \frac{e^2}{\kappa\ell}\int dq\,|J_{nm}(q)|^2 F_{ab}(q). \tag{8}$$

If $M = \max(m,n)$ and $N = \min(m,n)$

$$J_{nm}(q) = \sqrt{\frac{N!}{M!}}\,e^{-\ell^2q^2/4}(\ell q/\sqrt{2})^{N-M} L_M^{(N-M)}(\ell^2q^2/2) \tag{9}$$

in terms of the associated Laguerre polymonial $L_n^{(\alpha)}$. The form factors of the subbands reflect the weakening of the Coulomb interactions as compared with the strictly two-dimensional limit:

$$F_{ab}(q) = \int dz dz'\,|\xi_a(z)|^2\,|\xi_b(z')|^2\,e^{-q|z-z'|}, \tag{10}$$

where the ξ denote subband envelope functions which are chosen to be Gaussians with half-widths which minimize the total Hartree energy [18], thus including band-bending effects. Explicit expressions for the integral in Eq. (8) have been given by Chu and Chang [32].

In the following the discussion will be restricted to the conventional experimental situation of weak excitation with a small number of photoexcited holes. The exciton condensation can then certainly be neglected. The absorption spectra are easily obtained by setting the number of holes to zero. The luminescence is more subtle, however, and to the best of my knowledge has not been treated properly before. For general electron and hole densities the chemical potentials in Eqs. (2) and (5) have to be determined taking into account the presence of bound states, including the carrier sink of bound pairs with finite momentum [30,33]. Fortunately, the electron chemical potential will not be affected by a sufficiently small number of holes, which in turn may be treated classically ($\beta\mu_h \to -\infty$). The luminescence intensity is proportional to the density ρ_h of optically active holes. The quantity which is discussed in Section 3 is therefore the normalized spectrum

$$\lim_{\rho_h \to 0} I^{\mathrm{lum}}(\hbar\omega)/\rho_h, \tag{11}$$

where

$$\lim_{\beta\mu_h \to -\infty} \rho_h = \sum_n \{e^{-\beta(\varepsilon_h(n) - \mu_h)} - e^{-\beta(E(n) - \mu_e - \mu_h)}\} + \sum_i e^{-\beta(\lambda_i - \mu_e - \mu_h)}. \tag{12}$$

Eq. (12) is non-trivially obtained from the single-particle Green function of the holes as modified by the exciton effect. If λ_{ex} denotes the lowest eigenvalue which does not equal any $E(n)$, the hole ground state is of a single-particle nature when $\lambda_{ex} - \mu_e < \varepsilon_h(0)$, and an exciton otherwise, for any static interaction potential. At zero temperature and magnetic field (disregarding the state dependence of the self-energy) the exciton is thus occupied when its binding energy relative to the Fermi edge exceeds $\hbar^2 k_F^2/2m_h$, where k_F is the Fermi wave vector. This simple theorem defies conventional wisdom [19] and has unexpected consequences for the photoluminescence spectra, as discussed in the following section.

The equations for zero magnetic fields are recovered by replacing the Landau level indices by wave vectors, replacing the energies of a free particle a with effective mass m_a as

$$\frac{eB\hbar}{m_a}(n + \frac{1}{2}) \to \frac{\hbar^2 k^2}{2m_a}, \tag{13}$$

the Coulomb matrix elements Eq. (8) as

$$V_{ab}(n,m) \to \frac{2\pi e^2}{\kappa|\vec{k} - \vec{k}'|} F_{ab}(|\vec{k} - \vec{k}'|) \tag{14},$$

and deleting prefactors $1/(2\pi\ell^2)$. The situation is more complicated for finite-momentum pair states [29], which are optically inactive, however, and may be disregarded in the limit of vanishing hole number. The zero-field problem can be solved conveniently by assuming a small magnetic field which has no significant physical effect but transforms singular integral equations into easily solvable matrix equations.

3. Computational results

In numerical calculations the number of Landau level pair states in Eqs. (3) and (4) is finite and the convergence of the expansion must be carefully monitored, especially in the limit of small magnetic fields. At 0.2 tesla a basis set of 1200 Landau levels is necessary for accurate results, but the calculations presented here have been done using 600 basis functions. While the energies are virtually unchanged, the oscillator strengths can be too small by up to 10 per cent for strongly bound states, which is considered acceptable for the present purposes. It has been checked explicitly that the additional approximation of replacing the Coulomb matrix elements in the self-energies V_{ee} and V_{hh} by V_{eh} does not introduce significant errors. The parameters for GaAs $m_e = 0.067$, $m_h = 0.3$, $\varepsilon = 12.5$ for well and barrier material and a valence/conduction band offset ratio of 0.3/0.7 are used.

The luminescence and absorption spectra of a 100Å quantum well are plotted in Fig. 1 for two different densities and different temperatures. The densities are just below and well above the transition from free particle to excitonic ground state. The figures clearly show the counterintuitive result that the photoluminescence at low temperatures can be dominated by the exciton, even when the actual band gap is smaller than the exciton energy. The finite temperature luminescence looks very similar to published ones for InGaAs/InP [2] and GaAs/AlGaAs [3] quantum wells. Though hole localization has to be invoked to explain the strong enhancement of the luminescence at the Fermi edge in Ref. 2, it appears to be not necessarily required at lower densities. Lee *et al.* [3] measured an exciton binding energy of 4 meV at a density of $6 \cdot 10^{11}$ cm^{-2}. The criterion for exciton formation (for the present effective masses 4.8 meV) is not yet fulfilled, but the experimental luminescence is already strongly enhanced at the Fermi edge. Some hole localization thus seems also to be present in the AlGaAs samples of Ref. 3.

Fig. 1 also indicates that the Stokes shift between luminescence and absorption may vanish at still sizable electron densities. A (at zero temperature) discontinuous appearance of a Stokes shift is predicted which reflects the sudden transition from exciton to free-particle-like behaviour. For the present set of parameters the critical electron density in units of 10^{11} cm^{-2} should be about equal to the exciton binding energy in meV. The results of Delalande *et al.* [21] and Yoshimura *et al.* [22] on the luminescence of gated structures support the present theory in so far as the absorption and luminescence frequencies appear to "stick" together up to rather high gate voltages which might be sufficient to inject a significant free carrier concentration. Interesting is also the conspicuous additional peak observed in Ref. 21 when the electron density is depleted under illumination, which looks like a reappearing exciton. Similar high-energy luminescence satellites observed by Pinczuk *et al.* [17] might also have an excitonic origin.

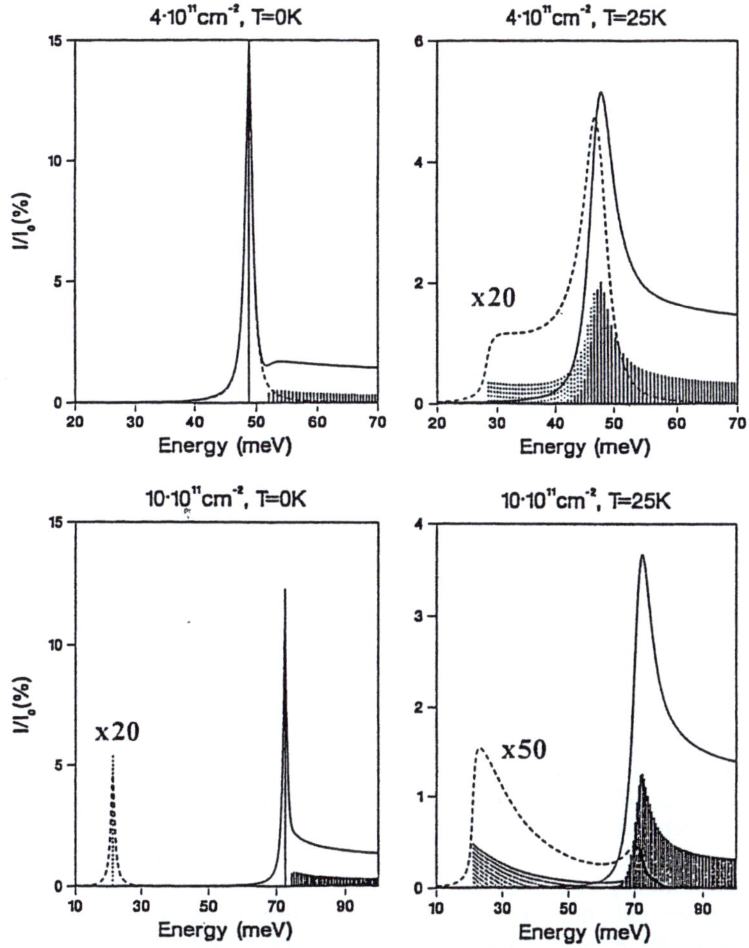

Fig. 1: Results of mean-field theory (Eqs.(1)-(3)) for the photoluminescence (dashed and dotted lines) and absorption spectra (full lines) of a 100 Å GaAs/$Al_{.3}Ga_{.7}As$ *symmetrically modulation-doped quantum well. The vertical lines are the results for the oscillator strengths (in percent of the oscillator strength I_0 of the strictly two-dimensional ground state exciton at zero field) in the presence of a magnetic field of 0.2 T normal to the interfaces. The envelopes are obtained by folding the line spectra with a Lorentzian of full width at half maximum of 1.5 meV.*

The effect of a magnetic field is shown in Fig. 2 for the low-density sample in Fig. 1. The main luminescence signal at all fields is due to exciton annihilation. The results in Fig. 2 do not agree with experiments of Yoshimura and Sakaki [11], as will be discussed below.

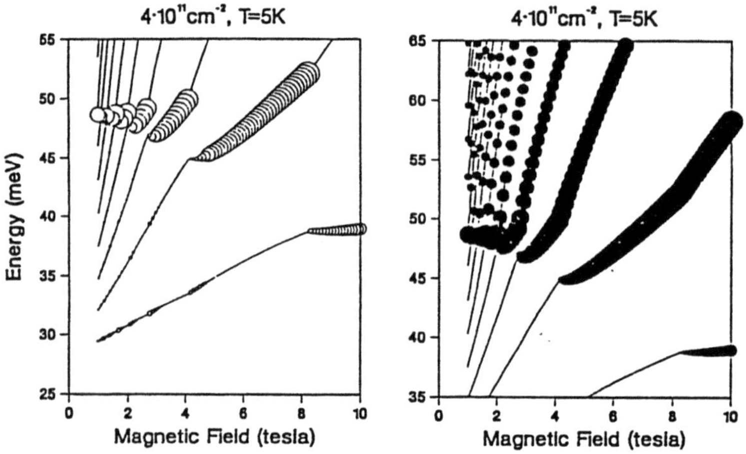

Fig. 2: Results of mean-field theory (Eqs.(1)-(3)) for the photoluminescence (left) and absorption (right) spectra of a 100 Å $GaAs/Al_{.3}Ga_{.7}As$ *symmetrically modulation-doped quantum well as a function of a magnetic field normal to the well. The area of the dots is proportional to the oscillator strength.*

4. Discussion

At zero magnetic field the main difference of the present theory from previous ones [20,23,25] is the explicit inclusion of bound excitonic states. Mahan [29] argued that in bulk systems bound states can be disregarded since their oscillator strengths can be estimated to be weak. We have seen that in quantum wells this is not necessarily the case. However, bound states are shown to dissociate quickly at higher densities and merge into the continuum to create what is often called a "Fermi edge singularity", a process which is accelerated by screening effects. Previous approaches [20,23,25] are thus valid at sufficiently high densities and temperatures.

At low magnetic fields and high electron densities excitons do not modify the band edge, which is therefore determined exclusively by the self-energies of the electrons and the holes (Eq. (7)) [18]. Here a criterion has been derived for the minimum density at which this theorem may be applied. At even lower but possibly still sizable densities the band gap is not located at the luminescence peak but in the low-energy tail. If the ground state is bound, an optical estimation of electron densities should be carried out with care.

Ohtaka and Tanabe [26] calculated the excitonic vertex correction for a strictly two-dimensional system at finite temperatures to all orders in a contact model potential. Because an infinite hole mass is assumed, a bound state is found at all densities and an unbinding transition does not occur. It is encouraging for the present theory that the optical spectra calculated by Ohtaka and Tanabe [26] in the mean-field "ladder" approximation appear to be quite similar to the exact result.

The weak point of the present theory is the neglect of screening. In three dimensions this would certainly be a bad approximation, but Schmitt-Rink *et al.* conjectured that in quasi-two-dimensional systems screening is relatively unimportant [16]. Kleinman [19] investigated the problem of excitons in doped quantum wells by a variational method for a statically screened electron-hole interaction. He derived a different criterion for the exciton binding and came to the conclusion that the transition occurs at much lower densities and should be continuous. By applying the present exciton unbinding criterion to his exciton binding energy curves the transition densities turn out a factor 2 to 3 smaller than the present ones. While the bare potential used here will without doubt overestimate exciton effects, a static approximation is expected to lead to an underestimation. Using both results it is possible to narrow down the real transition significantly.

While for zero (*i.e.* very small) magnetic fields the results presented in Figs. 1 can be reconciled with most experiments, problems are encountered when a magnetic field is applied (Figs. 2). While in Ref. 15 the calculated band gap has been shown to agree well with the experiments of Ref. 11, it is found here that at relevant densities the main luminescence should be excitonic for all magnetic fields. As discussed above, for a screened potential the ground state is not excitonic anymore for the density range under investigation and the conclusion in Ref. 15 that the experiments reflect a magnetoexcitonic effect should be reconsidered in view of the present results.

The effect of screening on the band gap of modulation-doped quantum wells in magnetic fields has been discussed by Katayama and Ando [12] and Uenoyama and Sham [13]. To avoid the singularities in the Landau level density of states and the screening, lifetime broadening due to impurity scattering had to be introduced into the formalism at the cost of an adjustable parameter. For small broadenings small but clear oscillations due to the self-consistent screening are reported for the band gap, in good agreement with some experiments [5,6]. These effects are clearly beyond a Hartree-Fock treatment which only produces slight cusps in the band gap as a function of the magnetic field. However, other experiments, *e.g.* [11], do not show these oscillations at higher filling factors. This might point to an essentially extrinsic mechanism [4,10], but could also be explained by an increased broadening [12] which washes the oscillations out. In the latter case an excitonic mechanism is required to explain the experiments of Ref. 11. On the other hand, Uenoyama and Sham [13] include excitonic corrections in their theory of the band gap, but found them to be rather unimportant. For densities above the exciton unbinding transition the band gap is probably quite insensitive to the static screening approximation (used by Ref. 13 in contrast to Ref. 12) because an underestimation of the exciton effect is compensated by an overestimation of the self-energies. It remains to explain the apparently too low critical magnetic fields in Ref. 11 as well as the strong excitonic effects in magnetoabsorption reported in Ref. 3, which indicate again that screening is unimportant, but which are not seen by other groups [9].

We have to conclude that more combined luminescence and absorption experiments on high-quality samples with a large range of electron densities are needed to get to grips with the fundamental problem of excitonic states in a fermion sea.

The support of Prof. M.F.H. Schuurmans is gratefully acknowledged.

References

[1] J.M. Worlock, A.C. Maciel, A. Petrou, C.H. Perry, R.L. Aggarwal, M.C. Smith, A.C. Gossard and W. Wiegmann, *Surf. Science* **142**, 486(1984).

[2] M.S. Skolnik, J.M. Rorison, K.J. Nash, D.J. Mowbray, P.R. Tapster, S.J. Bass, A.D. Pitt, *Phys. Rev. Lett.* **58**,2130(1987).

[3] J.S. Lee, Y. Iwasa, and N. Miura, *Semicond. Sci. Technol.* **2**, 675(1987); J.S. Lee, N. Miura and T. Ando, *J. Phys. Soc. Japan*, in press.

[4] J. Orgonasi, J.A. Brum, C. Delalande, G. Bastard, T. Rötger, J.C. Maan, G. Weimann, and W. Schlapp, *J. Phys.* (*Coll.*) **C5**, 407(1987).

[5] C.H. Perry, J.M. Worlock, M.C. Smith and A. Petrou, in *High Magnetic Fields in Semiconductor Physics*, edited by G. Landwehr, (Springer Verlag, Berlin, 1987).

[6] M.S. Skolnik, K.J. Nash, S.J. Bass, P.E. Simmonds, and M.J. Kane, *Solid State Commun.* **67**,637(1988).

[7] D. Heiman, B.B. Goldberg, A. Pinczuk, C.W. Tu, A.C. Gossard, and J.H. English, *Phys. Rev. Lett.* **61**,605(1988).

[8] I.V. Kukushkin, K. von Klitzing, and K. Ploog, *Phys. Rev.* **B37**,2021(1988).

[9] D.D. Goldberg, D. Heiman, M.J. Graf, D.A. Broido, A. Pinczuk, C.W. Tu, J.H. English and A.C. Gossard, *Phys. Rev.* **B38**,10131(1988).

[10] R. Stepniewski, W. Knop, A. Raymond, G. Martinez, T. Rötger, J.C. Maan and J.P. Andre, in: *High Magnetic Fields in Semiconductor Physics II*, edited by G. Landwehr, (Springer Verlag, Berlin, 1989).

[11] H. Yoshimura and H. Sakaki, *Phys. Rev.* **B39**,13024(1989).

[12] S. Katayama and T. Ando, *Solid State Commun.* **70**,97(1989).

[13] T. Uenoyama and L.J. Sham, *Phys. Rev.* **B39**,11044(1989).

[14] G.E.W. Bauer, *Proceedings of the* 8*th EP2DS*, Grenoble, 4-8 September 1989, *Surf. Science*, in press.

[15] G.E.W. Bauer, *Phys. Rev. Lett.* **64**,60(1990).

[16] For a review see: S. Schmitt-Rink, D.S. Chemla and D.A.B. Miller, *Adv. Physics* **38**,89(1989).

[17] A. Pinczuk, J. Shah, H.L. Störmer, R.C. Miller, A.C. Gossard and W. Wiegmann, *Surf. Science* **142**, 492(1984).

[18] G.E.W. Bauer and T. Ando, *Phys. Rev.* **B31**,8321(1985); *J. Phys.* **C19**,1553(1986).

[19] D.A. Kleinman, *Phys. Rev.* **B32**,3766(1985).

[20] J.M. Rorison, *J. Phys.C* **20**,*L*311(1987).

[21] C. Delalande, G. Bastard, J. Organasi, J.A. Brum, H.W. Liu, M. Voos, G. Weimann, and W. Schlapp, *Phys. Rev. Lett.* **59**, 2690(1987); C. Delalande, *Phys. Scr.* **T19**, 129(1987).

[22] H. Yoshimura, G.E.W. Bauer and H. Sakaki, *Phys.Rev.***B38**,10791(1988).

[23] G. Livescu, D.A.B. Miller, D.S. Chemla, M. Ramaswamy, T.Y. Chang, N. Sauer, A.C. Gossard, J.H. English, *IEEE J. Quantum. Electron.* **24**,1677(1988).

[24] S. Munnix, D. Bimberg, D.E. Mars, J.N. Miller, E.C. Larkins, and J.S. Harris Jr., in : *Proceedings of the* 19*th International Conference on the Physics of Semiconductors*, edited by W. Zawadzki, Institute of Physics, Polish Academy of Sciences, Warsaw, 1988.

[25] J.-W. Wu, *Phys.Rev.***B39**,7992(1989).
[26] K. Ohtaka and Y. Tanabe, *Phys.Rev.***B39**,3054(1989).
[27] H. Kalt, K. Leo, R. Cingolani, and K. Ploog, *Phys.Rev.* **B40**,12017(1989).
[28] G.E.W. Bauer, in: *Proceedings of the NATO – ARW on Condensed Systems of Low Dimensionality*, Marmaris, Turkey, April 1990, (Plenum Press, New York), to be published.
[29] G.D. Mahan, *Phys.Rev.***153**,882(1967).
[30] H. Haug and S. Schmitt-Rink, *Prog. Quantum Electron.* **9**,3(1984).
[31] H. Stolz, *Einführung in die Vielelektronentheorie der Kristalle*, (Akademie Verlag, Berlin, 1974)
[32] H. Chu and Y.-C. Chang, *Phys. Rev.* **B40**,5497(1989).
[33] P. Nozieres and S. Schmitt-Rink, *J. Low Temp.* **Phys. 59**,195(1985).
[34] C. Kallin and B.I. Halperin, *Phys. Rev.* **B30**,5655(1984); A.H. MacDonald, *J. Phys.* **C18**,1003(1984).

Pseudomorphic $In_{0.23}Ga_{0.77}As/GaAs$ Quantum Wells: Correlation of Anisotropic Lattice Relaxation and Degradation of Optical Properties

M. Grundmann[1], *U. Lienert*[1], *J. Christen*[1], *D. Bimberg*[1], *A. Fischer-Colbrie*[2], *and J.N. Miller*[2]

[1]Technische Universität Berlin, Institut für Festkörperphysik I, Hardenbergstr. 36, D-1000 Berlin 12, Germany
[2]Hewlett Packard Laboratories, High Speed Devices Laboratories, 3500 Deer Creek Road, Palo Alto, CA94304, USA

Abstract. The structural degradation of pseudomorphic quantum wells via misfit dislocations is correlated with the consequent modification of optical properties. The dislocation density is *asymmetric* in the (001) interface plane as directly visualized by scanning cathodoluminescence imaging. This leads to a change from tetragonal to monoclinic crystal symmetry for partially relaxed layers and to anisotropic tilt of lattice planes along the dislocations as revealed by means of double-crystal X-ray diffraction. The local drop of quantum efficiency and carrier recombination times around the misfit dislocations causes a macroscopic decrease of the total quantum efficiency and a change of the character of the recombination dynamics from purely radiative to dominantly nonradiative. This behavior is modelled and explained in the frame of a diffusion model.

1. Introduction

The inclusion of heterostructures and quantum wells (QWs) consisting of layers *non-lattice-matched* to each other into device concepts [1] has added a new dimension to semiconductor physics. By means of the $In_xGa_{1-x}As/GaAs$ material system the performance of high electron mobility transistors (HEMT) [2-4] or hot electron devices [5] has been dramatically improved. Additionally efficient photonic devices like lasers [6-8], detectors [9,10] and solar cells [11] in the near infrared were demonstrated.

If the quantum well width L_z is smaller than a certain critical thickness L_c, the resulting strain causes a tetragonal distortion of the unit cell in connection with a drastic change of the bandstructure. If L_z exceeds L_c, part of the strain is plastically relieved by generation of misfit dislocation lines across the heterointerfaces.

Much work has been done to derive the important design parameter *critical thickness* L_c theoretically [12-20] and to determine L_c experimentally [21-38]. It is the purpose of this paper to correlate the lattice relaxation and the optical degradation in terms of the influence of misfit dislocations on the recombination of excess charge carriers. It will turn out that the onset of optical degradation (manifested in a bandgap shift, increase of inhomogeneous spectral broadening, decrease of quantum efficiency and a change of the recombination dynamics) occurs delayed with respect to L_c.

2. Experimental Procedures

The subjects of our experiments were molecular beam epitaxy (MBE) grown samples, each one containing a *single* $In_{0.23}Ga_{0.77}As/GaAs$ QW. The various samples differ only by the QW

Springer Series in Solid-State Sciences, Vol. 97
Localization and Confinement of Electrons in Semiconductors
Editors: F. Kuchar · H. Heinrich · G. Bauer

thickness L_z=16.2, 21.6, 27.0, 32.4 and 37.8nm (samples B1309, B1310, ..., B1313). The growth temperature of the $In_{0.23}Ga_{0.76}As$ layer was 515°C [29]. On top a 50nm cap layer was grown. Details about the growth procedure may be found in [34].

The samples were investigated with plan view transmission electron microscopy (TEM), double-crystal X-ray diffraction (DXD), cathodoluminescence (CL) and photoluminescence (PL).

In our DXD setup a GaAs crystal is used as the first crystal and we work non-dispersively in the (+,–)-mode with the Cu K_{α_1} line. To obtain structural information on the samples, the experimental X-ray rocking curves are compared to theoretical ones derived from dynamical diffraction theory [39]. The *azimuthal* orientation ϕ of the sample is defined as a counter-clockwise rotation around the [001]-axis; at ϕ=0 the projections of the incident and the diffracted beam lie along the [110]-direction. Very recently the method of varying ϕ systematically has found increasing interest [40-42].

Our setup for detection of CL in the visible and near infrared regime described in detail elsewhere [43] was extended further to the infrared regime by attaching a Ge-*p*-*i*-*n*-diode (North Coast E0-817L, $\lambda \leq 1.8\mu m$) using a lock-in technique for non-time-resolved CLI. A Si-APD (Avalanche Photodiode, RCA C30902S, $\lambda \leq 1100nm$) adapted to our CL setup allows single photon counting in the near infrared regime with a total time-resolution of 350ps [44].

3. Structural Properties

We combined integral cathodoluminescence imaging (ICLI) and transmission electron microscopy to determine the linear dislocation density n_L [34,45,46]. These two methods complement each other.

The lateral distribution of the quantum efficiency is shown for a 130 x 73 μm^2 area of the interface for all samples in Fig.1. The L_z=22nm QW shows the misfit dislocations as single distinct straight lines of reduced quantum efficiency η, which are preferentially orientated along [110]. The L_z=16 nm QW shows no dislocations on this scale but ICLIs taken at lower magnification, imaging a larger area at once, indicate a line misfit dislocation density of $n_L=4\cdot10^3m^{-1}$. In QWs with $L_z \geq 27$nm single misfit dislocations are no longer observed and the CLI consists of dark line defects at an average distance saturating at about 2μm corresponding to groups of dislocations [45,46]. For the determination of n_L at higher dislocation densities, plan-view two-beam TEM experiments are used. The ICLI and TEM results are summarized in Fig.2. While the dark line defect density determined by ICLI saturates, the misfit dislocation density obtained from TEM still increases. The TEM value for the 16nm QW represents the detection limit of TEM due to the finite scanning area. The solid line in Fig.2 connects the valid experimental values for the misfit dislocation density of each sample.

All samples contain misfit dislocations; this implies that the critical thickness L_c for the start of generation of misfit dislocations is somewhat less than 16nm. Therefore the energy balance theory of People and Bean (P&B) [14], predicting L_c=37nm, does not agree with our experimental result. The theory of mechanical equilibrium of Matthews and Blakeslee (M&B) [13], recently improved by Dodson and Tsao [16], describes the situation very well with the theoretical value L_c=13nm. In all samples the misfit dislocations are preferentially orientated along [110] , while the density of dislocations along [1$\bar{1}$0] is substantially smaller. This explains the striped nature of the ICLIs at high dislocation density and is in agreement with previous work [47-49]. The *asymmetric* dislocation density causes several anisotropic strain relaxation phenomena as revealed by X-ray diffraction, reported in the following.

Fig.3 shows DXD rocking curves of the 38nm QW (B1313). This sample with the thickest well is the only one showing a partial relaxation (of 7%) of the lattice constant d_z in the growth direction leading to a larger in-plane lattice constant than that of bulk GaAs $a_{0,GaAs}$ at the upper interface. This causes in turn a strain of the GaAs cap-layer resulting in an additional peak shifted some arc minutes from the GaAs-substrate peak towards larger angles (see inset of

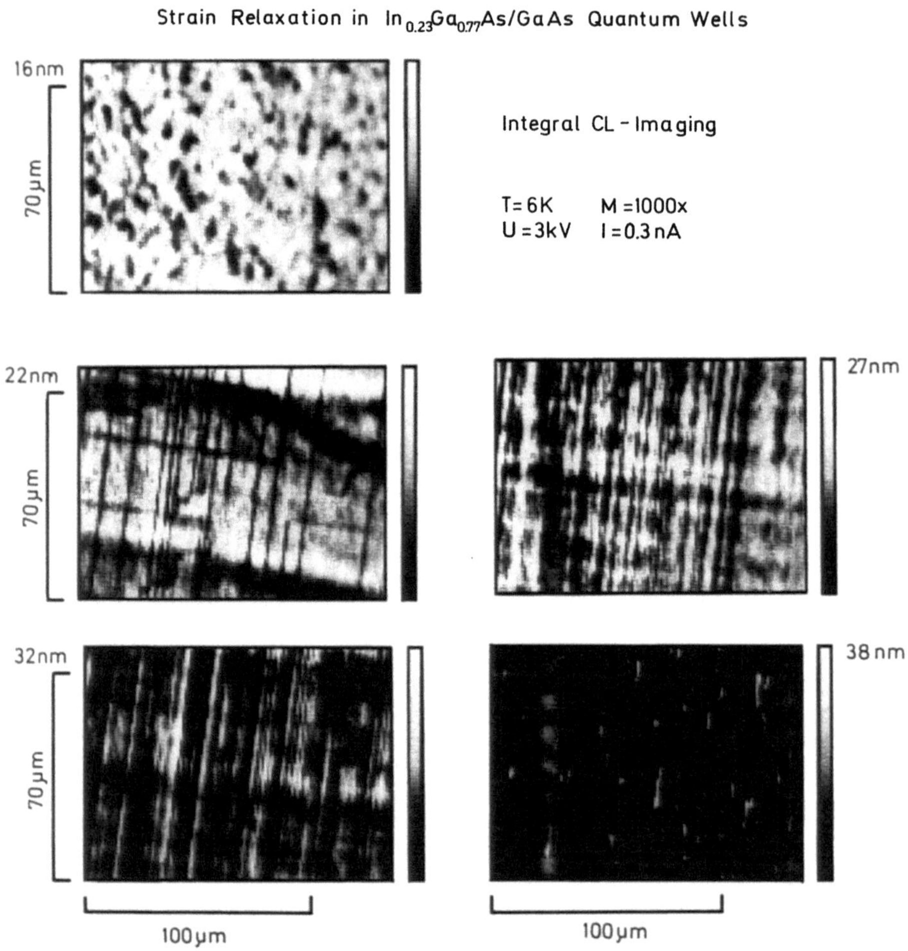

Fig.1: Integral cathodoluminescence images of $In_{0.23}Ga_{0.77}As/GaAs$ single QWs with different well width L_z. Increasing thickness leads to increasing density of dark line defects.

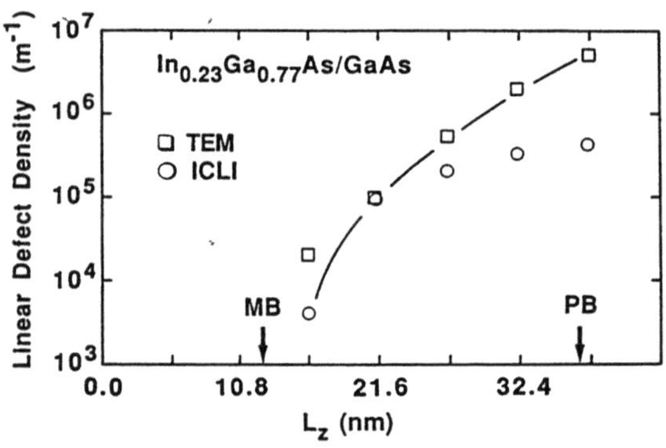

Fig.2: Linear defect density n_L as a function of QW width L_z measured with TEM (squares) and ICLI (circles). Arrows indicate critical thickness according to Matthews and Blakeslee (MB) and People and Bean (PB).

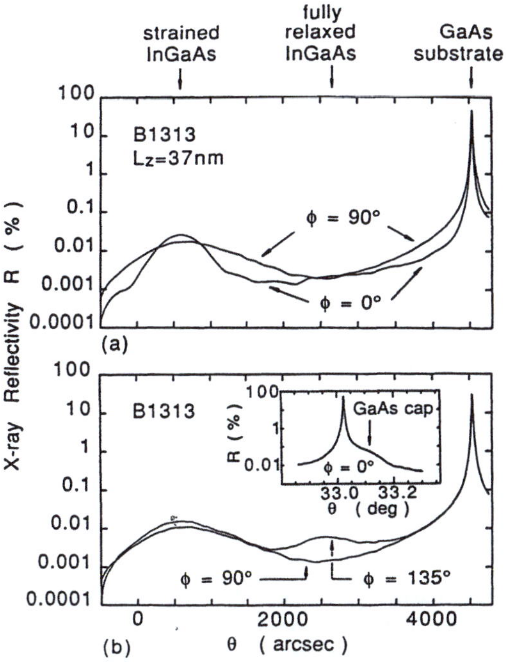

Fig.3: DXD rocking curves for different azimuthal orientations ϕ of sample B1313 showing (a) variation of FWHM of the InGaAs epilayer peak, (b) different response from fully relaxed InGaAs.

Fig.3b), indicating GaAs cap-layer lattice planes with decreased separation in the [001]-direction. Such strain in the cap-layer has been recently revealed by photoluminescence [50].

The relaxation of the L_z=38nm QW is studied in more detail by using asymmetric reflections sensitive to the in-plane lattice constants. From the (404) and $(0\bar{4}4)$ reflections we find that the lattice constants d_x along [100] and d_y [010] are equal, $d_x=d_y>a_{0,GaAs}$, both relaxed by about 7%. From the (224), (444), $(2\bar{2}4)$ and $(4\bar{4}4)$ reflections we obtain the lattice constants d_{110} along [110] and $d_{1\bar{1}0}$ along $[1\bar{1}0]$. The remarkable feature is now that these lattice constants are no longer equal, $d_{110}=\sqrt{2}a_{0,GaAs}$ and $d_{1\bar{1}0}$ relaxed by 14%. The relations of the lattice constants are shown schematically in Fig. 4. This anisotropy is caused by the asymmetric density of dislocations along the [110]- and $[1\bar{1}0]$-directions, which leads to a significant lattice relaxation along $[1\bar{1}0]$ but *not* along [110]. The inequality of the angle bisectors $d_{110}<\sqrt{2}d_x$ and $d_{1\bar{1}0}>\sqrt{2}d_x$ implies that the projection of the unit cell onto the (001) plane is no longer a square but an equilateral parallelogram. The inner angles differ by 0.16° from 90°. The crystal symmetry of the $In_xGa_{1-x}As$ QW is therefore no longer tetragonal as in the fully strained case but *monoclinic* ! The unit cells of the GaAs cap layer are found to have monoclinic symmetry too. This result is quite general for III-V- as well as for II-VI-compounds, since the defect formation in the zincblende lattice without a center of inversion is known to be anisotropic [51-54].

In quantum wells with dislocation densities $n_L>10^6m^{-1}$ (typical samples are B1312 and B1313) the FWHM of the $In_xGa_{1-x}As$-layer peak shows an oscillation with a periodicity of 180° as a function of the azimuthal orientation ϕ of the sample. Fig.3a shows that the two rocking curves of this sample for minimum and maximum FWHM (corresponding to the sample orientations ϕ=0° and ϕ=90°) are remarkably different. Fig.5 displays the experimental values of FWHM as a function of ϕ.

The anisotropy of FWHM has its origin in the asymmetric density of dislocations along the [110]- and $[1\bar{1}0]$-directions. Along [110] the lattice planes are somewhat tilted towards $[1\bar{1}0]$-

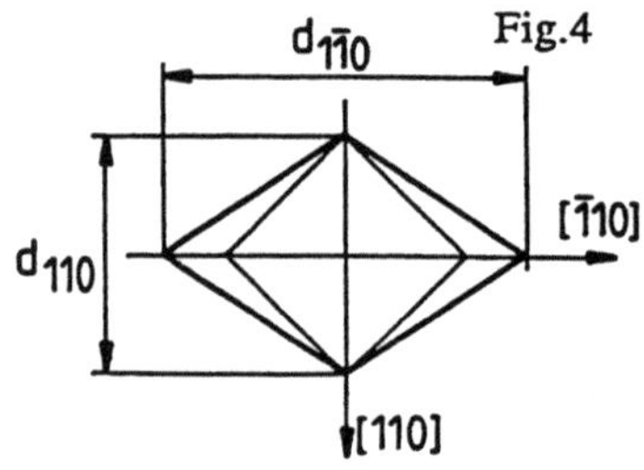

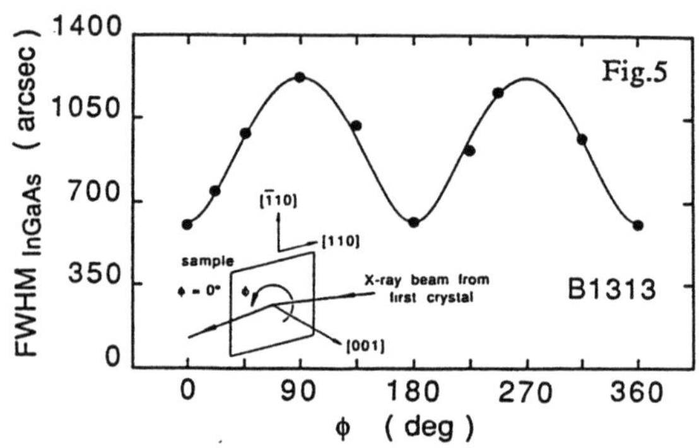

Fig.4: Monoclinic distortion of the unit cell (projection onto the (001) plane).

Fig.5: Full width at half maximum of InGaAs layer X-ray reflection as function of sample orientation. Dots are experimental values, solid line is theoretical fit (*equ.1*). The inset shows the schematic set-up for the diffraction geometry with varying azimuthal sample orientation ϕ.

and $[\bar{1}10]$-directions; along $[1\bar{1}0]$ they are tilted towards [110] and $[\bar{1}\bar{1}0]$. The anisotropy and periodicity of the FWHM are explained as follows: The plane orientations are distributed with Gaussians with standard deviations $\sigma_{1\bar{1}0}$ and σ_{110} for tilting angles along $[1\bar{1}0]$ and along [110] leading to a statistical broadening of the layer peak by convolution with both distributions. Adding the intrinsic width σ_0 caused by finite layer thickness, we find for σ^2 of the epilayer peak

$$\sigma^2(\phi) = \sigma_0^2 + \sigma_{110}^2 + (\sigma_{1\bar{1}0}^2 - \sigma_{110}^2) \cos^2(\phi - 90°) \quad . \qquad (1)$$

It is obvious that for $\sigma_{110} = \sigma_{1\bar{1}0}$ no oscillation of the FWHM of the epilayer will occur; in our case, however, $\sigma_{1\bar{1}0} > \sigma_{110}$. Taking for σ_0 the theoretical value σ_0=196" of an ideal layer, we deduce from the fit (solid line in Fig.4) to the experimental halfwidths the values $\sigma_{1\bar{1}0}$ = 480" and σ_{110} = 174". The standard deviation σ of the distribution of tilting angles is of the order of magnitude b/d, the ratio of the length of the Burgers vector (assuming 60° type) and the average dislocation distance. We note that the tilting of lattice planes directly proves that the dominating part of the dislocations is of 60° type because 90° type dislocations would not cause a tilt of (001) planes.

In addition to a broadening of the layer peak we observe a broadening of the base of the substrate peak similar to what has been reported before [54]. The broadening of the substrate peak has the same angular dependence as the broadening of the $In_xGa_{1-x}As$ peak indicating anisotropic tilting of GaAs lattice planes near the interfaces. We wish to emphasize that we do not observe an overall tilt of the QW with respect to the substrate.

Furthermore we observe reflection from fully relaxed unit cells having the bulk $In_{0.23}Ga_{0.77}As$ lattice constant (see Fig.3b). This indicates heterogeneous strain relaxation and coexistence of strained and fully relaxed domains. The ϕ-dependence of this effect proves that we observe a combined mechanism of lattice relaxation and lattice plane tilting. This effect is discussed in more detail elsewhere [55].

4. Optical Properties

The structural relaxation is followed by a change of a variety of optical properties. With increasing L_Z, i.e. increasing dislocation density, we observe a doubling of the inhomogeneous spectral broadening for $L_Z \geq 27$nm. For $L_Z \geq 37$nm it is possible to detect a shrinkage of the bandgap exceeding the decrease caused by the reduction of the confinement effects with increasing L_Z. This indicates that the strain effect on the bandstructure is reduced. We note that the monoclinic distortion *couples* heavy and light holes at the center of the Brillouin zone. In our case the bandgap energy is calculated to be modified in the 1meV range, a value too small to be unambiguously extracted from the PL spectra. The monoclinic QW is optically uniaxial and exhibits linear birefringence, which will be subject of another publication. In the following we wish to focus on the deterioration of the quantum efficiency and the recombination time constants.

The quantum efficiency η is determined with calorimetric absorption spectroscopy in connection with a calorimetric transmission experiment [56] in full agreement with photoluminescence intensities. The experimental values are shown in Fig.6b. Near L_C the quantum efficiency is almost *not* affected but then shows a strong decrease from over 90% down to only a few percent. For L_Z=27nm we find $\eta \approx 50\%$.

With time resolved CL we measured low temperature transients. With increasing L_Z, i.e. with increasing dislocation density n_L, the decay becomes much faster. The transients can be almost perfectly fitted by a sum of two exponentials with time constants τ_1 and τ_2, which are plotted in Fig.6c versus L_Z. The high quantum efficiency of the 16 and 22nm QWs makes it straightforward to identify the decay constant τ_1 with the radiative excitonic (X(e,hh),n=1) lifetime τ_r. The excitonic lifetime in the thicker well is slightly larger, as expected from the theory of excitonic oscillator strength in QWs [57]. The values of about 2ns compare well to values found for lattice-matched systems like $Al_xGa_{1-x}As/GaAs$ [58,59] and $In_{0.53}Ga_{0.47}As/InP$ [60,61] at low temperatures. The time τ_2 constant is given by (e,A^0) recombination, as clearly observed in time delayed spectra [62].

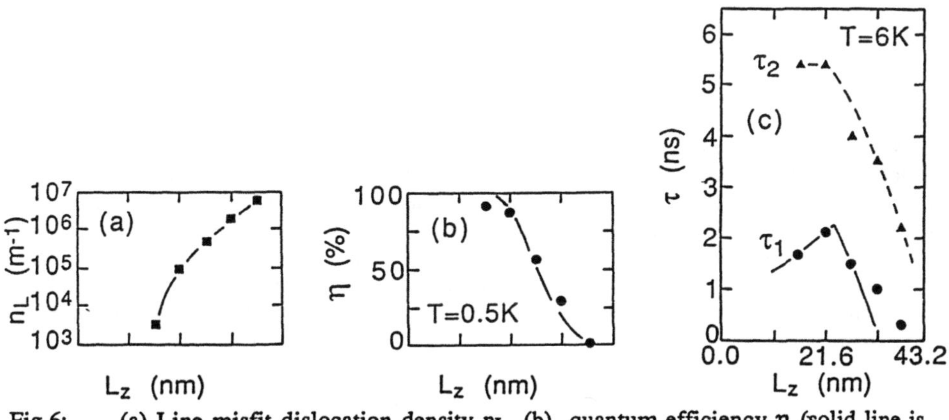

Fig.6: (a) Line misfit dislocation density n_L, (b) quantum efficiency η (solid line is theoretical curve from diffusion model, *equ.2*) and (c) decay constants of CL transients, solid line denotes theoretical description by diffusion model (*equ.4*), dashed line is guide to the eye.

5. Correlation of Structural and Optical Degradation

From the experimental point of view it is now absolutely clear that the degradation of optical properties occurs at much larger QW thickness than the critical thickness for the onset of dislocation generation, obtained with the single dislocation counting technique (ICLI).

The dislocations act as dark line defects, i.e. they are centers of locally reduced quantum efficiency. Additionally they are centers of locally reduced charge carrier lifetime [34]. With increasing dislocation density these local changes of optical properties affect the total quantum efficiency and the recombination dynamics as will be quantified in the following by means of a one-dimensional diffusion model. This diffusion model assumes misfit dislocations in an average distance $d=1/n_L$ acting as perfect sinks for excess charge carriers [34]. By solving the diffusion equation we find the theoretical dependence

$$\eta = 1 - 2Ln_L \tanh \{1/(2Ln_L)\} \quad , \qquad (2)$$

L being the ambipolar diffusion length. L is the only free parameter of the model. At low dislocation densities ($Ln_L<<1$) η is almost not affected, while at high dislocation densities ($Ln_L>>1$) the majority of the excess charge carriers reach a dislocation within their lifetime and recombine nonradiatively, resulting in low total quantum efficiency. An appropiate value of $L=0.5\mu m$ allows a perfect fit of the experimental values as is shown in Fig.6b.

To evaluate the influence of the misfit dislocations on the recombination dynamics we solve the rate equation including the diffusion term for the excess charge carrier density n

$$\partial n/\partial t = - n/\tau_{tot} + D\, \partial^2 n/\partial x^2 \quad , \qquad (3)$$

D being the diffusion coefficient and τ_{tot} the total recombination lifetime given by the radiative lifetime τ_r at high quantum efficiency. The recombination dynamics is well described by an approximate solution of equ.3 found by development into a Fourier series and taking only the leading term into account. The actually observed decay time τ is then given by

$$1/\tau = 1/\tau_r + 1/\tau_D = 1/\tau_r \{1+(\pi L n_L)^2\} \quad . \qquad (4)$$

$\tau_D = 1/\{D(\pi n_L)^2\}$ is a lifetime caused by the presence of misfit dislocations. Thus for low dislocation density ($Ln_L<<1$) the decay time is given by the radiative excitonic lifetime. At high dislocation densities ($Ln_L>>1$) the recombination dynamics is dominated by τ_D describing the diffusion into dislocations and fast nonradiative decay there. The theoretical L_Z-dependence of the dominating recombination time constant τ_1 including the variation of excitonic oscillator strength and the influence of the misfit dislocations is shown in Fig.6c as a solid line. The decay time changes its character from *radiative* ($n_L \leq 10^5 m^{-1}$) to increasingly *nonradiative* ($n_L \geq 10^6 m^{-1}$).

6. Conclusion

We have investigated the structural and optical degradation of pseudomorphic $In_xGa_{1-x}As/GaAs$ single quantum wells with TEM, CL, DXD and PL. The asymmetric strain relaxation leads to a change of crystal symmetry from tetragonal in unrelaxed QWs to monoclinic in partially relaxed QWs. The lattice planes show anisotropic tilting around misfit dislocations; strained and fully relaxed domains coexist.

The onset of dislocation generation (critical thickness L_C), determined with direct dislocation imaging techniques agrees with the theory of Matthews and Blakeslee. The degradation of optical properties occurs delayed with respect to L_C and is therefore in general not suitable to determine L_C. The decrease of quantum efficiency and the change of the character of recombination dynamics scale with the ratio of carrier diffusion length and average dislocation distance.

Acknowledgement

We are grateful to R.K.Bauer for generous experimental support, to J.Boehrer for CAS experiments and to R.Hull for TEM analysis. This work is funded by Deutsche Forschungsgemeinschaft (DFG) in the framework of Sonderforschungsbereich 6.

References

[1] G.C.Osbourne, Phys.Rev. B **27**, 5126 (1983)
[2] J.J.Rosenberg, M.Benlamri, P.D.Kirchner, J.M.Woodall, G.D.Petit, IEEE Electron.Dev.Lett. **EDL-6**, 491 (1985)
[3] L.D.Nguyen, D.C.Radulescu, M.C.Foisy, P.J.Tasker, L.F.Eastman, IEEE Transact. Electron.Dev. **36**, 833 (1989)
[4] J.M.Ballingall, P.Ho, G.J.Tessmer, P.A.Martin, N.Lewis, E.L.Hall, Appl. Phys. Lett. **54**, 2121 (1989)
[5] K.Seo, M.Heiblum, C.M.Knoedler, W.-P.Hong, P.Bhattacharya, Appl. Phys.Lett. **53**, 1946 (1988)
[6] P.K.York, K.J.Beernink, G.E.Fernández, J.J.Coleman, Appl. Phys Lett. **54**, 499 (1989)
[7] K.Y.Lau, S.Xin, W.I.Wang, N.Bar-Chaim, M.Mittelstein, Appl.Phys. Lett. **55**, 1173 (1989)
[8] A.Larsson, J.Cody, R.J.Lang, Appl.Phys.Lett. **55**, 2268 (1989)
[9] C.Jagannath, A.Siletti, A.N.M.M.Choudhury, B.Elman, P.Melman, Appl.Phys.Lett. **56**, 1892 (1990)
[10] A.F.-D.Chin, P.Battacharya, IEEE Transact. Electron Dev. **ED-36**, 2183 (1989)
[11] S.Katsumoto, A.Yamamoto, M.Yamaguchi, Jpn.J.Appl.Phys. **24**, 636 (1985)
[12] J.A.van der Merwe, J.Appl.Phys. **34**, 117 (1963)
[13] J.W.Matthews, A.E.Blakeslee, J.Cryst.Growth **27**, 118 (1974)
[14] R.People, J.C.Bean, Appl.Phys.Lett.**47**, 322 (1985)
[15] R.People, J.C.Bean, Erratum, Appl.Phys.Lett. **49**,229 (1986)
[16] B.W.Dodson, J.Y.Tsao, Appl.Phys.Lett.**51**, 1325 (1987)
B.W.Dodson, J.Y.Tsao, Phys.Rev.B **38**, 12383 (1988)
[17] B.W.Dodson, J.Y.Tsao, Appl.Phys.Lett. **55**, 1345 (1989)
[18] S.Luryi, E.Suhir, Appl.Phys.Lett. **49**, 140 (1986)
[19] P.M.J.Marée, J.C.Barbour, J.F.van der Veen, K.L.Kavanagh, C.W.T.Bulle-Lieuwma, M.P.A.Viegers, J.Appl.Phys. **62**, 4413 (1987)
[20] G.Wagner, V.Gottschalch, H.Rhan, P.Paufler, phys.stat.sol. (a) **113**, 71 (1989)
[21] I.J.Fritz, Appl.Phys.Lett. **51**, 1080 (1987)
[22] I.J.Fritz, P.L.Gourley, L.R.Dawson, Appl.Phys.Lett. **51**, 1004 (1987)
[23] P.J.Orders, B.F.Usher, Appl.Phys.Lett. **50**, 980 (1987)
[24] M.Gal, P.C.Taylor, B.F.Usher, P.J.Orders, J.Appl.Phys. **62**, 3898 (1987)
[25] T.G.Andersson, Z.G.Chen, V.D.Kulakovskii, A.Uddin, J.T.Vallin, Appl.Phys.Lett. **51**, 752 (1987)
[26] P.L.Gourley, I.J.Fritz, L.R.Dawson, Appl.Phys.Lett.**52**,377 (1988)
[27] G.L.Price, Appl.Phys.Lett. **53**, 1288 (1988)

[28] D.Morris, A.P.Roth, R.A.Masut, C.Lacelle, J.L.Brebner, J.Appl.Phys.**64**, 4135 (1988)
[29] A.Fischer-Colbrie, J.N.Miller, S.S.Laderman, S.J.Rosner, R.Hull, J.Vac.Sci.Technol. B **6**, 620 (1988)
[30] J.-P-Reithmaier, H.Cerva, R.Lösch, Appl.Phys.Lett. **54**, 48 (1989)
[31] J.Woodhead, Comment, Appl.Phys.Lett. **55**, 2147 (1989)
[32] G.L.Price, B.F.Usher, Appl.Phys.Lett. **55**, 1984 (1989)
[33] C.R.Wie, J.Appl.Phys. **65**, 2267 (1989)
[34] M.Grundmann, J.Christen, D.Bimberg, A.Fischer-Colbrie, R.Hull, J.Appl.Phys. **66**, 2214 (1989)
[35] C.Shieh, H.Lee, J.Mantz, D.Ackley, R.Engelmann, Electron. Lett. **25**, 1226 (1989)
[36] B.Elman, E.S.Koteles, P.Melman, C.Jagannath, D.Dugger, Proc. Int.Conf. on GaAs and Relat. Comp. 1989, Karuizawa, Japan, Inst. of Phys.Conf.Ser. **106**, 171 (1990)
[37] K.J.Beernink, P.K.York, J.J.Coleman, R.G.Waters, J.Kim, C.M.Wayman, Appl.Phys.Lett. **55**, 2167 (1989)
[38] S.L.Weng, J.Appl.Phys. **66**, 2217 (1989)
[39] W.J.Bartels, J.Hornstra, D.J.W.Lobeek, Acta Cryst. **A42**, 539 (1986)
[40] S.K.Gandhi, J.E.Ayers, Appl.Phys.Lett.**53**, 1204 (1988)
[41] A.Segmüller, Adv. X-ray Anal. **29**, 353 (1986)
[42] A.Segmüller, Mat.Res.Soc.Symp.Proc. **77**, 151 (1987)
[43] J.Christen, D.Bimberg, Oyo Buturi **57**, 69 (1988)
J.Christen, M.Grundmann, D.Bimberg, Appl.Surf.Sci. **41/42**, 329 (1989)
[44] M.Grundmann, J.Christen, D.Bimberg, Proc. Int.Conf. on the Study and Technol. of Defect Control in Semic. 1989, Yokohama, Japan
[45] E.A.Fitzgerald, G.D.Ast, P.D.Kirchner, G.D.Pettit, J.M.Woodall, J.Appl. Phys.**63**, 693 (1988)
[46] E.A.Fitzgerald, Y.Ashizawa, L.F.Eastman, G.D.Ast, J.Appl.Phys.**63**, 4925 (1988)
[47] M.S.Abrahams, L.R.Weisberg, C.J.Buiocchi, J.Blanc, J.Mater.Sci.**4**, 223 (1969)
[48] E.A.Fitzgerald, P.D.Kirchner, R.Proano, G.D.Pettit, J.M.Woodall, D.G.Ast, Appl.Phys.Lett.**52**, 1496 (1988)
[49] S.B.Qadri, B.T.Jonker, G.A.Prinz, J.J.Krebs, J.Vac.Sci.Technol. **A6**, 1526 (1988)
[50] M.Gal, P.J.Orders, B.F.Usher, M.J.Joyce, J.Tann, Appl.Phys.Lett.**53**, 113 (1988)
[51] W.J.Bartels, W.Nijman, J.Cryst.Growth **37**, 204 (1977)
[52] C.R.Wie, H.M.Kim, K.M.Lau, Proc. SPIE **877**, 41 (1988)
[53] K.L.Kavanagh, M.A.Capano, L.W.Hobbs, J.C.Barbour, P.M.J.Marée, W.Schaff, J.W.Mayer, D.Pettit, J.M.Woodall, J.A.Stroscio, R.M.Feenstra, J.Appl.Phys.**64**, 4843 (1988)
[54] P.D.Brown, G.J.Russell, J.Woods, J.Appl.Phys. **66**, 129 (1989)
[55] M.Grundmann, U.Lienert, D.Bimberg, A.Fischer-Colbrie, J.N.Miller, Appl.Phys.Lett. **55**, 1765 (1989)
[56] A.Juhl, D.Bimberg, J.Appl.Phys. **64**,303 (1988)
[57] M.Grundmann, D.Bimberg, Phys.Rev. B **38**, 13486 (1988),
[58] J.Christen, D.Bimberg, A.Steckenborn, G.Weimann, Appl.Phys. Lett. **44**, 84 (1984)
[59] J.Feldmann, G.Peter, E.O.Göbel, P.Dawson, K.Moore, C.Foxon, R.J.Elliott, Phys.Rev.Lett. **59**, 2337 (1987)
[60] M.Engel, R.K.Bauer, D.Bimberg, D.Grützmacher, H.Jürgensen, J.Cryst.Growth **93**, 359 (1988)
[61] U.Cebulla, G.Bacher, A.Forchel, G.Mayer, W.T.Tsang, Phys.Rev. B **39**, 6257 (1989)
[62] M.Grundmann, D.Bimberg, A.Fischer-Colbrie, J.N.Miller, Phys.Rev. B **41**, 10120 (1990)

Part VI

Impurities in Quantum Well Structures

Isolated Impurities, Impurity Bands and the Metal–Insulator Transition in Modulation-Doped Multiple-Quantum-Well Structures in Magnetic Fields

Y.-J. Wang [1], *B.D. McCombe* [1], *and W. Schaff* [2]

[1] Department of Physics and Astronomy, State University of New York at Buffalo, Buffalo, NY 14260, USA

[2] School of Electrical Engineering, Cornell University, Ithaca, NY 14853, USA

Abstract. Magnetotransport and magnetotransmission studies of several modulation-doped (Si) GaAs/$Al_{0.3}Ga_{0.7}As$ multiple-quantum-well samples are described. Samples are doped over a range of densities near the metal-insulator transition according to the Mott criterion at zero magnetic field. The magnetotransmission measurements show clear transitions associated with electrons in the GaAs wells bound to their parent donor ions in the barriers over the entire range of magnetic fields investigated (1 - 9T). At high fields (up to 9T) these absorption lines have the character of the isolated impurity transitions from the ground state (m = 0) to first excited state (m = +1) of hydrogenic donors, and the positions agree well with variational calculations for such transitions even for the most heavily doped sample studied (sheet doping density = 8 x 10^{10} cm^{-2}/well). At low fields for the most heavily doped sample the lines show marked broadening and a shift to <u>higher</u> energies compared with the isolated donor calculations. This sample shows activated conductivity even at very low fields with a "Hall dip" around 2T; at this field the conductivity appears metallic with density determined from the Hall voltage approximately equal to the donor doping density. At fields above this the longitudinal resistivity increases dramatically and the Hall resistivity shows more strongly activated behavior. Results are compared with current models of the metal-insulator transition in a magnetic field.

1. Introduction

The metal-insulator transition has been the subject of extensive study for many years in a number of bulk systems [1] and in quasi-two-dimensional systems formed by metal-oxide-semiconductor structures and single semiconductor heterostructures [2]. Recently, considerable experimental work on the magnetic-field-induced metal-insulator (MI) transition in doped bulk semiconductors [3 - 5], modulation-doped single GaAs/AlGaAs heterostructures [6], and delta-doped GaAs [7] has been carried out. The modulation-doped GaAs/AlGaAs system is of particular interest because of the extremely high mobility metallic electron gas that can be created in this system and the intense effort directed at studies at very high magnetic fields and very low temperatures to elucidate the properties of the condensed state that exists in the Fractional Quantum Hall Effect Hall Effect regime and the possibility of Wigner Crystallization at very high magnetic fields and very low temperatures.

Rather little attention has been paid to the modulation-doped multiple-quantum-well (MQW) system, even though these structures posseses a number of advantages for such studies compared with either bulk doped semiconductors or single heterostructures [8]. A major disadvantage of the single heterostructure is the dependence of the confining potential on the doping density in the barrier, which obviates the possibility of systematic studies of the MI transition for constant confinement. It is also not possible to vary the extent of the wave function in the plane in a

Springer Series in Solid-State Sciences, Vol. 97
Localization and Confinement of Electrons in Semiconductors
Editors: F. Kuchar · H. Heinrich · G. Bauer

systematic manner, nor to measure the far infrared (IR) optical properties of the impurity system in the dilute limit. By contrast, with MQW structures, since the confining potential can be well defined and made essentially independent of the doping density, and sufficient repetitions of the basic structure can be produced to yield adequate sensitivity for the optical measurements, all of the above systematic studies become possible. In addition, the dilute (isolated impurity) limit has been investigated both theoretically [9, 10] and experimentally [11], and this limiting case is now reasonably well understood.

The present work concentrates on a study of the magnetic field and temperature dependence of the transport and far infrared optical properties of modulation-doped GaAs/AlGaAs MQW samples with doping densities chosen to be above, but close to, the Mott criterion for the MI transition for these structures at zero magnetic field. The following section provides a brief outline of the models of the MI transition and a summary of the current understanding of the effects of confinement on the energy states of isolated barier-doped impurities in MQW structures. Section 3 gives experimental details, and in Section 4 the results are presented. Finally, the present results are discussed in the context of the models and other experimental work, and conclusions and future work are outlined in Section 5.

2. Background and Models

The MI transition for impurities in semiconductors is usually described in terms of one or a combination of the following simple models.

The Mott-Hubbard model considers an idealized perfect lattice of impurities having the average separation and a Coulomb interaction among the electrons (one electron per impurity) which is a constant energy U, whose magnitude is the difference between the ground state energies of the neutral hydrogenic impurity and the negatively charged impurity with two electrons of opposite spin (the D^- ion). This model leads to a picture in which the overlap of the impurity wavefunctions broadens the s-like, N-fold degenerate discrete ground state of the N isolated impurities out into a band whose width is proportional to the overlap integral between adjacent sites. The number of available states in this band, the lower Hubbard (LH) band, is N (not 2N). In addition, there is a higher-lying band of states centered at an energy U above the center of the lower Hubbard band which results from the overlap of the D^- wave functions on adjacent sites. This is referred to as the upper Hubbard (UH) band; for a given average impurity separation it is considerably broader than the LH band due to the larger extent of the D^- wave function. In this model the MI transition occurs at an impurity separation that is less than, but close to that corresponding to overlapping of the LH and UH bands. For densities less than this value the system is an insulator since the LH band is completely filled and there is a gap between the LH and UH bands.

The Anderson model is a strictly single particle model in which the Coulomb repulsion is neglected. This model considers individual electrons to be bound in potential wells located on a regular array of sites, but whose strengths and thus ground states are randomly distributed over a range W. The density of states is considered to be constant over this range and zero outside. All states in this band are localized until the overlap of wave functions of electrons on adjacent sites, which increases with increasing density of wells (impurities), gives rise to an overlap integral with magnitude I greater than some substantial fraction of the band width W. At values of density greater than some critical density the wave function spreads out throughout the crystal, and the system becomes metallic. The value of I/W at which the MI transition is predicted to take place is not precise. In the Anderson model the bandwidth due to disorder and the overlap integral are independent.

A related model in which the bandwidth associated with disorder and the overlap integral are intimately connected is that due to Lifshitz. In this model the potential wells of each impurity are taken to be identical

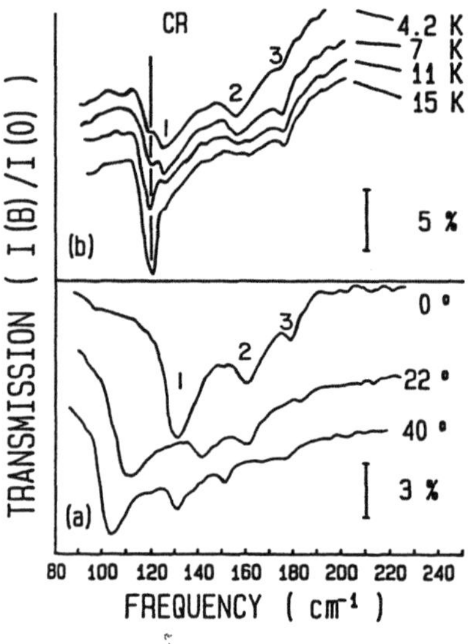

Fig. 1 Magnetotransmission spectra for a center-well and center-barrier-doped MQW sample at 9T. (a) Spectra at three angles of the magnetic field with respect to the sample surface normal at 4.2K.
(b) Dependence of the spectra on temperature with the field normal to the surface.

and the randomness is imposed in the form of topological disorder, often called off-diagonal disorder in the literature. This is a more complex situation due to the interplay between the bandwidth and the overlap integral, and it is thus more difficult to ascertain the condition for the MI transition. The condition is again determined in general by the ratio of the overlap integral to the bandwidth [1]. Topological disorder is closer to the experimental situation obtained in a doped semiconductor sample with very small compensation. In a realistic situation a combination of the above models is most appropriate; thus quantitative predictions are difficult to obtain in practical cases.

A limiting case for which quantitative calculations are available is that of the isolated impurity. The particular case of interest for the present work is that of isolated impurities in the AlGaAs barriers of GaAs/AlGaAs MQW structures. Theoretical calculations for isolated impurities [10], and far IR experiments [12, 13] on relatively lightly doped samples have been carried out on quantum well structures that are doped in the barriers (modulation doped). Representative experimental results are shown in Fig. 1 for a sample doped both over the central one third of the 210 A wells at 1×10^{16} cm^{-3} and over the central one third of the 150 A barriers at 4×10^{16} cm^{-3} with Si donors. The general configuration of the MQW samples is shown schematically in Fig. 2. Panel (a) of Fig. 1 shows transmission spectra taken at several angles of the magnetic field with respect to the normal to the sample surface. Three absorption lines are observed; the position of each scales with the normal component of the applied magnetic field, thus demonstrating the confined nature of all three. Line 3 is the 1s - 2p(m=+1) (standard hydrogen atom notation) transition of donors located at the centers of the GaAs wells [11], and lines 1 and 2 are associated with electrons in the wells bound to their parent donor ions in the barrier centers. The dominant line, line 1, is due to ground state (m = 0) to first excited state (m = +1) transitions [12,13]; the origin of the weaker line 2 is still uncertain. Since it is not of direct concern to the present study, it will not be discussed further. The upper panel (b) shows the temperature dependence of the spectra with the field at normal incidence. The important point here is that line 1 rapidly decreases in intensity with increasing temperature and is replaced by a sharp free electron cyclotron resonance at slightly lower frequency. This demonstrates the localized nature of the barrier impurity transition as well as the small binding energy of the ground state which

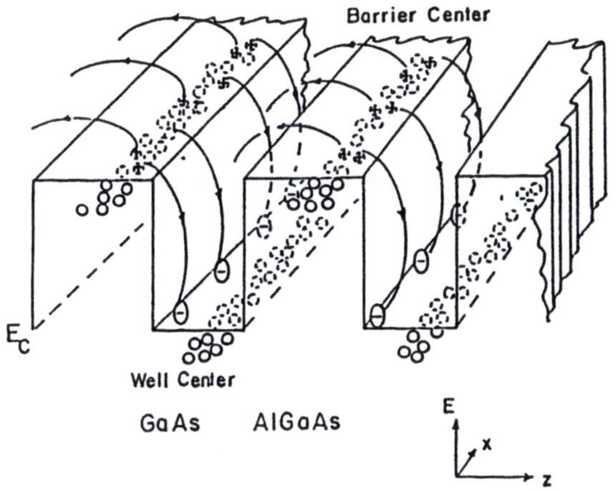

Fig. 2 Schematic energy diagram of the MQW structure and doping.

results from the separation of the electron from the positive donor ion. The magnetic field dependence of the position of line 1 agrees very well with that calculated for isolated impurities [10, 12]; thus at fields above about 2T these samples are well described by an isolated impurity model (or more appropriately by narrow, well separated impurity bands for the ground and first excited states). This model can be used as a starting point for understanding the evolution of the transport measurements and the far infrared spectra with doping densities near that of the MI transition.

3. Experimental Details

The samples for the present study were grown by molecular beam epitaxy. The general structure consisted of a set of 20 240 A GaAs wells separated by 240 Å $Al_{0.3}Ga_{0.7}As$ barriers clad on both sides by approximately 1500 Å of $Al_{0.3}Ga_{0.7}As$. The entire structure was grown on a 2000 Å undoped GaAs buffer layer on semi-insulating GaAs substrates and capped by 100 Å of GaAs. Two samples were investigated in detail in the present work. Both were doped over the central one third of the AlGaAs barriers with Si donors: Sample 1; $N_D = 4 \times 10^{16}$ cm^{-3} :Sample 2;$N_D = 1 \times 10^{17}$ cm^{-3}. The corresponding sheet densities are 3.2×10^{10} and 8×10^{10} cm^{-2}, respectively. Hall bar samples for the electrical transport measurements with dimensions 1.6 mm by 7 mm were cleaved out of the wafers, and ohmic contacts were fabricated by alloying In dots at 400 C for 12 minutes. For the far IR measurements samples with dimensions approximately 5 mm by 5mm were cleaved from the wafers, and the substrate side was wedged at a 6 degree angle to minimize multiple reflection effects.

Far IR transmission measurements were carried out in a helium exchange gas/light-pipe system in a 9T superconducting magnet system in conjuction with a BOMEM Fourier Transform Interferometric Spectrometer or a SPECAC polarizing Fourier Transform Interferometric Spectrometer. For measurements above 80 cm^{-1} and below 80 cm^{-1} a Ge:Ga photoconductive detector and Infrared Laboratories composite Si bolometer were used, respectively. Temperature dependent measurements were carried out with a wire-wound heater/capacitance thermometer combination mounted on the brass sample holder in helium exchange gas.

Standard dc electrical measurements were made as a function of temperature in pumped liquid helium between 1.6 and 4.2 K. Temperature was determined from the helium vapor pressure measured with a capacitance manometer.

4. Results

4.1 High Magnetic Fields

Transmission spectra for Sample 1 are shown in Fig. 3 at 8T and 1.7K. The normal to the sample surface was oriented at a 20 degree angle to the magnetic field. The lowest trace (A) shows the transmission ratioed to a zero field background. The strong line with minimum at 107 cm^{-1} is due to transitions from the ground state (m = 0) to the first excited state (m =+1) of barrier impurities as discussed above. A weak line 2 is also observed at approximately 137 cm^{-1}. The localized nature of these transitions can be clearly seen from the spectrum obtained in the presence of weak optical pumping with a red LED shown in the middle trace (B). A sharp free electron cyclotron resonance line appears at 102 cm^{-1}; the upper trace (C) shows the ratio of LED-on to LED-off which brings out the optically-induced cyclotron resonance line. This line is actually instrumentally broadened at 1 cm^{-1} resolution. Higher resolution data yield a linewidth that corresponds to an effective mobility of 250,000 cm^2/V sec. Measurements of the temperature dependence of the spectra shows a gradual shift from the impurity line at the lowest temperatures to cyclotron resonance at temperatures above approximately 20K. At 9T and with the magnetic field normal to the sample surface, the cyclotron resonance line occurs at 120 cm^{-1}.

A plot of the position of the transmission minimum vs magnetic field (field normal) at 4.2K is given in Fig. 4(a). The field dependence of cyclotron resonance is shown for comparison by the dashed line, and the theoretically predicted transition frequency vs field for isolated impurities located at the barrier centers is shown by the solid line [10]. Over the entire range of fields investigated the observed transmission minima agree well with the calculation.

Transmission spectra for Sample 2 are shown in Fig. 5 as a function of temperature at 9T. As temperature is raised the line gradually shifts down in frequency, but even at 30K the transmission minimum still lies 2 -3 cm^{-1} above the position of free carrier cyclotron resonance measured in Sample 1. The lines are also somewhat broader than those in Sample 1. A ratio of LED-on to LED-off spectra at 4.2K similar to that in Fig. 3 shows a clear cyclotron resonance at a frequency of 121 cm^{-1}. As can be seen from Fig. 4(b) the frequency of the transmission minima at 4.2K again agree well with calculations for isolated impurities at high magnetic fields. Below about 4T there are substantial deviations that are discussed below. Thus

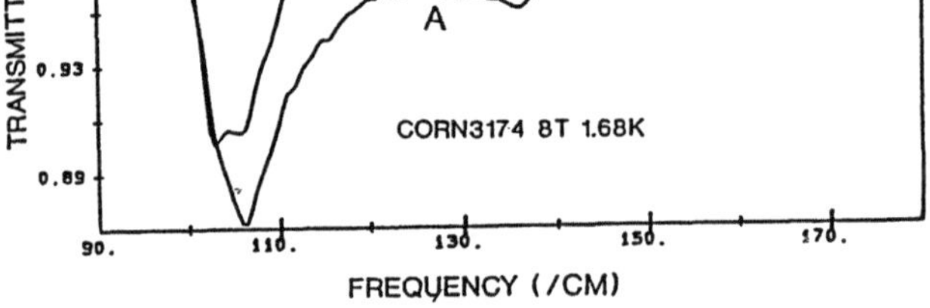

Fig. 3 Magnetotransmission spectra for Sample 1 at 1.68K with the sample tilted at 20 degrees. A - dark: B - LED on: C - ratio, B/A. Curves are displaced for clarity.

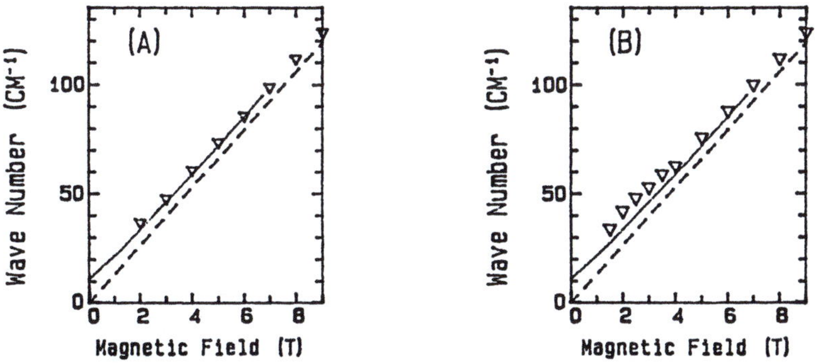

Fig. 4 Frequency vs magnetic field plots for the two samples. Solid lines: Variational calculations for the isolated center-barrier transition from ref. [10]. Dashed lines: cyclotron resonance. (A) Sample 1; (B) Sample 2.

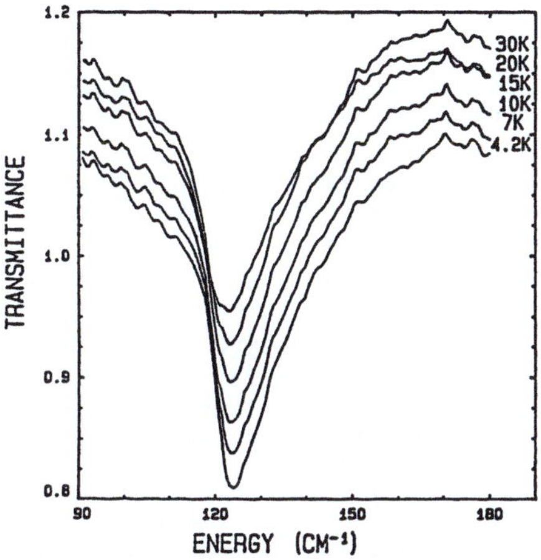

Fig. 5 Magnetotransmission spectra for Sample 2 at 9T and several temperatures.

overall, the behavior of the high field spectra in Sample 2 is similar to that in Sample 1, but quantitatively different in several respects.

The longitudinal resistance for both samples shows thermally activated behavior in the high field region with the magnetoresistance of Sample 2 somewhat larger than that of Sample 1.

4.2 Low and Intermediate Magnetic Fields

Low field magnetotransmission spectra for Sample 2 are shown in Fig. 6. As field is reduced from 4T the line rapidly broadens and loses intensity. Over this region of field the position of the transmission minimum is shifted to higher frequencies than predicted by the isolated impurity model (see Fig. 4(b)). At the lowest magnetic fields field shrinkage of the wave functions is minimal, and the full influence of the broadened impurity bands is evident in the spectra. As mentioned above similar data for Sample 1, although exhibiting some broadening of the line as B is reduced, do not show any clear shift from the position predicted by the isolated impurity calculation outside experimental error.

The Hall coefficient for Sample 2 exhibits very interesting behavior in the low to intermediate field region. The three dimensional carrier density as calculated from the Hall voltage and the effective thickness of the sample (taken to be the GaAs well thickness times the number of wells) is shown in Fig. 7 as a function of 1/T for several values of magnetic

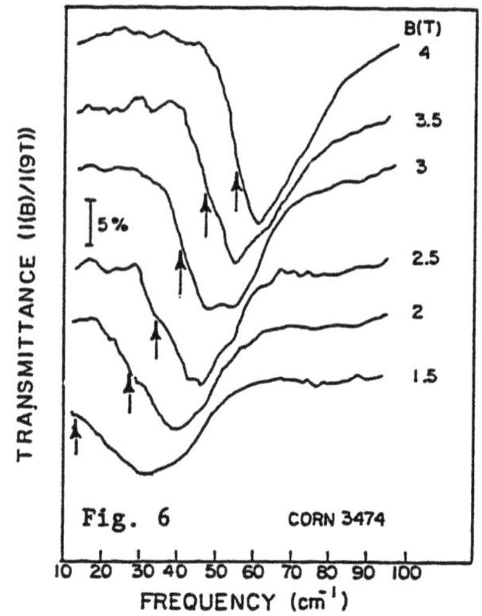

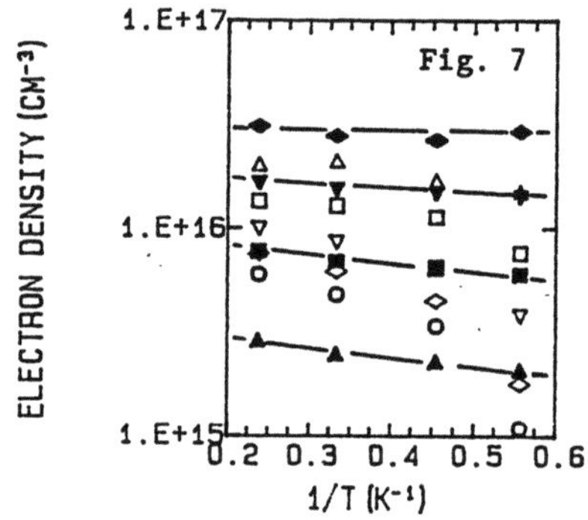

Fig. 6 Magnetotransmission spectra at 4.2K and several magnetic fields in the low-intermediate field region. The arrows indicate the positions of cyclotron resonance.

Fig. 7 Semilogarithmic plot of electron density vs 1/T for Sample 2 at several magnetic fields. ▲ - 0.2T: - 0.5T: ▼- 1T: ◆- 2T: Δ - 3T: □- 4T: ◇- 5T: ▽- 6T: o - 7T.

field. At the lowest fields the carrier density is thermally activated with a very small activation energy (approximately 0.08 meV). As field is increased the activation energy <u>decreases</u> and the carrier density <u>increases</u>. At a field of about 2T the activation energy is zero within experimental error, and the sample is clearly metallic with a carrier density approximately equal to that expected with all dopant impurities contributing. Above this field the Hall voltage becomes thermally activated again. This behavior is somewhat similar to the "Hall dip" data reported for bulk samples [5]. Hall results for Sample 1 are inconclusive at this time due to contact problems, but the Hall conductivity appears to be thermally activated at all values of magnetic field investigated.

5. Discussion and Conclusions

The criterion for the MI transition is not precisely defined. For the present situation of confined impurities the important dimensionless parameter is the product of the square root of the sheet doping density per well and the radius of the ground state wavefunction in the plane of the wells, $[N_{2D}]^{1/2} a_o^* = M$. An approximate criterion is $[N_{2D}]^{1/2} a_o^* > 0.3$ for the metallic side of the transition. An estimate of the extent of the wavefunction for barrier impurities at zero magnetic field with the parameters of the present samples yields a_o^* approximately 200 Å [14]. Thus M is approximately 0.35 for Sample 1 and 0.57 for Sample 2. The primary effect of the magnetic field is to shrink the wavefunction in the plane normal to the magnetic field. In the high field limit, $\hbar\omega_c/2Ry^* \gg 1$, where ω_c is the cyclotron frequency, and Ry^* is the effective Rydberg for the confined impurities, the length playing the role of a_o^* at zero field is the magnetic length, $\ell = [c\hbar/eB]^{1/2}$. For the present samples Ry^* is approximately 2.5 meV, $\hbar\omega_c/2Ry^* = 3$, and $\ell = 86$ Å at 9T. Thus over the range studied the magnetic field has a substantial effect on the impurity wave functions and their overlap. For example, replacing a_o^* by the magnetic length at 9T yields a value of M(B) = 0.24, below the (somewhat arbitrary) criterion given above for the MI transition. Without detailed calculations of the magnetic field dependence of the barrier impurity wavefunctions and a more precise definition of the criterion for the MI transition it is not possible to make other than these rather qualitative statements. Nevertheless, such estimates can be used as a guide in interpreting the far IR and electrical transport measurements.

5.1 High Magnetic Fields

Given the arguments above it is not surprising that the magnetotransmission measurements at high fields yield spectra whose main features are reasonably well explained by the isolated impurity calculations. The increase of the linewidth from Sample 1 to Sample 2 is consistent with a broader impurity band for the more heavily doped Sample 2. It should be emphasized that the absorption lineshapes, in general, result from a complicated average over the ground state and first excited state impurity bandwidths; the particular form of the average depends on the details of the mechanism leading to the impurity bands. It is not clear why at higher temperatures, or under optical excitation, the free carrier cyclotron resonance for the two samples should occur at slightly different energies. It is possible that a larger well in Sample 2 leads to decreased nonparabolicity and thus a slightly larger dydlotron frequency. The thermally activated behavior of the conductivity for both samples is also consistent with both being on the insulating side of the MI transition at high fields.

5.2 Low and Intermediate Magnetic Fields

Sample 1 appears to be below, but very close to the MI transition even at zero magnetic field, consistent with estimates based on the approximate criterion given above. The observation of activated conductivity over the whole range of fields, as well as impurity ground state to first excited state transitions that agree with the isolated impurity calculations down to the lowest field studied are in agreement with this assertion. The broadening of the absorption line at the lowest fields indicates that there is a substantial impurity band that is narrowed appreciably by the magnetic field.

The most interesting results concern Sample 2. The behavior of the Hall conductivity at fields up to about 2T is particularly striking and reminiscent of recent work in bulk semiconductors doped with donors [5]. In this work arguments based on exclusion of volume around isolated donors and large metallic clusters were presented to explain a so-called "Hall dip" as a function of magnetic field. This model explains the increasing metallic behavior with increasing magnetic field as due to increasing numbers of <u>isolated</u> donors (those with essentially no overlap with nearest neighbors) and concomitant removal of volume resulting from the field induced decrease in overlap. The excluded volume per isolated impurity is greater than the average volume occupied by an impurity, and thus the remaining impurities for which there <u>is</u> overlap have an increased value of M, i.e., they are more metallic. The present results are at least consistent with such an explanation; however the data is too sparse at present to make more positive statements. In any case, above about 3T the magnetic field begins to cause the sample to become insulating, apparently due to the reduction in overlap of the wavefunctions of a sufficiently large fraction of the impurities to remove the possibility of large metallic clusters with percolation pathways.

In this region where there is a MI transition observable in the transport measurements, there is no dramatic change in the magnetotransmission spectra; the line simply broadens continuously as field is reduced. This is an indication that the transition is predominantly a Mott-Anderson or Mott-Lifshitz transition resulting from an overlap (or the removal thereof) of the LH and UH bands. The MI transitions occurs in the impurity bands. Complete elucidation of the details of this transition at low densities requires more extensive work, particularly in the temperature dependence of the transport measurements.

In conclusion, the present work demonstrates the promise of the GaAs/AlGaAs MQW system for the study of the MI transition in confined structures. It will be particularly interesting to examine in detail the evolution of the transition that occurs from impurity band metallic behavior to conduction band metallic behavior at higher doping densities, and to examine how the magnetic field affects this transition at very high fields and very low temperatures.

6. Acknowledgements

The work at SUNY at Buffalo was supported in part by the Office of Naval Research and by the ONR/SDIO MFEL program. YJW and BDM are grateful to L. Cooper for continuing support and encouragement, and to R. Ranganathan, S.Holmes, W.J.Li, J-P. Cheng, X. Liu and A. Petrou for experimental assistance, helpful discussions, and photoluminescence and reflectivity characterization of the samples. They know how to sort this out.

References

1. See e.g., B.I. Shklovskii and A.I. Efros, "Electronic Properties of Doped Semiconductors", Springer Series in Solid State Sciences (Springer-Verlag, Berlin, 1984).
2. See e.g., T. Ando, A.B. Fowler and F. Stern, Rev. Mod. Phys. 54, 437 (1982).
3. M-W. Lee, D. Romero, H. D. Drew, M. Shayegan and B.S. Elman, Solid State Commun. 66, 23 (1988).
4. M. Shayegan, V.J. Goldman and H.D. Drew, Phys. Rev.B 38, 5585 (1988).
5. V.J. Goldman, M.,Shayegan and H.D. Drew, Phys.,Rev. Letters 56, 1056 (1986).
6. J.L. Robert, A. Raymond, L.Konczewicz, C. Bousquet, W. Zawadzki, F. Alexandre, I.M. Masson, J.P. Andre and P.M. Frijlink, Phys. Rev. B 33, 5935 (1986).
7. A. Zrenner, F. Koch and K. Ploog, Surface Sci. 196, 671 (1988).
8. H. Sigg, K. v. Klitzing, M. Hauser and K. Ploog, "Shallow Impurities in Semiconductors 1988", Edited by B. Monemar (Inst. of Physics, Bristol, 1988) pp 11 -20.
9. R.L. Greene and K.K. Bajaj, Phys. Rev. B 34, 951 (1986).
10. R.L. Greene and P.Lane, Phys. Rev. B 34, 8639 (1986).
11. See e.g., A.A. Reeder, J-M. Mercy and B.D. McCombe, IEEE J. Quant. Elec. QE-24, 1690 (1988).
12. J-M. Mercy, B.D. McCombe W. Beard, J.Ralston and G. Wicks, SurfacemSci. 196, 334 (1988).
13. E. Glaser, B.V. Shanabrook, R.L. Hawkins, W. Beard, J-M. Mercy, B.D. McCombe and D. Musser, Phys. Rev. B 36, 8185 (1987).
14. M. Stopa and S. DasSarma, Phys. Rev. B 40, 8466 (1989).

Localized Conduction States due to Magneto-Acceptors in 2D Systems

M. Kubisa and W. Zawadzki

Institute of Physics, Polish Academy of Sciences, PL-02 668 Warsaw, Poland

It is shown that in the 2D systems with an external magnetic field perpendicular to the interfaces, acceptors produce bound (localized) states also in the conduction band. Binding energies of magneto-acceptors are compared to those of magneto-donors in similar conditions. Implications for the broadening of Landau levels in 2D systems and for the ratio of the free states to the bound states in a magnetic field are discussed.

1. Introduction

Donors and acceptors in semiconductors have been for many years the subject of experimental and theoretical studies. Shallow donors produce bound electron states just below the bottom of the conduction band, while shallow acceptors produce bound electron states just above the top of the valence band. This is directly related to the sign of the potential energy: negative for the donors and positive for the acceptors, and the sign of the electron effective mass: positive for the conduction band and negative for the valence band. It is clear, for example, that in the bulk semiconductor a conduction electron is not bound to an acceptor (negative) charge. Being repulsed by the Coulomb interaction it runs away.

However, if for any reason the conduction electron is confined to the proximity of an acceptor, it will form states having energies in the range of the conduction band. One of such systems is provided by the commonly investigated 2D semiconductor structures with an external magnetic field perpendicular to interfaces. In this situation the conduction electron can not run away from the acceptor because it is confined in the z direction by the potential well, and in the x-y plane by the Lorentz force of the magnetic field. Below we describe theoretically this case. One should mention here the work of Laughlin [1] who, using the first-order perturbation theory, showed that a positive potential in such a system gives a discrete electron energy above the conduction Landau level.

It is clear from the above reasoning, that one can also expect in 2D systems subjected to magnetic field bound states of valence electrons related to donor potentials having energies in the range of the valence band.

2. Perfectly two-dimensional magneto-impurities

We consider first a perfectly 2D conduction electron in an external magnetic field perpendicular to the 2D plane. Neglecting spin the initial Hamiltonian reads (cylindrical coordinates, atomic units),

$$H = H_0 \pm \frac{2}{\rho} \quad , \tag{1}$$

Springer Series in Solid-State Sciences, Vol. 97
Localization and Confinement of Electrons in Semiconductors
Editors: F. Kuchar · H. Heinrich · G. Bauer

where the minus sign describes the donor case and the plus sign the acceptor case. The kinetic energy is [2]

$$H_0 = -\frac{1}{\rho}\frac{\delta}{\delta\rho}\left(\rho\frac{\delta}{\delta\rho}\right) - \frac{1}{\rho^2}\frac{\delta^2}{\delta\varphi^2} - i\gamma\frac{\delta}{\delta\varphi} + \frac{\gamma^2\rho^2}{4} , \tag{2}$$

in which

$$\gamma = \frac{\hbar\omega_c}{2Ry^*} \tag{3}$$

characterizes relative strengths of magnetic and Coulomb interactions. The cyclotron frequency is $\omega_c = eB/m^*$, the effective Rydberg is $Ry^* = m^*e^4/2\hbar^2\kappa^2$, where κ is the static dielectric constant. The energies in Eqs. (1) and (2) are counted in Ry* and the lengths in the effective Bohr radia $a_B^* = \kappa\hbar^2/m^*e^2$.

Solutions of the hamiltonian H_0 (free 2D electron in a magnetic field) are characterized by two quantum members: N=0,1,2,...., and a value of projection of the angular momentum on the magnetic field direction M=...-2,-1,0,1,2,.... The energies are then

$$E_{NM} = 2\gamma\left(N + \frac{M+|M|+1}{2}\right) . \tag{4}$$

On the other hand, in absence of the magnetic field ($\gamma=0$) the donor energies are

$$E_l = -\frac{1}{\left(l+\frac{1}{2}\right)^2} , \quad l = 0,1,2\ldots \tag{5}$$

M is a good quantum number in the general case (1), reflecting the cylindrical symmetry of the physical situation.

We solve the eigenenergy problem for the Hamiltonian (1) variationally. For the states with N=0 we take the trial function in the form

$$\Phi_{OM}(\rho,\varphi) = \frac{1}{\sqrt{2\pi}}\exp(iM\varphi)C_{OM}\rho^m\exp\left(-\frac{1}{4}\alpha_0^2\rho^2 - \beta_0\rho\right) , \tag{6}$$

where $m=|M|$, and the normalization coefficient is $C_{OM} = (\Delta_{2m+2})^{-1/2}$, in which

$$\Delta_i = \frac{(i-1)!}{\alpha_0^i}\exp\left(\frac{\beta_0^2}{\alpha_0^2}\right)D_{-i}\left(2\frac{\beta_0}{\alpha_0}\right) , \tag{7}$$

and $D_{-i}(t)$ is the parabolic cylinder function

$$D_{-i}(t) = \frac{1}{(i-1)!}\exp\left(-\frac{t^2}{4}\right)\int_0 x^{i-1}\exp\left(-\frac{1}{2}x^2 - tx\right)dx . \tag{8}$$

α_0 abd β_0 are variational parameters. The trial function (6) combines atomic and magnetic features.

The variational average of the energy is

$$\bar{E}_{OM} = M\gamma - \beta_0^2 + \frac{1}{4}(\gamma^2+\alpha_0^4)\frac{\Delta_{2m+4}}{\Delta_{2m+2}} + (2m+1)\beta_0\frac{\Delta_{2m+1}}{\Delta_{2m+2}} \pm 2\frac{\Delta_{2m+1}}{\Delta_{2m+2}} . \tag{9}$$

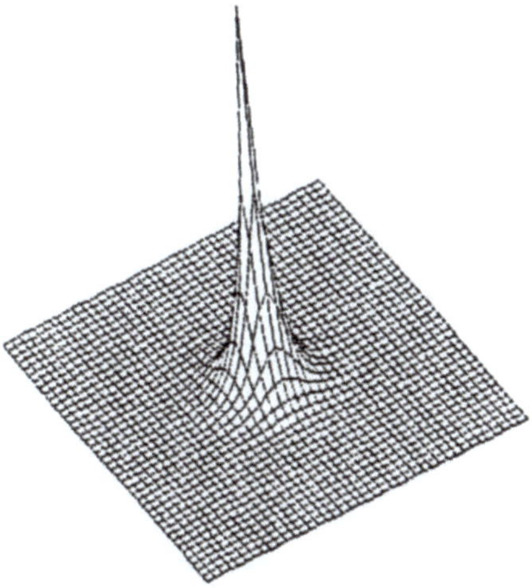

Fig. 1 Square of the wavefunction $|\Phi_{00}(\rho,\varphi)|^2$ for a perfectly two-dimensional magneto-donor, as calculated for γ=1.

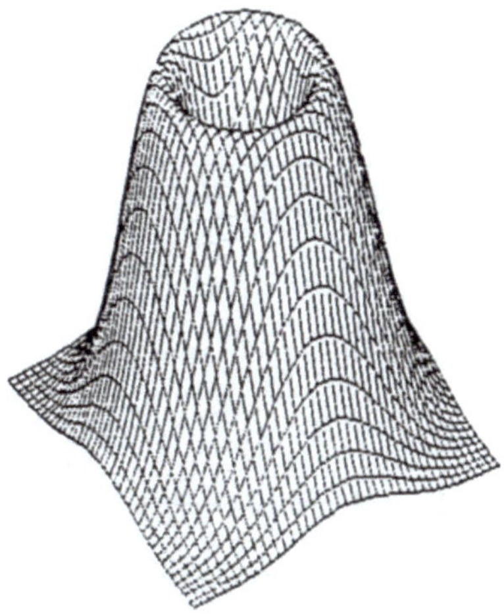

Fig. 2 Square of the wavefunction $|\Phi_{00}(\rho,\varphi)|^2$ for a perfectly two-dimensional magneto-acceptor, as calculated for γ=1. The same x-y area is shown as in Fig. 1, but the height of the donor wave-function in Fig. 1 is much larger than that of the acceptor.

Looking for the minimum of the trial average (9), given by the conditions $\delta\bar{E}/\delta\alpha_0=0$ and $\delta\bar{E}/\delta\beta_0=0$, we calculate variational energies of magneto-acceptor and magneto-donor. It turns out that for the repulsive (acceptor) potential the minimizing value of β_0 is negative, so that a maximum of the wavefunction (6) occurs at some distance from the acceptor. the wavefunctions for the states (MN)=(00) in the two cases are shown in Figs 1 and 2.

Having calculated the variational absolute energy one can determine an energy distance from the free electron value $E_{00}=\gamma$. The results are shown in Fig. 3. It can be seen that, with the use of our trial function (6), the exact binding energy $E_b=4$ is calculated for the 2D ground donor state at B=0 (cf. Eq(5)).

The magneto-acceptor levels occur above the free-electron Landau level. At γ=0 these levels coalesce with the free-electron energy. This indicates, in agreement with intuitive expectations, that it is the magnetic field which keeps the electron in a proximity of the acceptor. In absence of the field the electron runs away from the acceptor and the Coulomb energy vanishes. For larger states (with M≠0) the Coulomb interaction is weaker; this corresponds to higher donor and lower acceptor energies. Since the variational procedure gives an upper bound to the exact energy, the true energies of the donor and acceptor states can be somewhat lower than the ones shown in Fig. 3. However, the two-parameter wavefunctions (6) are believed to give very good results.

In order to describe possible cyclotron-resonance-type transitions for magneto-acceptors we have also calculated variational energies of the

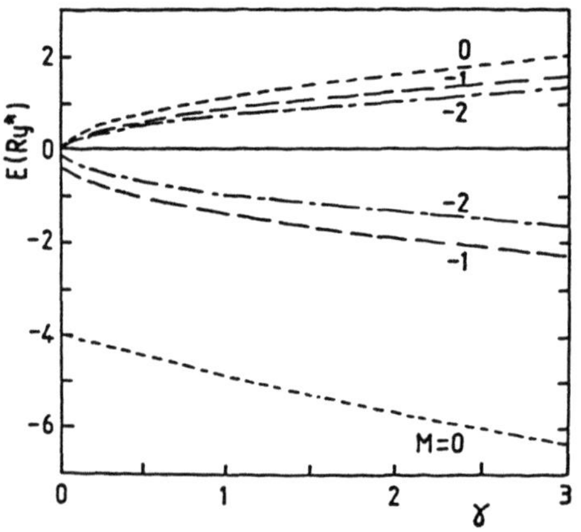

Fig.3 Variational energies of the three perfectly 2D magneto-acceptor (OM) states (positive values) and magneto-donor (OM) states (negative values), as counted from the lowest conduction Landau level (horizontal solid line) versus γ. The quantum numbers M are indicated. At γ=0 the donor states have finite binding energies, while the acceptor energies coalesce with the free-electron energy.

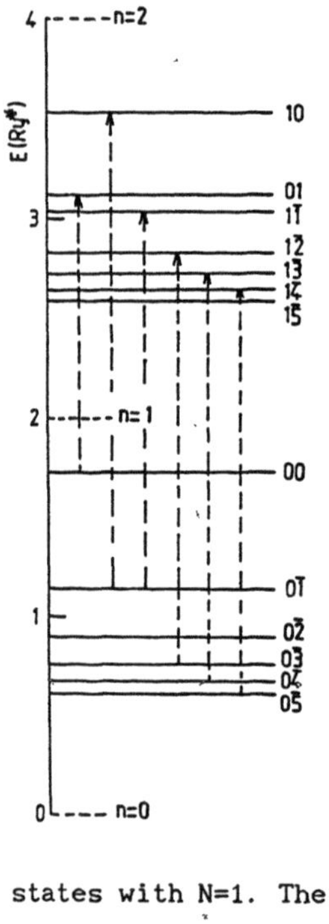

Fig.4 Energies of perfectly 2D magneto-acceptor states related to n=0 and n=1 Landau levels, as calculated variationally for γ=1. Optical cyclotron-resonance-type transitions obeying the selection rules ΔM=1, Δn=1, are indicated.

states with N=1. The trial function has been chosen in the form

$$\Phi_{1M} = \frac{1}{\sqrt{2\pi}} \exp(iM\varphi) C_{1M} (\alpha_1\rho)^m (m+1-\beta\rho-\frac{1}{2}\alpha^2\rho^2) \exp(-\frac{1}{4}\alpha_1^2\rho^2-\beta_1\rho) \quad , \qquad (10)$$

where $\alpha^2=(\alpha_0^2+\alpha_1^2)/2$ and $\beta=(\beta_0+\beta_1)/2$, in which α_1 and β_1 are variational parameters and α_0 and β_0 have been defined above. The (1M) states are orthogonal to the (OM) states. Using the trial function (10) the variational energy has been calculated in terms of the parabolic cylinder functions and minimized with respect to α_1 and β_1 (for the fixed minimizing values of α_0 and β_0). The calculated magneto-acceptor energies at $\gamma=1$ are shown in Fig. 4.

The optical selection rules for the left circular polarized radiation σ_L are: $\Delta M=+1$ and $\Delta n=+1$, where $n=N+(M+|M|)/2$ is the Landau quantum number [3]. The corresponding transitions are indicated in Fig. 4. The energy of (00)$\rightarrow$(01) transition is smaller than the free-electron cyclotron energy $\hbar\omega_c$, all the other transitions have energies larger than $\hbar\omega_c$.

3. Magneto-impurities in GaAs-GaAlAs heterojunctions

We shall now consider magneto-impurities in a quasi-2D situation, provided by an electric potential U(z) of a heterojunction. To this end we use a method first proposed for rectangular quantum wells by Brum et al. [3] and employed for heterojunctions by Zawadzki et al. [4]. The method is based on two approximations. 1) The trial function is separated in the form: $\Psi(x,y,z)=\Phi(x,y)f(z)$. 2) It is assumed that the envelope f(z) of the bound electron (in the presence of U(z) and the Coulomb potential V(x,y,z) is identical to the envelope $f_0(z)$ of the electron in the electric subband (i.e. without the Coulomb potential). These simplifications allow one to reduce the 3D problem of an electron in V(x,y,z)+U(z) potentials to the 2D problem

$$H = H_0 + V_{ef}(\rho) \quad , \tag{11}$$

with an effective potential

$$V_{ef}(\rho) = \int_{-}^{+} V(\rho,z)f_0^2(z)dz \quad , \tag{12}$$

where H_0 is given in Eq. (2). We take $V(\rho,z)$ in the Coulomb form

$$V(\rho,z) = \pm \frac{2}{[\rho^2+(z-z_0)^2]^{1/2}} \quad . \tag{13}$$

Taking a two-dimensional Fourier transform

$$V(\vec{\rho},z) = \int V(\vec{q},z)\exp(i\vec{q}.\vec{\rho})d^2\vec{q} \quad , \tag{14}$$

one gets

$$V(q,z) = \pm\frac{1}{\pi q}\exp(-q|z-z_0|) \quad , \tag{15}$$

where $\vec{q}=(q_x,q_y)$ and $q=(q_x^2+q_y^2)^{1/2}$. Defining

$$W(q) = 2\pi\int_{-}^{+} V(q,z)f_0^2(z)dz \quad , \tag{16}$$

we obtain

$$V_{ef}(\rho) = \int_0 W(q)J_0(\rho q)qdq \quad , \tag{17}$$

where J_0 is the Bessel function of the first kind.

If $\Phi(x,y)=\Phi_{0M}(\rho,\varphi)$ is the trial function describing the electron motion in the (x,y) plane, the trial potential energy is

$$U_M = \langle\Phi_{0M}|V_{ef}|\Phi_{0M}\rangle = \int_0 W(q)\chi(q)qdq \quad , \tag{18}$$

where

$$\chi_M(q) = \int|\Phi_{0M}|^2 J_0(\rho q)d^2\vec{\rho} \tag{19}$$

depends only on the trial function. The complete information on the 2D structure and the impurity potential is contained in W(q) of Eq.(16). The effective potential $V_{ef}(\rho)$ of Eq.(17) depends on the shape of the potential well through $f_0(z)$ and on the impurity position z_0. It may differ considerably from the Coulomb potential.

Considering a GaAs-GaAlAs heterojunction we take the simplest Fang-Howard trial function for the lowest electric subband,

$$f_0(z) = (b^3/2)^{1/2}z\exp(-bz/2) \quad , \qquad z\geq 0 \quad , \tag{20}$$

$$= 0 \quad , \qquad z<0 \quad ,$$

where $b=(12m^*eF_{eff}/\hbar^2)^{1/3}$ and $F_{eff}=(4\pi e/\kappa)(N_{depl}+11N_s/32)$ in the standard notation. Using the above trial function and taking the impurity at the interface ($z_0=0$), we calculate

$$W(q) = \pm\frac{2}{q(1+q/b)^3} \quad . \tag{21}$$

We have performed variational calculations for a GaAs-GaAlAs heterojunction taking $m^*=0.067m_0$, $\kappa=13.1$, $N_s=5\times10^{11}cm^{-2}$, $N_{depl}=6\times10^{10}cm^{-2}$. For the states N=0, M=1,2,3 the trial function (6) has been used. The results are shown in Fig.5. The qualitative picture is similar to that of Fig.3 for perfectly 2D magneto-impurities. However, the effective two-dimensional impurity potential in a heterojunction does not have a singularity at the impurity site, which has a large weakening effect on the energy of the ground donor state.

At magnetic fields typically used in magneto-transport investigations on GaAs-GaAlAs heterojunctions one has $\gamma\geq 1$. It can be seen from Fig.5 that in this range of γ values the acceptor and donor energies become almost symmetric around the free-electron Landau level. This means that a broadening of Landau levels due to randomly distributed donors and acceptors should be symmetric below and above the level, in agreement with the observations.

The presented theory considers only one impurity. In a real structure many impurities are present, their wave functions may overlap, and sharp localized states cease to exist. This corresponds to a metal-nonmetal transition, çf.Ref.[4]. Magneto-impurity states with higher negative M values are progressively larger. This means that at a given magnetic field the wave functions of the states closer to the Landau level (which are larger) may still overlap, while those further from the Landau level (which are smaller) do not overlap anymore, i.e. they become localized. This serves as a model for the mobility edges within broadened Landau levels, also observed experimentally.

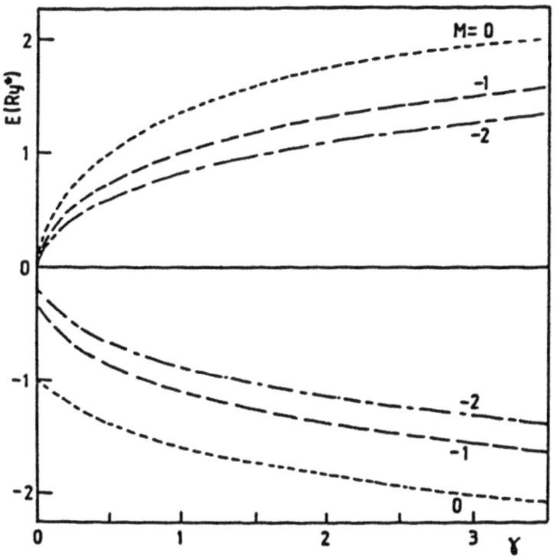

Fig.5 Variational energies of magneto-donor (negative values) and magneto-acceptor (positive values) in a GaAs-GaAlAs heterojunction, as counted from the lowest conduction Landau level (horizontal solid line) versus γ.

References

1. R.B.Laughlin, Phys.Rev. B27, 3383 (1983).
2. Y.Yafet, R.W.Keyes, and E.N.Adams, J.Phys.Chem.Solids 1, 137 (1956).
3. W.Zawadzki and J.Wlasak, in "Theoretical Aspects and New Developments in Magneto-Optics" (Editor J.T.Devreese), Plenum Press (New York, 1980), p.347.
4. J.A.Brum, G.Bastard, and L.Guillemont, Phys.Rev. B30, 905 (1984).
5. W.Zawadzki, M.Kubisa, A.Raymond, J.L.Robert, and J.P.Andre, Phys.Rev. B36, 9297 (1987).

Part VII

Optical and Magnetic Properties of Dilute Magnetic Semiconductor Quantum Well Structures

Ultrafast Magnetic and Electronic Spectroscopy in Dilute Magnetic Semiconductor Quantum Wells

D.D. Awschalom, M.R. Freeman, J.M. Hong, and L.L. Chang

IBM Research, T.J. Watson Research Center, P.O. Box 218, Yorktown Heights, NY 10598, USA

Abstract. A parallel set of two low-temperature optical experiments have been performed in order to explore the connection between the magnetic and electronic properties in $Cd_{1-x}Mn_xTe$ diluted magnetic semiconductor quantum structures. Optically-induced magnetic spectroscopy using integrated dc SQUID susceptometers has revealed both the static and time-resolved *magnetic* signature of carrier localization in multiple quantum wells. In addition, the electronic spin scattering of optically excited charge carriers has been directly observed in time scales short compared to the characteristic magnetic interaction times with femtosecond spin-polarized luminescence measurements. Unlike three dimensional systems, quantum confinement results in unexpectedly rapid carrier spin relaxation and a weak polarization of the magnetic ions which subsequently decay independently of the electronic behavior.

1. Introduction

The field of optically induced magnetization in semiconductors began in 1967 with the work of Lampel on optical pumping in silicon [1]. Optically excited electrons and holes acted as intermediaries to transfer a certain amount of the angular momentum of incoming circularly polarized light to the nuclear system via the hyperfine interaction. The change in polarization of the nuclei was detected using nuclear magnetic resonance techniques. More recently, a resurgence of interest in diluted magnetic semiconductor (DMS) systems has been driven by the discovery of very large magneto-optical effects and by progress in epitaxial growth of the materials [2]. Charge carriers in these systems are coupled to the d-electrons of localized magnetic ions by a strong spin-spin exchange interaction and can give rise to an optically pumped magnetization which has been seen directly with Faraday rotation [3], and Superconducting Quantum Interference Device (SQUID) magnetometers [4] and susceptometers [5]. In particular, these experiments have developed into a genuine spectroscopy [5,6] in which one may directly observe the energy-dependent response of the magnetic species. In this article we discuss studies of quantum well structures of the II-VI DMS $Cd_{1-x}Mn_xTe$ in which carrier confinement effects are clearly found in

Springer Series in Solid-State Sciences, Vol. 97
Localization and Confinement of Electrons in Semiconductors
Editors: F. Kuchar · H. Heinrich · G. Bauer

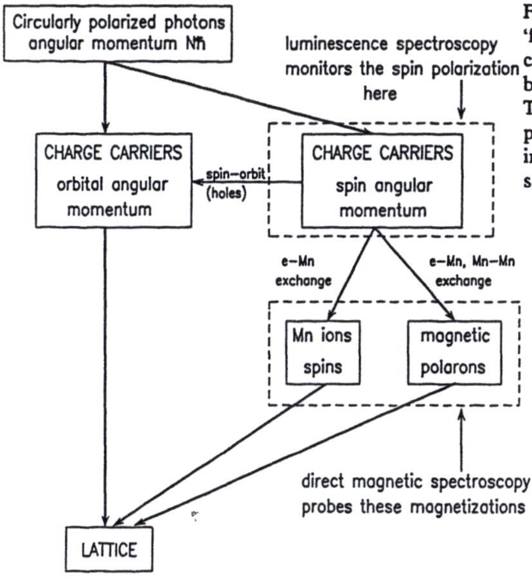

Figure 1. A chart to schematically illustrate the 'flow' of angular momentum initiated by a circularly polarized optical excitation above the bandgap of a diluted magnetic semiconductor. The charge carrier spin orientations relax primarily through manifestations of the spin-orbit interaction in the band structure or by spin-flip scattering with the magnetic ions.

the magnetic measurements [6], using a susceptometer operating near the quantum noise limit to overcome the relatively small size of the signals. The high photoluminescence efficiency of the quantum structures enables us to monitor in addition the link between the incoming light and the magnetic ions, namely the initially spin-polarized electron-hole population. The temporal decay of spin alignment of these charge carriers is observed in a femtosecond time-resolved luminescence polarization experiment [7] and can be studied as a function of energy and degree of carrier confinement. The importance of having more than one means of probing the system is illustrated by the angular momentum 'flowchart' in Fig. 1, which schematizes the chain of events initiated by a circularly polarized excitation. In concert, these two studies enable one to monitor the evolution of angular momentum in the quantum structure.

2. Diluted Magnetic Semiconductor Quantum Wells

The quantum well structures used in this work are grown by molecular epitaxy and have been described in detail elsewhere [8]. Layer thicknesses range from 9 Å to 200 Å, with Mn concentrations from 0 to 38%, where the latter is approximately the upper limit for well-controlled crystalline growth. The Mn substitution for Cd results in a contraction of the lattice and a concomitant upward shift in the bandgap which is linear in the Mn concentration. Quasi-two dimensional layers for carrier confinement are produced by varying the Mn content during growth. The structures are typically characterized by x-ray diffraction studies and low-temperature

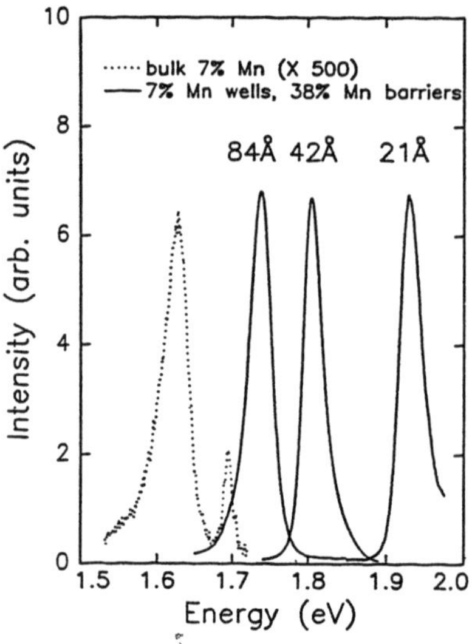

Figure 2. Photoluminescence spectra of three $Cd_{1-x}Mn_xTe$-$Cd_{1-y}Mn_yTe$ quantum structures with $x=0.065$ wells and $y=0.38$ barriers, at T=5K. The barrier widths are 86 Å and the well widths are indicated on the plot. A measurement for an MBE-grown thick film having the same Mn concentration as the wells is shown for comparison.

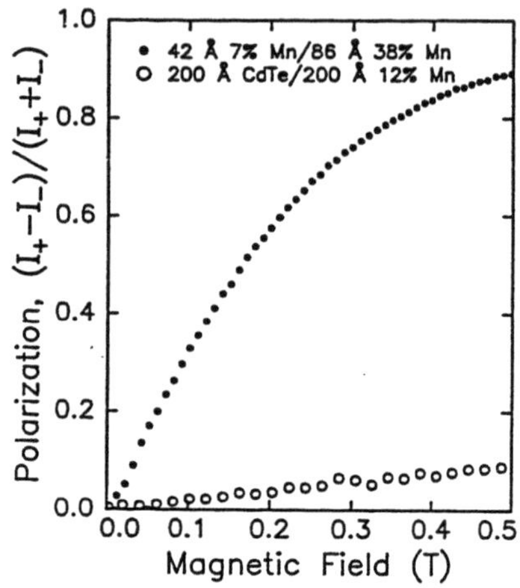

Figure 3. The circular polarization of luminescence as a function of applied magnetic field at T=4K for two quantum structures, one of which has pure CdTe wells (open circles). The luminescence is detected perpendicular to the layer planes, and parallel to the field (Faraday configuration). The excitations are linearly polarized.

photoluminescence measurements. Liquid helium temperature luminescence spectra of a variety of $Cd_{1-x}Mn_xTe$-$Cd_{1-y}Mn_yTe$ structures all having 86 Å, 38% Mn barrier layers are shown in Fig. 2. The luminescence peak positions agree to 1% with the lowest subband energies estimated in a simple Kronig-Penney calculation assuming a small (~10 meV) valence band offset. The lower intensity luminescence from an epitaxial thick film is shown for comparison, and is seen to be dominated by recombination of deeply bound excitons.

The strong spin-spin exchange interaction appears as a large effective g-factor (~1000) for the charge carriers at low temperatures. This is seen in measurements of the luminescence polarization in magnetic fields perpendicular to the layer planes, such as those in Fig. 3. The influence of the exchange interaction on the carriers is proportional to the number of unpaired Mn ions in the system, due to the antiferromagnetic Mn-Mn interaction. Another factor in the quantum well system, unlike a bulk crystal or a uniform thin film, is the relative probability, as determined by the carrier wavefunction, of finding the particle in the barrier or in the well. Thus the observed polarization is much larger in Fig. 3 for the Mn-containing well in comparison to the wide non-magnetic well. In the latter there is relatively little overlap of the wavefunction with the barrier. This demonstrates that the carriers are in fact predominantly confined to the wells and not, for example, bound to the interfaces because of poor interfacial quality. The data of Fig. 3 are measured using a linearly polarized excitation, and the luminescence polarization in finite magnetic fields arises as the carrier spins settle into the equilibrium distribution appropriate to the Zeeman splitting of the energy levels. Under such conditions one can also measure an increase in magnetization of the sample due to unpolarized illumination [9].

3. Direct Magnetic Measurements

A wealth of spectroscopic information can be obtained from both static and time-resolved measurements of the optically induced magnetization. The power of this technique stems from the fact that the magnetic response is determined by the *local* environment of the polarized carriers rather than by a spatial average over the whole system in a uniform magnetic field as in traditional susceptibility measurements. Some of the electron spin alignment is transferred to the Mn ions through spin-flip scattering, and in addition the electronic polarization, before it decays, represents a small bias field tending to orient the ion spins. These are short range interactions, making the experiments particularly sensitive to the difference in the magnetic environments of the wells and barriers. The magnetization is therefore sensitive to the degree of carrier confinement both through the well width and through the excitation energy (the initial carrier energy is important because we have found the carrier spin scattering times and energy loss times to be comparable). In Fig. 4 we show the static magnetization induced in several quantum structures as a function of exciting photon energy. These magnetic excitation spectra display peaks at the subband energies in the same manner as traditional photoluminescence excitation spectra. Here the oscillator strength for optical absorption is convolved with the spin relaxation rates and the spatial extent of the carrier wavefunction to produce the observed signal. The peak positions nevertheless are in good agreement with the subband energies predicted by the Kronig-Penney model. The carrier magnetization itself is not seen in these measurements, because the short spin relaxation time of the carriers leads to a small time-integrated magnetization. The picosecond bursts of carrier magnetization are also attenuated by a high frequency roll-off of the input circuit coupling the flux from the sample into the SQUID.

A two pulse pump-probe technique is used for the time-resolved magnetic measurements, with the SQUID susceptometer acting as an integrating detector in a boxcar averaging mode. The transient magnetization generated in the sample by the initial (pump) pulse is perturbed at a time δt later by the weaker probe pulse. The response to the probe pulse depends on the instantaneous magnetization, allowing a

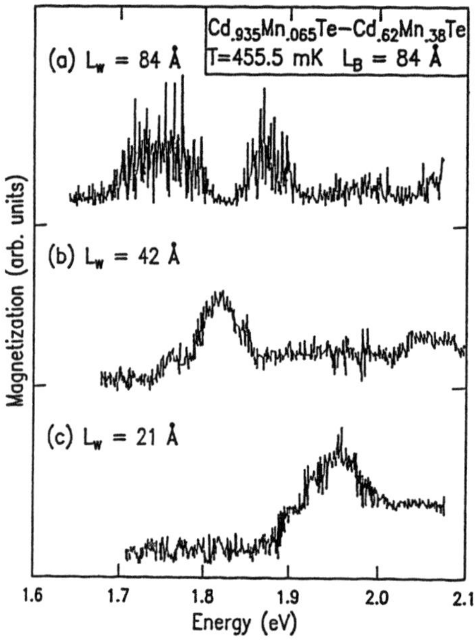

Figure 4. Direct magnetic spectroscopy of the three multiple quantum wells of Fig. 2, with fixed barrier thickness and varying well widths as shown. The total number of periods in the structures are (a) 30, (b) 60 and (c) 120.

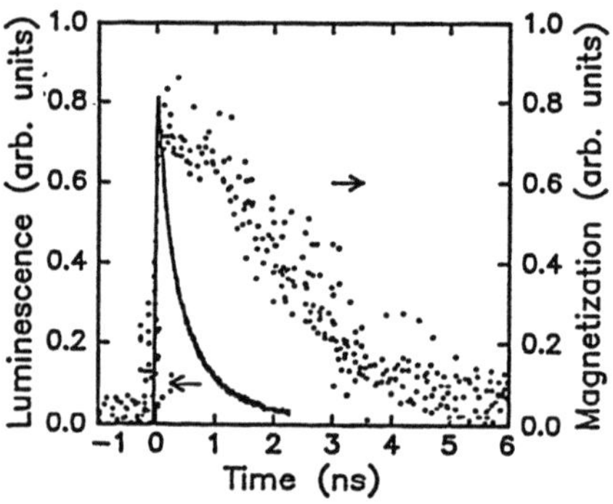

Figure 5. The temporal decays of energy-integrated luminescence intensity (solid line) and of optically induced magnetization (points) for the 30 period multiple well of Fig. 4(a). The luminescence is excited at 1.97 eV, and the magnetic signal induced using 1.75 eV photons.

temporal profile of the pump-induced magnetization to be produced by ramping the delay time of the probe pulse. An example of Mn spin relaxation measured in this manner for a quantum well is shown in Fig. 5. Note that in this case the magnetic signal persists well beyond the carrier lifetime as seen in time-resolved luminescence, conclusively demonstrating it to be a measure of magnetization transferred to the Mn ions. The observed decay time is in good agreement with estimates for the spin-lattice relaxation time of bulk $Cd_{1-x}Mn_xTe$ [5,6].

The ionic spin relaxation time is very sensitive to concentration [10], falling from tens of nanoseconds at a few percent to tens of picoseconds at tens of percent. This is manifested in the time-resolved magnetic measurements as a decrease in the decay time with decreasing well thickness (and hence increasing overlap of the carrier wavefunctions and barriers). Figure 6 shows the dramatic variation in time scales among a series of samples of differing widths, even though the carrier lifetimes are similar for all the samples (~400 ps). The magnetic dynamics are clearly seen to be independent of the carrier lifetimes, and vary a great deal depending on the initial carrier energy and degree of confinement. One may isolate the magnetic behavior from the electronic dynamics, by changing either the

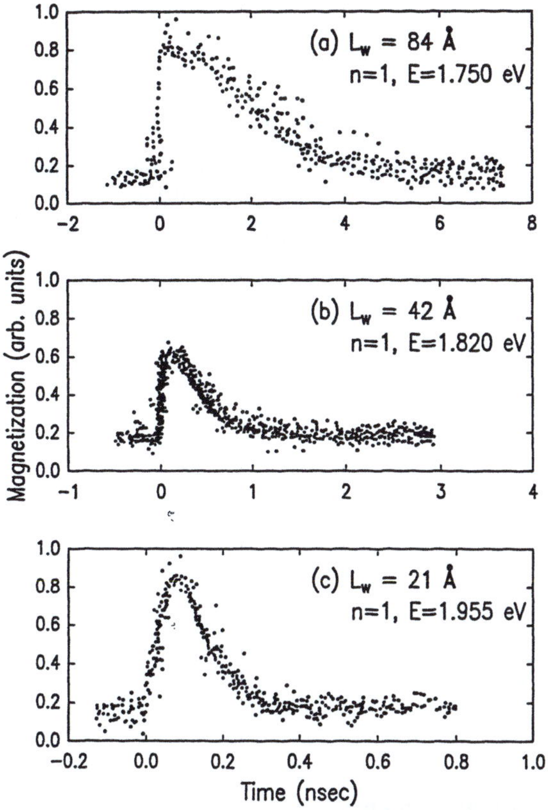

Figure 6. Time-resolved magnetic spectroscopy signals for the same series of samples as in Figs. 2 and 4. The excitation energy in each case is at the n=1 level, corresponding to the peak of the DC luminescence curve.

photon energy or the geometry of a quantum well structure. The physical location of the induced magnetization is controlled by the shape of the carrier wavefunction. The carrier wavefunction is clearly a useful localized probe of magnetic behavior in microscopic structures. Given this tool, and given the fact that we have seen dimensional effects on the electronic properties, including manifestations of these effects in magnetic studies, we can now explore the effects of lower dimension on the ionic spin interactions. Susceptibility experiments have shown dimensional crossover effects for thin layers of these materials in the spin glass regime [11]. With carriers as magnetic probes, a natural question arises as to the effect of confinement on magnetic polaron formation [3]. The data of Fig. 7 demonstrate the occurrence of bound magnetic polarons in 40 Å wide layers of 28% Mn material. The thickness of these barriers is comparable to the diameter of the fully formed polaron in the bulk material (~30Å).

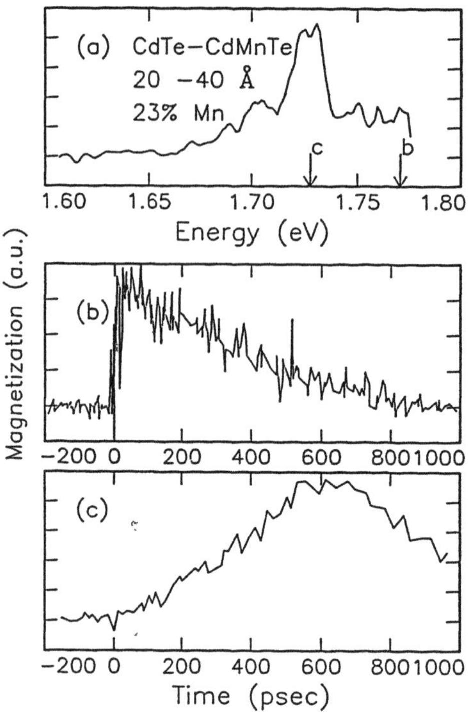

Figure 7. Magnetic spectroscopy of a non-magnetic well superlattice structure. In this case all of the magnetization arises from carrier wavefunction overlap with the barriers. (a). The time-integrated magnetic signal. The arrows mark the excitation energies for the time-resolved signals. (b). Time-resolved magnetic signal for a high energy excitation, showing the spin-lattice relaxation of the Mn magnetization. (c). Time-resolved signal at a lower excitation energy for which oriented carriers become trapped on magnetic fluctuations and form polarons. The polaron formation and decay is governed by the spin-lattice time.

4. Femtosecond Spin Scattering Measurements

The absence of a carrier magnetization signal in the susceptometer measurements suggests that the carrier spin relaxation is much faster than that of the Mn ions. To measure the carrier spin orientation decay, we perform a luminescence polarization experiment. When the spin relaxation time is comparable to the carrier lifetime this information can reasonably be extracted from DC luminescence spectra, but in situations such as those encountered here a time-resolved measurement must be performed. Spin scattering can in fact be fast compared to the luminescence risetime, and the luminescence decay is strongly energy dependent [7].

In Fig. 8 we show the decay of luminescence polarization after optical pumping of an 86 Å 7% Mn/86 Å 38% Mn multiple quantum well structure. The luminescence intensity is measured for both circular components, and the difference divided by the sum of the two is plotted as the polarization. Here the excitation is circularly polarized and at 2.0 eV, and the luminescence is measured at 1.88 eV. The luminescence intensity shows a decay time of 10 ps at 1.88 eV and grows with decreasing energy to 450 ps in the ground state (1.75 eV). The relative strengths of the selection rule-allowed transitions are such that the initial electron spin polarization is 50%. Hole 'spin scattering' is effectively much faster than

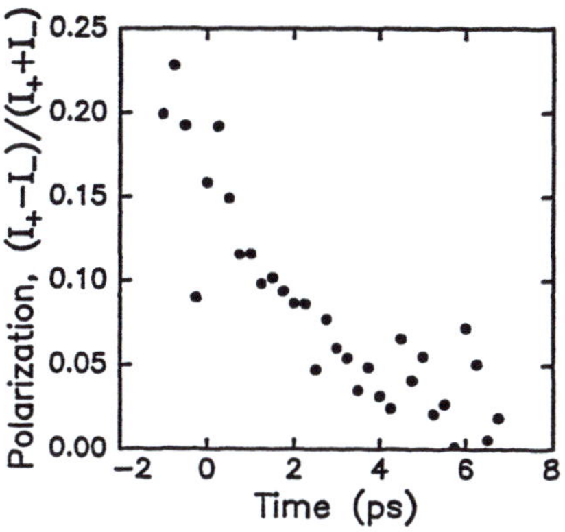

Figure 8. The temporal decay of circular luminescence polarization for the sample of Fig. 5, following circularly polarized excitation at 2.0 eV. The detected luminescence energy is 1.88 eV.

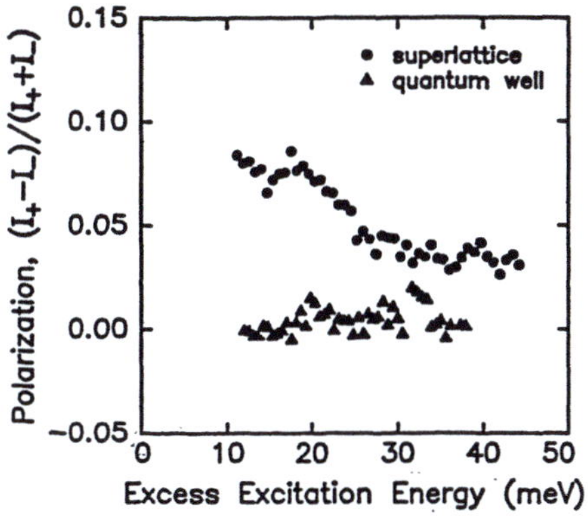

Figure 9. The time-integrated luminescence polarization seen in optical pumping as a function of excitation energy above the detection point. Circles: superlattice of 18Å, 15% Mn barrier regions interrupting CdTe with a 90Å period. Triangles: 86Å, 7% Mn/86Å, 38% Mn quantum wells.

the times recorded here -- hole spin states are not well defined because of the strong spin-orbit interaction in the valence band -- so that the luminescence polarization equals the electron spin polarization divided by two. The polarization decay time constant of a few ps is in good agreement with theoretical estimates [12] for this structure of conduction electron spin-flip scattering due to spin-spin exchange with the Mn ions. However, we have observed equally rapid electron spin relaxations in structures with no Mn in the wells [7]. Therefore another fast scattering channel, independent of the Mn, is also present and has been confirmed by studies of $GaAs/Ga_{1-x}Al_xAs$ quantum wells. The new channel has been ascribed to electron-hole exchange, which is enhanced by quantum confinement [13] due to stronger overlap of the carrier wavefunctions. This discovery is consistent with the fact that the efficiency of optical generation of magnetization in quantum structures is small relative to bulk material of similar composition [5,6].

Further evidence of a confinement effect on spin scattering comes from the steady state luminescence polarizations in optical pumping shown in Fig. 9 as a function of exciting photon energy. The luminescence is detected 5 meV above the peak of the line in each case, and the abscissa is the excitation energy relative to the detection point. The multiple quantum well is the same sample as in Fig. 8, for which there is very little overlap of carrier wavefunctions in neighboring wells. The superlattice sample is more bulk-like in that the wavefunctions are extended. This result is in excellent agreement with optical pumping studies of bulk $Cd_{1-x}Mn_xTe$ [14]. Electron spin relaxation is clearly more rapid in the quantum wells (the two samples have similar luminescence lifetimes).

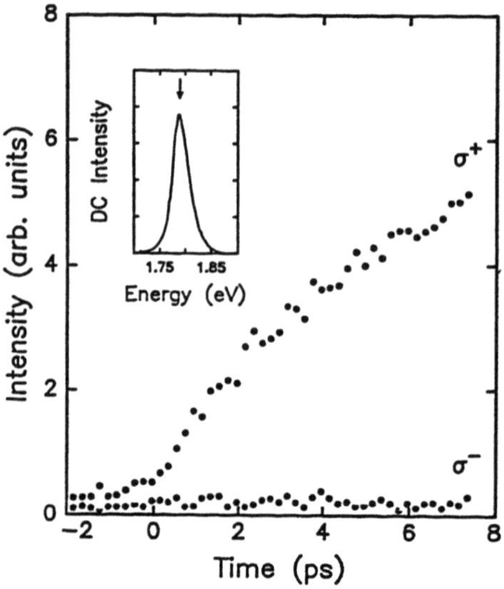

Figure 10. The time-dependence of left- and right-circularly polarized components of luminescence for an 86Å, 13% Mn well, 86Å, 38% Mn barrier multiple quantum well. The sample is at 4K and in a 0.75 T magnetic field to polarize the luminescence. The detection energy is indicated by the arrow in the inset, and the excitation pulse consists of linearly polarized, 1.99 eV photons.

Finally we demonstrate that hole orientation dynamics may also be examined in these materials through the application of a magnetic field. The $j=3/2$ character of the valence band leads to two allowed transitions for electrons of either spin orientation, and consequently the luminescence cannot be 100% polarized unless the holes are 'polarized'. The data of Fig. 10 are measured under the modest field and temperature conditions necessary to accomplish this (because of the large g-factors). In this case the initial nonequilibrium population is not spin-polarized and is generated by linearly polarized light. The measurements place an upper limit of 150 fs on the orientation time of the holes. It is interesting to note that a large magnetization change due to unpolarized excitation should be observable in this case, because all the carriers become oriented and remain so throughout their lifetime.

5. Conclusion

We have described direct measurements of photoinduced magnetization and electron spin scattering which have been made as a function of photon (carrier) energy in a variety of DMS quantum well structures. Magneto-optic spectroscopy allows the magnetic observation of carrier localization and spin relaxation in these systems, with very low noise and picosecond temporal resolution. Although the carrier lifetimes are found to be similar within a variety of geometries, the magnetic behavior is seen to be widely adjustable by varying the confinement and thereby the spin "imprint" upon the quantum wells and barriers. The subsequent magnetic dynamics proceed independently from the electronic structure of the material. Femtosecond luminescence studies enable observations of charge carrier spin scattering within time scales short compared to the carrier-ion spin exchange, yielding the fastest direct measurements of spin relaxation to date. Compared to three dimensional materials, the confined carriers are seen to have unexpectedly fast scattering times which are remarkably independent of the strong electron-ion exchange in DMS. We attribute this behavior to a consequence of quantum confinement and the enhanced electron-hole correlations in a two dimensional system.

References

1. G. Lampel, Phys. Rev. Lett. **20**, 491 (1968).
2. O. Goede and W. Heimbrodt, Phys. Stat. Sol. (b) **146**, 11 (1988).
3. D.D. Awschalom, J.-M. Halbout, S. von Molnar, T. Siegrist and F. Holtzberg, Phys. Rev. Lett. **55**, 1128 (1985).
4. H. Krenn, W. Zawadzki and G. Bauer, Phys. Rev. Lett. **55**, 1510 (1985); H. Krenn, K. Kaltenegger, T. Dietl, J. Spalek, and G. Bauer, Phys. Rev. B39, 10918 (1989).
5. D.D. Awschalom, J. Warnock and S. von Molnar, Phys. Rev. Lett. **58**, 812 (1987).
6. D.D. Awschalom, J. Warnock, J.M. Hong, L.L. Chang, M.B. Ketchen and W.J. Gallagher, Phys. Rev. Lett. **62**, 199 (1989).
7. M.R. Freeman, D.D. Awschalom, C. Hsu, J.M. Hong and L.L. Chang, in Proceedings of the 34th Conference on Magnetism and Magnetic Materials, Boston, November 1989. J. Appl. Phys, in press.
8. J. M. Hong, D.D. Awschalom, L.L. Chang and Armin Segmuller, J. Appl. Phys. **63**, 3285 (1988).
9. S.M. Ryabchenko, Yu. G. Semenov and O.V. Terletskii, Zh. Eksp. Teor. Fiz. 82, 951 (1982) [Sov. Phys. JETP **55**, 557 (1982)].
10. E.A. Harris and K.S. Yngvesson, J. Phys. **C1**, 990 (1968).
11. D.D. Awschalom, J.M. Hong, L.L. Chang and G. Grinstein, Phys. Rev. Lett. **59**, 1733 (1987).
12. G. Bastard and L.L. Chang, in press.
13. Y. Chen, B. Gil, P. Lefebvre, H. Mathieu, T. Fukunaga and H. Nakashima, in *Excitons in Confined Systems*, edited by R. Del Sole et al., Springer-Verlag, Berlin (1988), p. 200.
14. J. Warnock, R.N. Kershaw, D. Ridgley, K. Dwight, A. Wold, and R.R. Galazka, Sol. Stat. Comm. **54**, 215 (1985).

Light-Induced Magnetization in Dilute Magnetic PbTe/PbMnTe Quantum Well Structures

H. Krenn

Institut für Physik, Montanuniversität Leoben, A-8700 Leoben, Austria

Abstract: We have investigated the evolution of confinement effects from 3D to 2D behaviour in diamagnetic/paramagnetic heterosystems $PbTe/Pb_{1-x}Mn_xTe$ ($x<0.038$) using CW photomagnetization (PM) and photoluminescence (PL) in the excitation (PM) and probing (PL) energy range 215 ... 245 meV at T = 2K and at zero external magnetic field. PM comes about from exchange interaction between Mn ions and either spin-polarized free carriers in PbMnTe barriers or spinpolarized confined carriers within PbTe wells which couple into the PbMnTe barriers. Epilayers of PbTe and PbMnTe, as well as single and multiple quantum wells with spatial confinement t(PbTe) = 100 ... 350 Å between PbMnTe barriers ($x<0.038$) are investigated. The formation of minibands is proved for a 346 Å- and 202 Å-period superlattice having PbTe wells of 58 Å and 76 Å, respectively.

1. Introduction

Heterostructures of IV-VI semiconductors $PbTe/Pb_{1-x}Mn_xTe$ form a new class of homoepitaxial systems of alternate diamagnetic PbTe wells and paramagnetic $Pb_{1-x}Mn_xTe$ barriers. In spite of the rather complicated energy band scheme [1] due to the many-valley behaviour of holes and electrons, the bands of PbTe and its derived ternary compound $Pb_{1-x}Mn_xTe$ ($x<0.1$) share many properties of the related narrow gap II-VI semiconductors. The direct energy gap lies at the L-point of the Brillouin zone and increases rapidly with Mn substitution. Another difference to the common II-VI compounds is the almost mirrorlike conduction and valence band energy dispersion in PbTe and $Pb_{1-x}Mn_xTe$. The magneto-optical properties of $Pb_{1-x}Mn_xTe$ are governed [2] by the Heisenberg exchange interaction between the mobile (electrons and holes) and localized (Mn^{2+}) spins. The band parameters include four exchange integrals in addition to the momentum matrix elements $P_{||}$ and $P_{\perp}$ due to the principal axes of the rotation ellipsoids of constant energy surfaces at the L-point of the Brillouin zone [2] .
In this paper the method of light-induced magnetization [3,4] is used to orient localized Mn spins in the $Pb_{1-x}Mn_xTe$ barriers, whereas spin-polarized confined carriers (electrons and holes) are excited into the energy levels in diamagnetic PbTe wells. Interion transitions of Mn^{2+} cannot be excited directly by the midinfrared (215 ... 245 meV) circularly polarized pump beam. Thus the angular momentum of photons is transferred initially into the spin-orientation of confined

Springer Series in Solid-State Sciences, Vol. 97
Localization and Confinement of Electrons in Semiconductors
Editors: F. Kuchar · H. Heinrich · G. Bauer

carriers within the PbTe wells. Afterwards, in dependence on their confinement energy, the polarized spins couple through overlapping wavefunctions with the localized Mn-ions at the interface and within a certain penetration depth of the $Pb_{1-x}Mn_xTe$ barrier. This Heisenberg exchange coupling can be viewed as an effective molecular field which originates from the spin-polarized confined carriers and which acts on the Mn spins in the barrier. The extension of the molecular field can be controlled by varying the well and barrier thicknesses, depending on the degree of confinement of spin-polarized carriers. Thus the behaviour of isolated confined levels up to a formation of minibands (delocalized carriers) can be studied. Is is worthwhile noting that the molecular fields produced by spin-polarized carriers replace an external magnetic field. The polarization mechanism is a consequence of the direct spin-spin exchange and takes place at zero magnetic field. Photomagnetization and spin dynamics in wide-gap $Cd_{1-x}Mn_xTe$/CdTe quantum wells were investigated by Awschalom et al. [3] . In this paper photomagnetization (PM) and photoluminescence (PL) spectra are related to each other in the excitation (PM) and probing (PL) energy range of 215 - 245 meV. The experiments were performed in the CW mode as already published in the pioneering work Ref. 4. After an estimate of the expected PM signal in units of pico-Tesla, monolayers of PbTe and $Pb_{1-x}Mn_xTe$, single and multiple quantum wells, and superlattices of PbTe/$Pb_{1-x}Mn_xTe$ ($x < 0.038$) are investigated. The PM and PL spectra are compared with envelope function calculations [5] .

2. Experimental Setup

A conventional 20 MHz radiofrequency SQUID with an ultimate flux sensitivity of $2x10^{-4}$ $\Phi_0/(Hz)^{1/2}$ is used within a signal bandwidth of 40 kHz. To avoid thermal instability in the sample and in the SQUID, the pickup-loop (diameter 5 mm) of the flux-transformer and the sample incorporated are immersed in superfluid helium at 2.2 K. The SQUID is hermetically sealed against the superfluid helium (otherwise the hybrid Josephson junction is deteriorated) maintaining a good thermal coupling to the helium bath. The specified noise figure of the SQUID is slightly improved at 2.2 K with respect to 4.2 K. The flux-transformer consists of a single pickup coil. The sample area is shielded against stray fields using Pb foils, a copper can and a superconducting niobium housing. The infrared beam of the CO laser (215 ... 245 meV, $\lambda = 5.0$ ... 5.8 μm) is fed through a photoelastic ZnSe modulator, which alternates the polarization state from left to right circularity at 37 kHz chop rate, holding the beam intensity constant. A small pin hole in the niobium housing serves as the entrance "slit" for the pump beam. The polarization state is checked for each measurement, and the proof of "direct" photomagnetism in the samples is simply done by turning a $\lambda/2$-plate over a full circle which yields two maxima and minima and four zero crossings in the photomagnetization signal. The flux sensitivity of the SQUID can also be related to sample dimensions and calibrated in units of magnetic induction ΔB = $4pT/(Hz)^{1/2}$ or as total magnetic moment with a resolution $\Delta m = 6.6x10^{-12}$ $emu/(Hz)^{1/2}$.

3. Estimate of Photomagnetization Signals

The constant energy ellipsoids at the L-point of the Brillouin zone are oriented either one parallel to the [111]-growth direction of an epitaxial film (longitudinal or l-valley) or three valleys oriented $[\bar{1}11]$, $[1\bar{1}1]$, $[11\bar{1}]$ oblique (o) to the growth direction. The density of states (DOS) masses for both sets of valleys are given in Fig. 1 for the 3D case of bulk crystals (Fig. 1a) and for the 2D case of confined carriers in multiple quantum wells (Fig. 1b). In the latter case the projected constant energy contours in the x= $[1\bar{1}0]$, y= $[11\bar{2}]$ - plane are ellipses for the oblique valleys with an average DOS mass $m_o = 0.063\ m_0$, whereas $m_t = 0.021\ m_0$ remains equal to the 3D-mass of a longitudinal valley [6]. For the calculation of the Fermi level in 2D confined systems the modified 2D-DOS masses must be used.

The degree of spin polarization of carriers excited in interband transitions depends on

i) the valley type,

ii) the strength of spin orbit interaction, given by parameters $\Theta_{+/-}$, Ref.1,

iii) the polarization state of the light wave,

iv) the recombinative and spin lifetimes due to various kinds of spin relaxation (the Elliott Yafet spin relaxation is assumed to dominate in PbTe due to the narrow gap).

For circularly polarized excitation the initial spin polarization (neglecting recombination and relaxation) can be calculated from the wavefunctions given in Ref. 1 up to a Fermi level of about 30 meV within the bands:

$$P_i = (n_+ - n_-) / (n_+ + n_-) = \begin{cases} 1 \text{ .... longitudinal valley,} \\ 3 / (5 + 4P^2_{||} / P^2_{\perp}) = 0.56 \text{ ....oblique valleys.} \end{cases} \tag{1}$$

$P^2_{||} / P^2_{\perp} = 0.0877$ is the ratio of the squares of momentum matrix elements due to the principal axes of the energy ellipsoids. n_+ , n_- are stationary electron densities for spin(+/-) and are replaced for holes by p_+, p_-. The final excess density of spins depends on the total electron concentration $n_+ + n_-$, on the spin

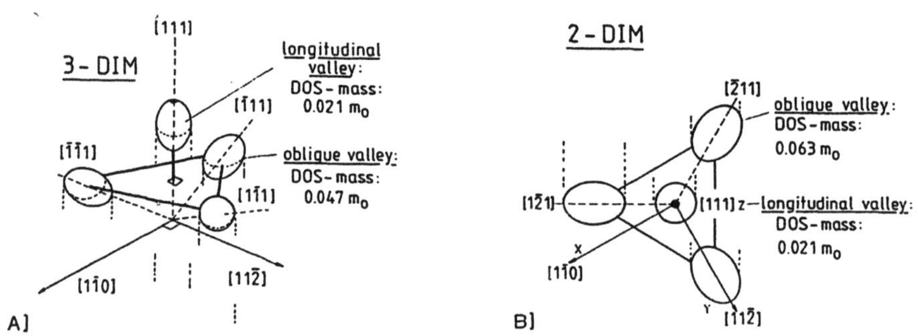

Fig. 1: Constant energy surfaces in bulk (a) and layered (b) PbTe.

relaxation time T_1 and on electron lifetime τ in the following way:

$$n_+ - n_- = P_i \frac{T_1}{T_1 + \tau} (n_+ + n_-) . \quad (2)$$

For an estimate of the upper boundary of photomagnetization, just the initial polarization in the longitudinal valley is considered and $T_1 = 2x10^{-9}$s and $\tau = 3x10^{-8}$s are taken, [7] for an irradiation of 5 mW and a photon energy 240 meV and an illumination area of 0.2 cm^2. The mean fields of spin-polarized electrons (e) and holes (h) acting on the Mn spins in $Pb_{1-x}Mn_xTe$ (for longitudinal valley) are given by

$$B_e^* = -(n_+ - n_-) / (2g\mu_B) (\beta_{||}\cos^2\Theta^- - \beta_\perp \sin^2 \Theta^-) = -(n_+ - n_-) \cdot 7.8x10^{-23} ,$$

$$B_h^* = -(p_+ - p_-) / (2g\mu_B) (\alpha \cos^2\Theta^+ - \delta \sin^2\Theta^+) = +(p_+ - p_-) \cdot 1.2x10^{-22}$$

$$[cm^{-3} eV cm^3] . \quad (3)$$

$\alpha, \beta_{||}, \beta_\perp, \delta$ are the four exchange parameters [2] due to Heisenberg exchange interaction between spinpolarized carriers and Mn ions, using the wavefunctions of Ref.1. Eqs. (3) are derived in the framework of a mean-field approach and are analogous to formula (15) of Ref. 4b. The mean fields $B_e^* = 1.5x10^{-5}$T and $B_h^* = 2.3x10^{-5}$T have opposite signs. For $Pb_{1-x}Mn_xTe$ ($x < 0.038$) the paramagnetic susceptibility is $\chi = 10^{-4}$ emu (at 2K) [8]. Thus the photomagnetization M in units of 10^{-12} T (pT) is estimated according to a spin density of $n_+ - n_- = p_+ - p_- = 4.6x10^{13}cm^{-3}$:

$$M = \chi(B_e^* + B_h^*) < 800 \text{ pT}. \quad (4)$$

The maximum signal observed in a variety of our $Pb_{1-x}Mn_xTe$ samples gives the correct order of magnitude. Clearly, Eq.(4) is an overestimate due to spin relaxation data taken for "pure" PbTe.

4. Samples, Results and Discussions

The samples grown by molecular beam epitaxy were specially tailored for our rather small wavelength range of excitation, $\lambda = 5.0 ... 5.8$ µm. The Mn content (x) in the $Pb_{1-x}Mn_xTe$ layers has a lower limit of $x_{min} = 0.02$ in order to prevent direct excitation in these layers which should act as barriers in the multiple quantum well (MQW) and superlattice (SL) samples. The sample parameters are given in Table I. In the epilayer sample ML#231 and single-quantum-well sample (SQW) the manganese content was chosen to be $x < 0.02$ in order to extract the onset of interband transitions from the photomagnetization signal within the $Pb_{1-x}Mn_xTe$ layers themselves. The thickness of the well width in the SQW sample was 104 Å to recover photomagnetization also from the wavefunction overlap of confined carriers within the PbTe-wells. We investigated first a diamagnetic PbTe epilayer ML#65. Except for a spurious background no evaluable signal was detected from

TABLE I. Sample Parameters

Sample ML = monolayer SQW = single quantum well MQW = multiple quantum well SL = superlattice		Geometry t = thickness w = well width b = barrier width	Carrier-concentration (cm^{-3})
ML#65	PbTe (d)*	t = 1.76 μm	$n = 1.3x10^{17}$
ML#231	$Pb_{1-x}Mn_xTe$ (p)** x = 0.0134	t = 1.25 μm	$n = 3.5x10^{17}$
SQW# 226	PbMnTe/ PbTe/ PbMnTe (p/d/p) x = 0.017)	t = 1.65 μm w = 104 Å b = 3060 Å	$n = 2.1x10^{17}$
MQW#154	PbTe/ PbMnTe/ (p/d/p/d/ ... x20) x = 0.028	t = 3.56 μm w = 350 Å b = 680 Å	$n = 4x10^{17}$
SL#114	PbTe/ PbMnTe/ PbTe/ (p/d/p/d ... x20) x = 0.027	t = 0.84 μm w = 58 Å b = 288 Å	$n = 2x10^{17}$
SL#144	PbTe/ PbMnTe/ PbTe/ (p/d/p/d ... x20) x = 0.037	t = 1.0 μm w = 76 Å b = 126 Å	$n = 2x10^{17}$

* d ..... diamagnetic

** p ..... paramagnetic

spin-polarized free carriers in PbTe. (Fig. 3, at the bottom, compares this "null" experiment with a useful signal from a $Pb_{1-x}Mn_xTe$/PbTe SQW sample).
In Fig. 2 an epitaxial layer of $Pb_{1-x}Mn_xTe$ (x = 0.0134, t = 1.25 μm) was illuminated near the fundamental absorption gap. The photomagnetization (at its largest observable level) unambiguously shows the onset of interband transitions at 230 meV, but only for circular polarization of the excitation beam. In Fig. 3 an un-

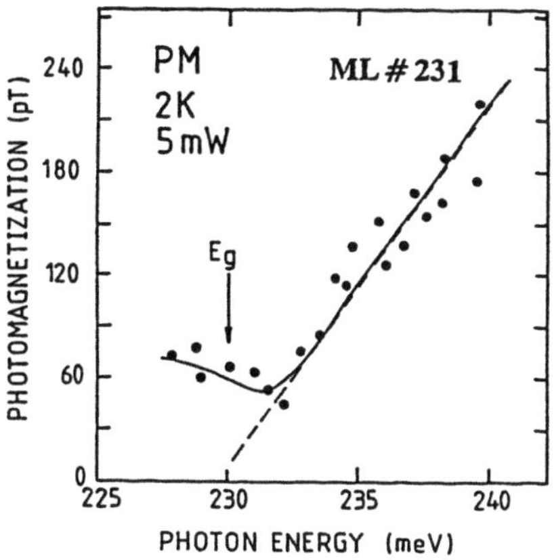

Fig. 2: Photomagnetization in PbMnTe epilayer (x=0.0134).

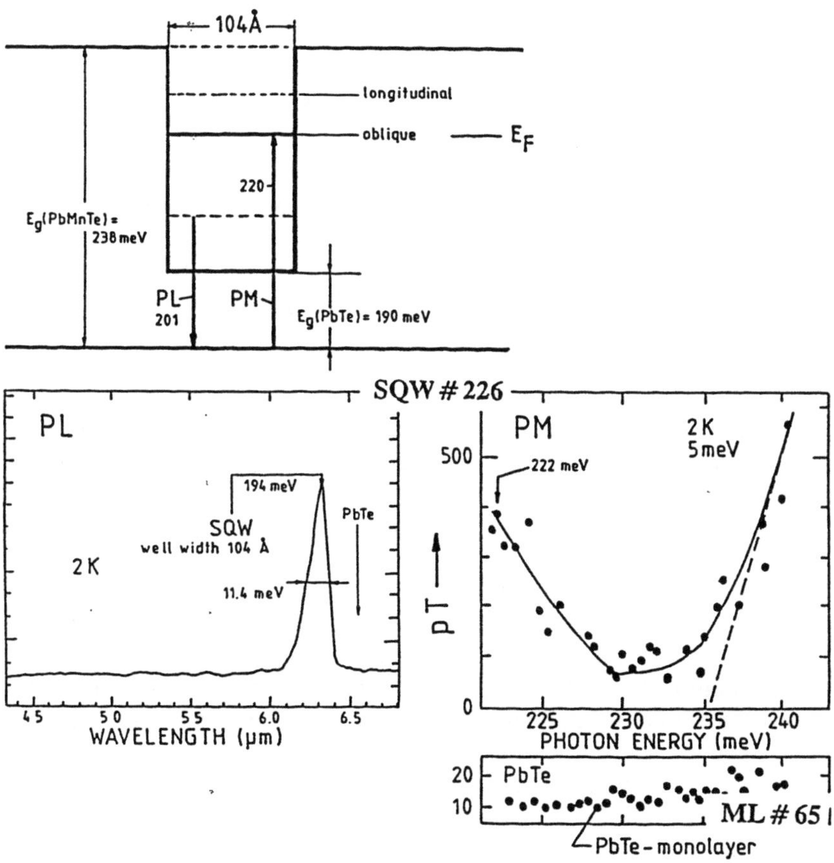

Fig. 3: Photoluminescence (PL) and photomagnetization (PM) in SQW#226 (PbTe well 104 Å). The "null" PM signal for the diamagnetic PbTe epilayer ML#65 is given at the bottom.

strained PbTe single quantum well SQW#226 of w=104 Å is embedded between (x=0.017) PbMnTe barriers of b=3060 Å. The top layer is transmissible for $\hbar\omega < 238$ meV and keeps away surface effects from the active well. The confined levels are sketched for the oblique and longitudinal valleys, if a rectangular well is assumed, and the band offset is (arbitrarily) located at the conduction band edge (flat valence band condition). The Fermi level masks any photomagnetization (PM) in the excitation energy range $\hbar\omega < E_F$. Transitions to the 2D longitudinal valley level above E_F within the well have much lower probability from DOS arguments. On the other hand, photoluminescence (PL) is observed from the masked (and confinement shifted) levels below E_F. The PL peak at 194 meV is above the PbTe energy gap, but is at lower energy than expected from the calculated oblique valley PL=201 meV, since interdiffusion deteriorates the rectangular shape of the well band offset.
A PM signal - neither PbMnTe-like nor PbTe-like - is observed for $190\text{meV} < \hbar\omega < 238\text{meV}$, which comes about from the tails of the wavefunction penetrating into the PbMnTe barrier. The PL intensity is due to recombination from the bottom-like level in the well with nonequilibrium holes, excited through the rather intense "hot" excitation into the barrier continuum (Nd:YAG-laser, $\lambda = 1.06$ µm). For comparison the absence of any PM signal in the diamagnetic PbTe- epilayer ML#65 is shown.
To enhance the PM signal, a (x20) multiple quantum well of a well thickness w=350 Å between 680 Å $Pb_{1-x}Mn_xTe$ (x=0.028) was investigated, Fig.4, MQW#154. Envelope function calculations [5] simulating an abrupt band offset ($\Delta E_v = 0$) delivered 2D confined levels within the well (dispersionless for $k_z \| [111]$, and with bulk-like dispersion for $k_x \| [1\bar{1}0]$ and $k_y \| [11\bar{2}]$, hatched area). The calculated Fermi level corresponds to the carrier concentration of $n = 4\text{x}10^{17}\text{cm}^{-3}$, taken from the Hall data.
The transitions are assigned numbers 1, 2, 3 for PL mainly from oblique valley levels, and 4, 5 for PM and excitation within the wells. For x=0.028 the PbMnTe-barrier photomagnetization sets in at a threshold 268 meV far away from our excitation energy range.
Envelope function calculations [5] have shown that the formation of minibands (finite dispersion $E(k_z)$) starts for superlattice periods D<400 Å in the case of sample SL#114, Fig. 5, at least for oblique valleys, whereas the longitudinal valleys maintain their 2D character. The transition 2D→3D broadens the miniband $E(k_z)$ as the barrier thickness decreases. The strong k_x, k_y-dispersion shifts the carrier continuum. Thus the PM signal should not peak for $\hbar\omega = E_F$, since the miniband wavefunction changes continuously to a behaviour of a PbMnTe barrier wavefunction for complete delocalization. An observed PM signal of 238 meV is in favourable agreement with the calculated 245 meV. Electron-hole recombination from the bottom of the miniband at $k_x = k_y = 0$ yields the slightly shifted PL signal of $\hbar\omega$ = 230 meV (calculated 238meV).
To separate the miniband more clearly from the barrier continuum, a superlattice SL#144, Fig. 6, was grown with a period of 202 Å and thus a broader miniband. Calculation and experiments are in even better agreement, because for x=0.037

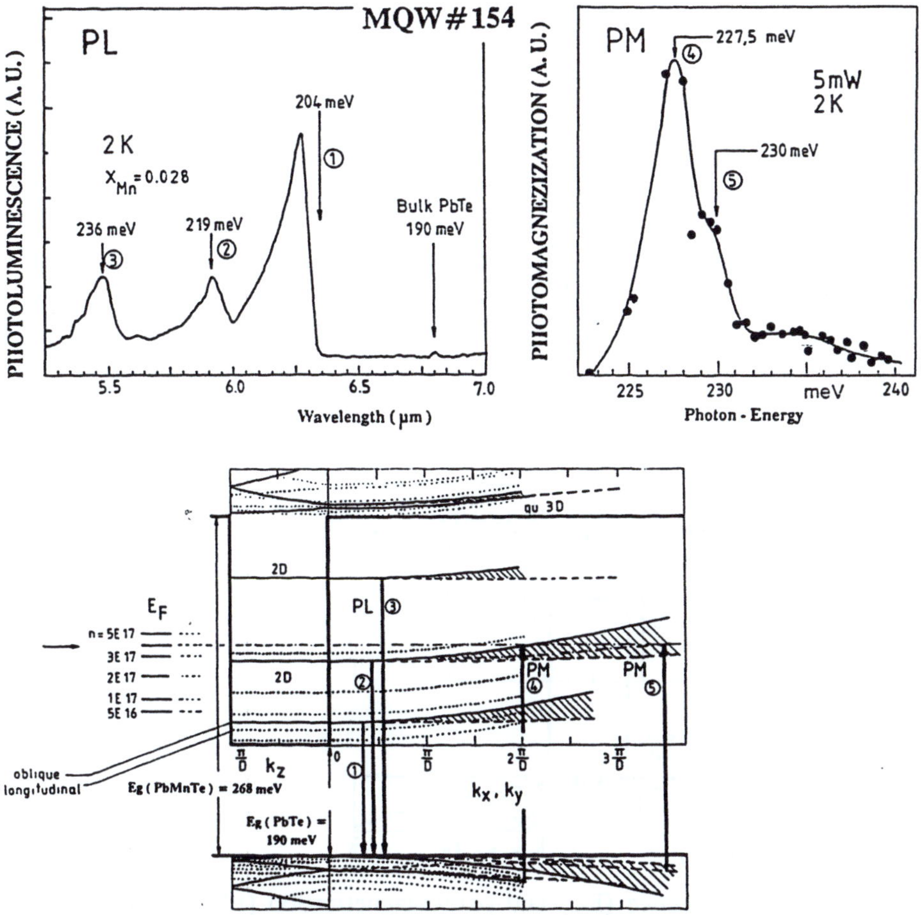

Fig. 4: Photoluminescence (PL) and photomagnetization (PM) for MQW#154 (20 periods); well: 350 Å, barrier: 680 Å. Envelope function calculation (left) gives the k_x, k_y, k_z-dispersion of energy bands indicating a 2D behaviour of confined carriers within the wells. The circled numbers correspond to transitions as described in the text.

the larger band offset with respect to the PbTe energy gap rather justifies the rectangular box approximation for the 76 Å PbTe well.

To date, using MBE growth, it is difficult to change the overall doping level from n- to p-type samples without deteriorating the transport properties of the samples (e.g. mobility and spin relaxation times). We believe from the comparison of PM and PL spectra that a decision about the location of the hetero-band-offset (ΔE_c, ΔE_v) is hardly possible from our experiments. However, envelope function calculations and modelling of the carrier concentration enables us to find the tendency of a shift PM-PL in dependence on the band-offset location. Fig. 7 shows the de-

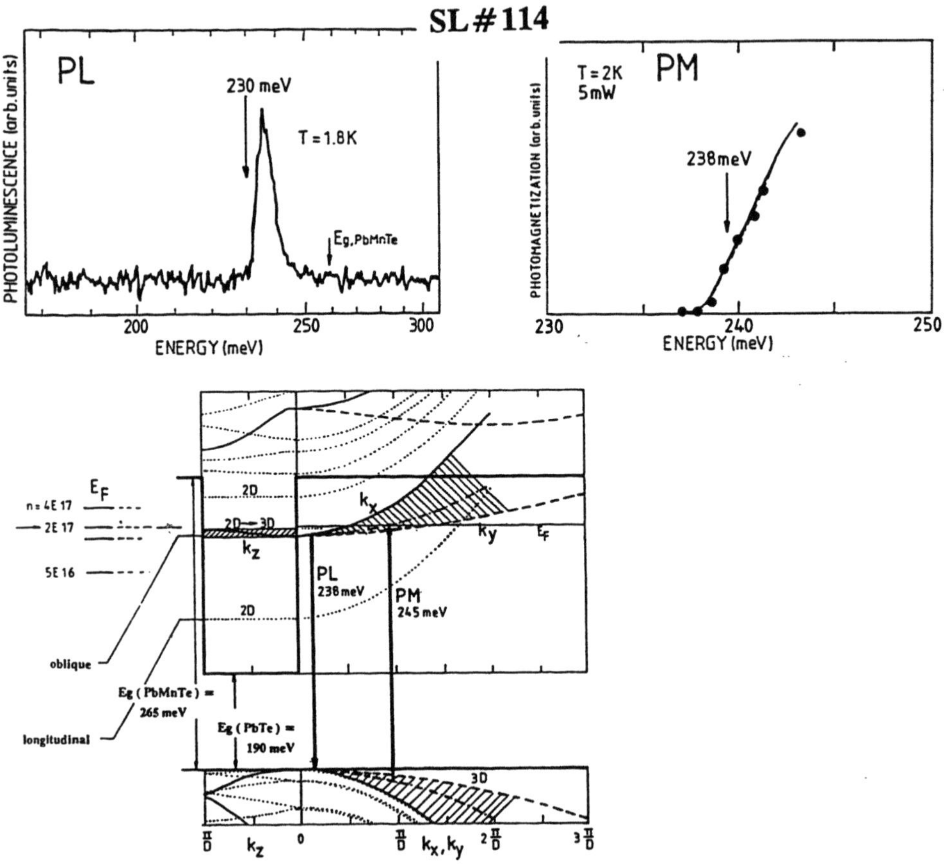

Fig. 5: Photoluminescence (PL) and photomagnetization (PM) for superlattice Sl#114, period 346 Å, wells 58 Å, grown on PbMnTe buffer with x = 0.027. Note the miniband formation for the oblique valley (hatched) and the different behaviour of PL and PM signal vs energy.

pendence PM-PL (energy shift of PM and PL peaks) on the carrier concentration: The results of the experiments with SL#114 favour a band-offset location in the conduction band. $\Delta E_c = 0$ could not explain the increase in energy PM-PL, which is not observed in our entire measurements for $n < 2x10^{17} cm^{-3}$. Also a symmetric partition of the band offsets $\Delta E_c = \Delta E_v$ is not conclusive from our measurements.

In summary, photomagnetization and photoluminescence spectra in diamagnetic/paramagnetic IV-VI semiconductor structures were presented for interband transitions in the range 215 ... 245 meV. These data were compared with envelope function calculations for which the total band offset was assumed to be located at the conduction band edge. Photomagnetization (PM) experiments are distinct from photoluminescence (PL) measurements since PM emerges from polarized

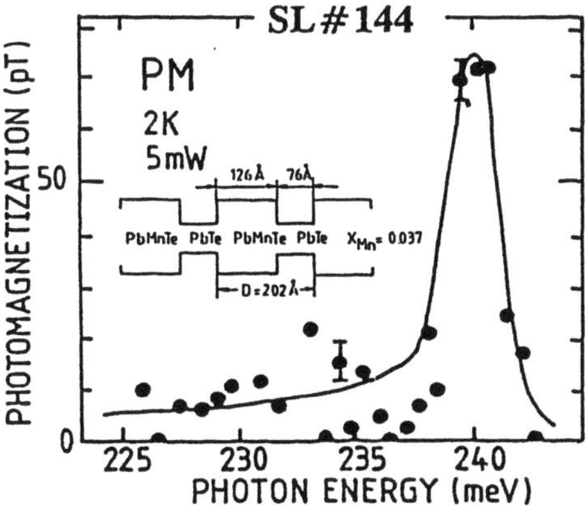

Fig. 6: Photomagnetization (PM) in superlattice SL#144 for 20 periods of 76 Å PbTe wells and 126 Å $Pb_{1-x}Mn_xTe$ (x=0.037) barriers.

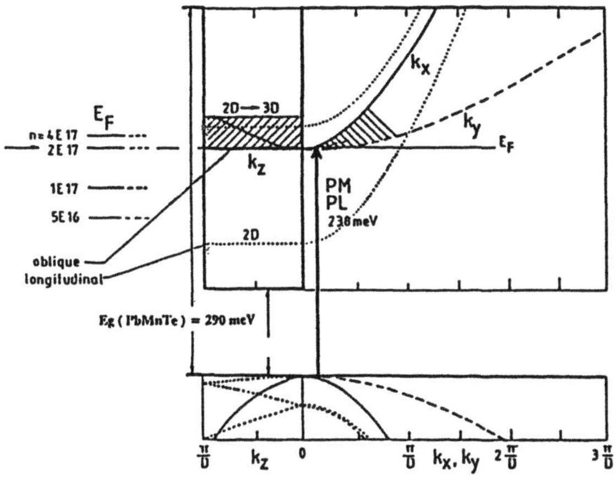

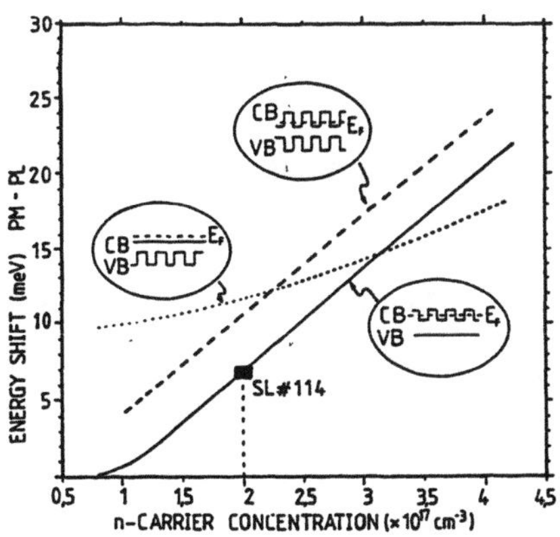

Fig. 7:
Energy shift between photomagnetization and photoluminescence vs electron concentration.

Mn ions in the paramagnetic barriers at rather high initial excitation energies, whereas PL mainly originates from thermalized carriers within the diamagnetic wells. The barrier Mn ions cannot be directly polarized by the off-resonance radiation, which generates spin-polarized confined carriers within the diamagnetic wells. In spite of the fact that our data are analysed preferentially in the wavelength scale, the evolution of photomagnetization from bulk-like behaviour of epilayers, quantum confined structures up to the formation of minibands in superlattices has been demonstrated as a new spectroscopic tool.

Acknowledgements

I am grateful to a number of coworkers who were involved in various aspects of this work: K. Kaltenegger for PM experiments, H. Pascher for PL measurements, M. Kriechbaum for envelope function calculations, H. Clemens, A. Voiticek and N. Frank for sample growth. Special thanks go to G. Bauer for his continuous interest in the success of this work and to the Fonds zur Förderung der wissenschaftlichen Forschung for financial support under grant P6476Phys.

References

1. D.L. Mitchell and R.F. Wallis, Phys. Rev. 151, 581 (1966).

2. H. Pascher, P. Röthlein, G. Bauer, and M. von Ortenberg, Phys. Rev. B40, 10469 (1989-II).

3. D.D. Awschalom et al., in this volume.

4. H. Krenn, W. Zawadzki, and G. Bauer, Phys. Rev. Lett. 55, 1510 (1985).

   H. Krenn, K. Kaltenegger, T. Dietl, J. Spalek, and G. Bauer, Phys. Rev. B39, 10918 (1989-II).

5. M. Kriechbaum, K.E. Ambrosch, E.J. Fantner, H. Clemens, and G. Bauer, Phys. Rev. B30, 3394 (1984).

6. P. Pichler, G. Bauer, and H. Clemens, Z. Phys. B - Condensed matter 67, 475 (1987).

7. B. Kaufmann, Phys. Rev. B30, 4640 (1984).

8. M. Gorska and J.R. Anderson, Phys. Rev. B38, 9120 (1988).

Index of Contributors

Springer Series in Solid-State Sciences

Editors: M. Cardona P. Fulde K. von Klitzing H.-J. Queisser

Springer Series in Solid-State Sciences

Editors: M. Cardona P. Fulde K. von Klitzing H.-J. Queisser

MIX
Papier aus verantwortungsvollen Quellen
Paper from responsible sources
FSC® C105338

If you have any concerns about our products,
you can contact us on
ProductSafety@springernature.com

In case Publisher is established outside the EU,
the EU authorized representative is:
Springer Nature Customer Service Center GmbH
Europaplatz 3, 69115 Heidelberg, Germany

Printed by Libri Plureos GmbH
in Hamburg, Germany